REDUCTIONS IN ORGANIC CHEMISTRY

ELLIS HORWOOD BOOKS IN ORGANIC CHEMISTRY

SYNTHETIC REAGENTS
Edited by J. S. PIZEY, University of Aston in Birmingham

Volume 1: Dimethyl Formamide — Lithium Aluminium Hydride — Mercuric Oxide — Thionyl Chloride

Volume 2: N-Bromosuccinide — Diazomethane — Manganese Dioxide — Raney Nickel

Volume 3: Diborane — 2,3-Dichloro-5,6-Dicyanobenzoquinone — Iodine — Lead tetraacetate

Volume 4: Mercuric Acetate — Periodic Acid and Periodates — Sulphuryl Chloride

Volume 5: Ammonia — Iodine Monochloride — Thallium (III) Acetate and Trifluoroacetate

LITHIUM ALUMINIUM HYDRIDE
Reprinted from Synthetic Reagents Volume 1

CHEMISTRY OF ORGANIC FLUORINE COMPOUNDS: A Laboratory Manual
M. HUDLICKÝ, Virginia Polytechnic Institute and State University

BIOSYNTHESIS OF NATURAL PRODUCTS
P. MANITTO, Professor of Organic Chemistry, University of Milan

FUNDAMENTALS OF PREPARATIVE ORGANIC CHEMISTRY
R. KEESE, University of Berne, R. K. MULLER, Hoffman-La Roche, Basel, and
T. P. TOUBE, Queen Mary College, University of London

CHEMISTRY OF HETEROCYCLIC FLAVOURING AND AROMA COMPOUNDS
G. VERNIN, National Centee of Scientific Research, Marseille, France

REDUCTIONS IN ORGANIC CHEMISTRY
M. HUDLICKÝ, Virginia Polytechnic Institute and State University

REDUCTIONS IN
ORGANIC CHEMISTRY

MILOŠ HUDLICKÝ
Professor of Chemistry
Virginia Polytechnic Institute and State University
USA

ELLIS HORWOOD LIMITED
Publishers · Chichester

Halsted Press: a division of
JOHN WILEY & SONS
New York · Chichester · Brisbane · Toronto

First published in 1984 by

ELLIS HORWOOD LIMITED
Market Cross House, Cooper Street, Chichester, West Sussex, PO19 1EB, England

The publisher's colophon is reproduced from James Gillison's drawing of the ancient Market Cross, Chichester.

Distributors:

Australia, New Zealand, South-east Asia:
Jacaranda-Wiley Ltd., Jacaranda Press,
JOHN WILEY & SONS INC.,
G.P.O. Box 859, Brisbane, Queensland 40001, Australia

Canada:
JOHN WILEY & SONS CANADA LIMITED
22 Worcester Road, Rexdale, Ontario, Canada.

Europe, Africa:
JOHN WILEY & SONS LIMITED
Baffins Lane, Chichester, West Sussex, England.

North and South America and the rest of the world:
Halsted Press: a division of
JOHN WILEY & SONS
605 Third Avenue, New York, N.Y. 10016, U.S.A.

©1984 M. Hudlický/Ellis Horwood Limited

British Library Cataloguing in Publication Data
Hudlický, Miloš
Reductions in organic chemistry. —
(Ellis Horwood series in chemical science)
1. Reduction, Chemical   2. Chemistry, Organic
I. Title
547'.23   QD281.R4

Library of Congress Card No. 84-3768

ISBN 0-85312-345-4 (Ellis Horwood Limited)
ISBN 0-470-20018-9 (Halsted Press)

Printed in Great Britain by Butler & Tanner, Frome, Somerset.

To whom it may concern

Sic ait, et dicto citius tumida aequora placat
collectasque fugat nubes solemque REDUCIT.

Publius Vergilius Maro, Aeneis, I, 142–3

All springs REDUCE their currents to mine eyes.
William Shakespeare, *The Tragedy of King Richard the Third*, II, 2, 68–70.

TABLE OF CONTENTS

PREFACE

Thirty years after my book *Reduction and Oxidation* was first published in Czech (1953) I am presenting this volume as an entirely new reference work.

Myriads of reductions reported in the literature have been surveyed in countless review articles and a respectable row of monographs, most of which are listed in the bibliography at the end of this book. With the exception of two volumes on reduction in Houben-Weyl's *Methoden der Organischen Chemie*, the majority of the monographs deal mainly with catalytic hydrogenation, reductions with hydrides and reductions with metals.

This book encompasses indiscriminately all the types of reductions and superimposes them over a matrix of types of compounds to be reduced. The manner of arrangement of the compounds is a somewhat modified Beilstein system and is explained in the introduction. Numerous tables summarize reducing agents and correlate them with the starting compounds and products of the reductions. Reaction conditions and yields of reductions are mentioned briefly in the text and demonstrated in 175 examples of reductions of simple types of compounds and in 50 experimental procedures.

The material for the book has been collected from original papers (till the end of 1982) after screening *Organic Syntheses*, Theilheimer's *Synthetic Methods*, Harrison and Harrison's *Compendium of Synthetic Methods*, Fieser and Fieser's *Reagents for Organic Synthesis*, and my own records of more or less systematic scanning of main organic chemistry journals. The book is far from being exhaustive but it represents a critical selection of methods which, in my view and according to my experience, give the best results based on yields. In including a reaction preference was given to simple rather than complicated compounds, to most completely described reaction conditions, to methodical studies rather than isolated examples, to synthetically rather than mechanistically oriented papers, and to generally accessible rather than rare journals. Patent literature was essentially omitted. Consequently many excellent chemical contributions could not be included, for which I apologize to their authors. Unfortunately this was the only way in which to turn in not an encyclopedia but rather a 'Pocket Dictionary of Reductions'.

In preparing the book I enjoyed the assistance and cooperation of many people whose work I would like to acknowledge. The students in my graduate courses in synthetic organic chemistry helped me in the literature search of some specific reactions. Thanks for that are due R. D. Allen, R. D. Boyer, D. M. Davis, S. C. Dillender, R. L. Eagan, C. C. Johnson, M. E. Krafft, N. F. Magri, K. J. Natalie, T. A. Perfetti, J. T. Roy, M. A. Schwalke, P. M. Sormani, J. Subrahmanyam, J. W. Wong and especially A. Mikailoff and D. W.

McCourt. I would like to thank Ms Tammy L. Henderson for secretarial help and (together with the word-processor of Virginia Polytechnic Institute and State University) for typing of the manuscript, and Ms Melba Amos for the superb drawings of 175 schemes and equations. To my colleague Dr H. M. Bell I owe thanks for reading some of the procedures, and to my son Dr Tomáš Hudlický I am grateful for reading my manuscript and for his critical remarks. My special gratitude is due to my wife Alena for her patience in 'holding the fort' while I was spending evenings and weekends immersed in work, for her efficient help in proof reading and indexing, and for her releasing me from my commitment never to write another book.

Finally, I would like to express my appreciation of the excellent editorial and graphic work involved in the publishing of my book. For them, my thanks are extended to Ellis Horwood and his publishing team.

March 16, 1983

Miloš Hudlický
Blacksburg, Virginia

INTRODUCTION

In the development of reductions in organic chemistry zinc, iron and hydrogen sulfide are among the oldest reducing agents, having been used since the forties of the last century. Two discoveries mark the most important milestones: catalytic hydrogenation (1897) and reduction with metal hydrides (1947). Each accounts for about one fourth of all reductions, the remaining half of the reductions being due to electroreductions and reductions with metals, metal salts and inorganic as well as organic compounds.

The astronomical number of reductions of organic compounds described in the literature makes an exhaustive survey impossible The most complete treatment is published in Volumes 4/1c and 4/1d of the Houben-Weyl compendium *Methoden der Organischen Chemie*. Other monographs dealing mainly with sections of this topic are listed in the bibliography (p. 257).

Most of the reviews on reductions deal with a particular method such as catalytic hydrogenation, or with particular reducing agents such as complex hydrides, metals, etc. This book gives a kind of a cross-section. Reductions are discussed according to what bond or functional group is reduced by different reagents. Special attention is paid to selective reductions which are suitable for the reduction of one particular type of bond or function without affecting another present in the same molecule. Where appropriate, stereochemistry of the reactions is mentioned.

Types of compounds are arranged according to the following system: hydrocarbons and basic heterocycles; hydroxy compounds and their ethers; mercapto compounds, sulfides, disulfides, sulfoxides and sulfones, sulfenic, sulfinic and sulfonic acids and their derivatives; amines, hydroxylamines, hydrazines, hydrazo and azo compounds; carbonyl compounds and their functional derivatives; carboxylic acids and their functional derivatives; and organometallics. In each chapter, halogen, nitroso, nitro, diazo and azido compounds follow the parent compounds as their substitution derivatives. More detail is indicated in the table of contents. In polyfunctional derivatives reduction of a particular function is mentioned in the place of the highest functionality. Reduction of acrylic acid, for example, is described in the chapter on acids rather than functionalized ethylene, and reduction of ethyl acetoacetate is discussed in the chapter on esters rather than in the chapter on ketones.

Systematic description of reductions of bonds and functions is preceded by discussion of methods, mechanisms, stereochemistry and scopes of reducing agents. Correlation tables (p. 177) show what reagents are suitable for conversion of individual types of compound to their reduction products. More detailed reaction conditions are indicated in schemes and equations, and selected laboratory procedures demonstrate the main reduction techniques (p. 201).

CATEGORIES OF REDUCTION

CATALYTIC HYDROGENATION

The first catalytic hydrogenation recorded in the literature is reduction of acetylene and of ethylene to ethane in the presence of platinum black (von *Wilde*, 1874) [*1*]. However, the true widespread use of catalytic hydrogenation did not start until 1897 when *Sabatier* and his coworkers developed the reaction between hydrogen and organic compounds to a universal reduction method (Nobel prize 1912) [*2*]. In the original work hydrogen and vapors of organic compounds were passed at 100–300° over copper or nickel catalysts. This method of carrying out the hydrogenation has now been almost completely abandoned, and the only instance of hydrogenation still carried out by passing hydrogen through a solution of a compound to be reduced is Rosenmund reduction (p. 144).

After having proceeded through several stages of development catalytic hydrogenation is now carried out essentially in two ways: low-temperature low-pressure hydrogenation in glass apparatus at temperatures up to about 100° and pressures of 1–4 atm, and high pressure processes at temperatures of 20–400° and pressures of a few to a few hundred (350) atm. (The first high-pressure catalytic hydrogenations were carried out over iron and nickel oxide by *Ipatieff* [*3*].) An apparatus for the low-pressure hydrogenation usually consists of a glass flask attached to a liquid-filled graduated container connected to a source of hydrogen and to a reservoir filled with a liquid (water, mercury) (Fig. 1). The progress of the hydrogenation is followed by measuring the volume of hydrogen used in the reaction (*Procedure 1*, p. 201).

A special self-contained glass apparatus was designed for catalytic hydrogenation using hydrogen evolved by decomposition of sodium borohydride (Fig. 2). It consists of two Erlenmeyer flasks connected in tandem. Hydrogen generated in the first flask by decomposition of an aqueous solution of sodium borohydride with acid is introduced into the second flask containing a soluble salt of a catalyst and a compound to be reduced in ethanolic solution. Hydrogen first reduces the salt to the elemental metal which catalyzes the hydrogenation. The hydrogenation can also be accomplished *in situ* in just one (the first) Erlenmeyer flask which, in this case, contains ethanolic solution of the catalytic salt to which an ethanolic solution of sodium borohydride is added followed by concentrated hydrochloric or acetic acid and the reactant [*4*] (*Procedure 2*, p. 202).

High-pressure hydrogenation requires rather sophisticated (and expensive) hydrogenation autoclaves which withstand pressures up to about 350 atm and which can be heated and rocked to insure mixing (Fig. 3). For medium-

Categories of Reduction

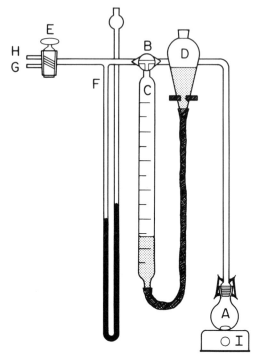

Fig. 1 Apparatus for hydrogenation at atmospheric pressure.

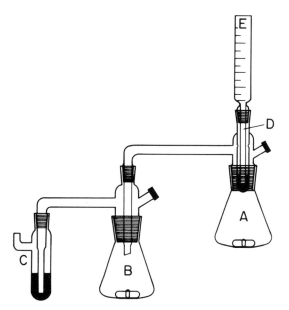

Fig. 2 Apparatus for hydrogenation with hydrogen generated from sodium boro-hydride.

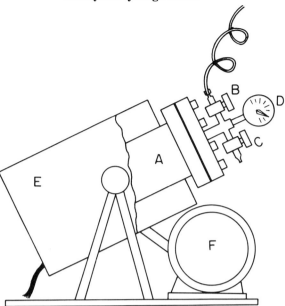

Fig. 3 Apparatus for high pressure hydrogenation. A, autoclave; B, hydrogen inlet valve; C, pressure release valve; D, pressure gauge; E, heating mantle; F rocking device.

pressure small-scale hydrogenations (and in the laboratories the really high-pressure hydrogenations are fairly rare nowadays) a simple apparatus may be assembled from stainless steel cylinders (available in sizes of 30 ml, 75 ml and 500 ml) attached to a hydrogen tank and a pressure gauge by means of copper or stainless steel tubing and swage-lock valves and unions (Fig. 4).

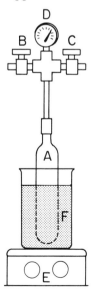

Fig. 4 Apparatus for hydrogenation at medium pressure.

Stirring is provided by magnetic stirring bars, and heating by means of oil baths. Such apparatus can withstand pressures up to about 125 atm. The progress of the reaction is controlled by measuring the decrease in the pressure of hydrogen (*Procedure 3*, p. 203).

Mechanism and Stereochemistry of Catalytic Hydrogenation

With a negligible number of exceptions such as reduction of organolithium compounds [5] elemental hydrogen does not react with organic compounds at temperatures below about 480°. The reaction between gaseous hydrogen and an organic compound takes place only at the surface of a catalyst which adsorbs both the hydrogen and the organic molecule and thus facilitates their contact. Even under these circumstances the reaction has an activation energy of some 6.5–16 kcal/mol, as has been measured for all nine transition elements of Group VIII for the reaction between hydrogen and propene [6]. For this particular reaction the catalytic activities of the nine metals decrease in the sequence shown and do not exactly match the activation energies (in kcal/mol).

[6] $Rh > Ir > Ru > Pt > Pd > Ni > Fe > Co > Os$ (1)

          13.0   15.0   6.5   16.0   11.0   13.0   10.0   8.1   7.4

Although all of the above elements catalyze hydrogenation, only platinum, palladium, rhodium, ruthenium and nickel are currently used. In addition some other elements and compounds were found useful for catalytic hydrogenation: copper (to a very limited extent), oxides of copper and zinc combined with chromium oxide, rhenium heptoxide, heptasulfide and heptaselenide, and sulfides of cobalt, molybdenum and tungsten.

It is believed that at the surface of a catalyst the organic compounds react with individual atoms of hydrogen which become attached successively through a half-hydrogenated intermediate [7]. Because the attachment takes place at a definite time interval reactions such as hydrogen–hydrogen (or hydrogen–deuterium) exchange, *cis–trans* isomerism and even allylic bond shift occur. Nevertheless the time interval between the attachment of the two hydrogen atoms must be extremely short since catalytic hydrogenation is often stereospecific and usually gives predominantly *cis* products (where feasible) resulting from the approach of hydrogen from the less hindered side of the molecule. However, different and sometimes even contradictory results were obtained over different catalysts and under different conditions [8, 9, 10, 11, 12] (p. 50).

Catalytic hydrogenation is subject to steric effects to the degree that it is faster with less crowded molecules. The rate of hydrogenation of alkenes decreases with increasing branching at the double bond, and so do the heats of hydrogenation [13]. Electronic effects do not affect catalytic hydrogenation too strongly, although they may play some role: hydrogenation of styrene is about three times as fast as that of 1-hexene [14].

(2)

[13]	$CH_2=CHC_3H_7$	$CH_2=CC_2H_5$	$CH_3C=CHCH_3$	$CH_3C=CCH_3$
		CH_3	CH_3	CH_3 CH_3
Relative rate	1.00	0.62	0.06	0.02
Kcal/mol	30.1	28.5	26.9	26.6

Catalytic hydrogenation giving optically active compounds was accomplished over palladium on poly(β-S-aspartate) and on poly(γ-S-glutamate) [15] and over rhodium on phosphine-containing substrates [16, 17, 18, 19, 20]. Optical yields up to 93% were reported [16].

More general statements about catalytic hydrogenation are difficult to make since the results are affected by many factors such as the catalyst, its supports, its activators or inhibitors, solvents [21], pH of the medium [21] (Auwers–Skita rule: acidic medium favors *cis* products; neutral or alkaline medium favors *trans* products [22]) and to a certain extent temperature and pressure [8].

Catalysts

Many catalysts, certainly those most widely used such as platinum, palladium, rhodium, ruthenium, nickel, Raney nickel, and catalysts for homogeneous hydrogenation such as tris(triphenylphosphine)rhodium chloride are now commercially available. Procedures for the preparation of catalysts are therefore described in detail only in the cases of the less common ones (p. 205). Guidelines for use and dosage of catalysts are given in Table 1.

PLATINUM

The original forms of platinum referred to in older literature (colloidal platinum or platinum sponge) are not used any more. Instead, catalysts of more reproducible activities gained ground. *Platinum oxide* (PtO_2) (*Adams' catalyst*) [23, 24] made by fusing chloroplatinic acid and sodium nitrate is considered among the most powerful catalysts. It is a brown, stable, non-pyrophoric powder which, in the process of hydrogenation, is first converted to black, very active platinum. Each molecule of this catalyst requires two molecules of hydrogen, and this amount must be subtracted from the final volume of hydrogen used in exact hydrogenations. In most hydrogenations the difference is minute since the amount of platinum oxide necessary for hydrogenation is very small (1–3%). Platinum oxide is suitable for almost all hydrogenations. It is subject to activation by some metal salts, especially stannous chloride and ferric chloride [25] (p. 98), and to deactivation by sulfur and other catalytic poisons [26] (p. 144). It withstands strong organic and mineral acids and is therefore suited for hydrogenation of aromatic rings and heterocycles.

Another form of very active elemental platinum is obtained by reduction

Table 1 Recommended reaction conditions for catalytic hydrogenation of selected types of compounds

Starting compound	Product	Catalyst	Catalyst/compound ratio in weight per cent	Temperature (°C)	Pressure (atm)
Alkene	Alkane	5% Pd (C)	5–10%	25	1–3
		PtO$_2$	0.5–3%	25	1–3
		Raney Ni	30–200%	25	1
			10%	25	50
Alkyne	Alkene	0.3% Pd(CaCO$_3$)	8%	25	1
		5% Pd(BaSO$_4$)	2% + 2% quinoline	20	1
		Lindlar catalyst	10% + 4% quinoline	25	1
	Alkane	PtO$_2$	3%	25	1
		Raney Ni	20%	25	1–4
Carbocyclic aromatic	Hydro-aromatic	PtO$_2$	6–20%, AcOH	25	1–3
		5% Rh(Al$_2$O$_3$)	40–60%	25	1–3
		Raney Ni	10%	75–100	70–100
Heterocyclic aromatic (pyridines)	Hydro-aromatic	PtO$_2$	4–7%, AcOH or HCl/MeOH	25	1–4
		5% Rh (C)	20%, HCl/MeOH		
		Raney Ni	2%	65–200	130
Aldehyde, ketone	Alcohol	PtO$_2$	2–4%	25	1
		5% Pd (C)	3–5%	25	1–4
		Raney Ni	30–100%	25	1
Halide	Hydro-carbon	5% Pd (C) or	1–15%, KOH	25	1
		5% Pd(BaSO$_4$)	30–100%, KOH	25	1
		Raney Ni	10–20%, KOH	25	1
Nitro compound, azide	Amine	PtO$_2$	1–5%	25	1
		Pd (C)	4–8%	25	1
		Raney Ni	10–80%	25	1–3
Oxime, nitrile	Amine	PtO$_2$	1–10%, AcOH or HCl/MeOH	25	1–3
		5% Pd (C)	5–15%, AcOH	25	1–3
		Raney Ni	3–30%	25	35–70

of chloroplatinic acid by sodium borohydride in ethanolic solution. Such platinum hydrogenated 1-octene 50% faster than did platinum oxide [27].

In order to increase the contact of a catalyst with hydrogen and the compounds to be hydrogenated platinum (or other metals) is (are) precipitated on materials having large surface areas such as activated charcoal, silica gel, alumina, calcium carbonate, barium sulfate and others. Such 'supported catalysts' are prepared by hydrogenation of solutions of the metal salts, e.g. chloroplatinic acid, in aqueous suspensions of activated charcoal or other solid substrates [28]. Supported catalysts which usually contain 5, 10 or 30 weight percent of platinum are very active, and frequently pyrophoric.

The support exerts a certain effect on the hydrogenation, especially its rate but sometimes even its selectivity. It is therefore advisable to use the same type of catalyst when duplicating experiments.

The platinum catalysts are universally used for hydrogenation of almost any type of compound at room temperature and atmospheric or slightly elevated pressures (1–4 atm). They are usually not used for hydrogenolysis of benzyl-type residues, and are completely ineffective in reductions of acids or esters to alcohols. At elevated pressures (70–210 atm) platinum oxide converts aromatics to perhydroaromatics at room temperature very rapidly [8].

PALLADIUM

Palladium catalysts resemble closely the platinum catalysts. *Palladium oxide* (PdO) is prepared from palladium chloride and sodium nitrate by fusion at 575–600° [29, 30]. Elemental palladium is obtained by reduction of palladium chloride with sodium borohydride [27, 31]. *Supported palladium catalysts* are prepared with the contents of 5% or 10% of palladium on charcoal, calcium carbonate and barium sulfate [32]. Sometimes a special support can increase the selectivity of palladium. Palladium on strontium carbonate (2%) was successfully used for reduction of just γ, δ-double bond in a system of α, β, γ, δ-unsaturated ketone [33].

Palladium catalysts are more often modified for special selectivities than platinum catalysts. Palladium prepared by reduction of palladium chloride with sodium borohydride (*Procedure* 4, p. 205) is suitable for the reduction of unsaturated aldehydes to saturated aldehydes [31]. Palladium on barium sulfate deactivated with sulfur compounds, most frequently the so-called quinoline-*S* obtained by boiling quinoline with sulfur [34], is suitable for the Rosenmund reduction [35] (p. 144). Palladium on calcium carbonate deactivated by lead acetate (*Lindlar's catalyst*) is used for partial hydrogenation of acetylenes to *cis*-alkenes [36] (p. 44).

Palladium catalysts can be used in strongly acidic media and in basic media, and are especially suited for hydrogenolyses such as cleavage of benzyl-type bonds (p. 151). They do not reduce carboxylic groups.

RHODIUM, RUTHENIUM AND RHENIUM

A very active elemental *rhodium* is obtained by reduction of rhodium chloride with sodium borohydride [27]. *Supported rhodium catalysts*, usually 5% on carbon or alumina, are especially suited for hydrogenation of aromatic systems [37]. A mixture of rhodium oxide and platinum oxide was also used for this purpose and proved better than platinum oxide alone [38, 39]. Unsaturated halides containing vinylic halogens are reduced at the double bond without hydrogenolysis of the halogen [40].

Use of *ruthenium catalysts* [41] and *rhenium heptoxide* [42] is rare. Their specialty is reduction of free carboxylic acids to alcohols.

NICKEL

Nickel catalysts are universal and are widely used not only in the laboratory but also in the industry. The supported form – *nickel* on *kieselguhr* or *infusorial earth* – is prepared by precipitation of nickel carbonate from a solution of nickel nitrate by sodium carbonate in the presence of infusorial earth and by reduction of the precipitate with hydrogen at 450° after drying at 110–120°. Such catalysts work at temperature of 100–200° and pressures of hydrogen of 100–250 atm [43].

Nickel catalysts of very high activity are obtained by the *Raney process*. An alloy containing 50% of nickel and 50% of aluminium is heated with 25–50% aqueous sodium hydroxide at 50–100°. Aluminum is dissolved and leaves nickel in the form of very fine particles. It is then washed with large amounts of distilled water and finally with ethanol. Depending on the temperature of dissolution, on the content of aluminum, and on the method of washing, Raney nickel of varied activity can be produced. It is categorized by the symbols W1–W8 [44]. The most active is Raney nickel W6 which contains 10–11% of aluminum, has been washed under hydrogen and has been stored under ethanol in a refrigerator. Its activity in hydrogenation is comparable to that of the noble metals and may decrease with time. Some hydrogenations can be carried out at room temperature and atmospheric pressure [45]. Many Raney nickel preparations are pyrophoric in the dry state. Some, like Raney nickel W6, may react extremely violently, especially when dioxane is used as a solvent, the temperature is higher than about 125°, and large quantities of the catalyst are used. If the ratio of the Raney nickel to the compound to be hydrogenated does not exceed 5% the hydrogenation at temperatures above 100° is considered safe [45]. It is to be kept in mind that the temperature in the autoclave may rise considerably because of heat of hydrogenation.

Raney nickel can be used for reduction of practically any functions. Many hydrogenations can be carried out at room temperature and atmospheric or slightly elevated pressure (1–3 atm) [45] (*Procedure 5*, p. 205). At high temperatures and pressures even the difficultly reducible acids and esters are reduced to alcohols [44] (p. 154). Raney nickel is not poisoned by sulfur and is used for desulfurization of sulfur-containing compounds [46] (p. 86) (*Procedure 6*, p. 205). Its disadvantages are difficulty in calculating dosage – it is usually measured as a suspension rather than weighed – and ferromagnetic properties which preclude the use of magnetic stirring.

Nickel of activity comparable to Raney nickel is obtained by reduction of nickel salts, e.g. nickel acetate, with 2 mol of sodium borohydride in an aqueous solution and by washing the precipitate with ethanol [13, 47] (*Procedure 7*, p. 205). Such preparations are designated *P-1* or *P-2* and can be conveniently prepared *in situ* in a special apparatus [4] (*Procedure 2*, p. 202). They contain a high percentage of nickel boride, are non-magnetic and non-pyrophoric and can be used for hydrogenations at room temperature and atmospheric pressure. Nickel P-2 is especially suitable for semihydrogenation of acetylenes to *cis*-alkenes [*47*] (p. 43, 44).

Nickel precipitated from aqueous solutions of nickel chloride by aluminum or zinc dust is referred to as *Urushibara catalyst* and resembles Raney nickel in its activity [*48*].

Another highly active non-pyrophoric nickel catalyst is prepared by reduction of nickel acetate in tetrahydrofuran by sodium hydride at 45° in the presence of *tert*-amyl alcohol (which acts as an activator). Such catalysts, referred to as *Nic catalysts*, compare with P nickel boride and are suitable for hydrogenations at room temperature and atmospheric pressure, and for partial reduction of acetylenes to *cis*-alkenes [*49*].

The Raney nickel process applied to alloys of aluminum with other metals produces *Raney iron*, *Raney cobalt*, *Raney copper* and *Raney zinc*, respectively. These catalysts are used very rarely and for special purposes only.

OTHER CATALYSTS

Catalysts suitable specifically for reduction of carbon–oxygen bonds are based on oxides of copper, zinc and chromium (*Adkins' catalysts*). The so-called copper chromite (which is not necessarily a stoichiometric compound) is prepared by thermal decomposition of ammonium chromate and copper nitrate [*50*]. Its activity and stability is improved if barium nitrate is added before the thermal decomposition [*51*]. Similarly prepared zinc chromite is suitable for reductions of unsaturated acids and esters to unsaturated alcohols [*52*]. These catalysts are used specifically for reduction of carbonyl- and carboxyl-containing compounds to alcohols. Aldehydes and ketones are reduced at 150–200° and 100–150 atm, whereas esters and acids require temperatures up to 300° and pressures up to 350 atm. Because such conditions require special equipment and because all reductions achievable with copper chromite catalysts can be accomplished by hydrides and complex hydrides the use of Adkins catalyst in the laboratory is very limited.

The same is true of *rhenium catalysts*: rhenium heptoxide [*42*], rhenium heptasulfide [*53*] and rhenium heptaselenide [*54*] all require temperatures of 100–300° and pressures of 100–300 atm. Rhenium heptasulfide is not sensitive to sulfur, and is more active than molybdenum and cobalt sulfides in hydrogenating oxygen-containing functions [*53*, *55*].

All the catalysts described so far are used in heterogeneous catalytic hydrogenations. In the past twenty years *homogeneous hydrogenation* has been developed, which is catalyzed by compounds soluble in organic solvents. Of a host of complexes of noble metals *tris(triphenylphosphine)rhodium chloride* is now used most frequently and is commercially available [*56*, *57*, *58*] (*Procedure 8*, p. 206). Homogeneous catalytic hydrogenation is often carried out at room temperature and atmospheric pressure and hardly ever at temperatures exceeding 100° and pressures higher than 100 atm. It is less effective and more selective than the heterogeneous hydrogenation and is therefore more suitable for selective reductions of polyfunctional compounds, for example for conversion of α,β-unsaturated aldehydes to saturated aldehydes [58] (*Procedure 8*, p. 206). It also causes less rearrangements and isotope exchange and is therefore convenient for deuteration. Because it is possible to prepare chiral catalysts homogeneous hydrogenation is useful for asymmetric reductions [20]. The disadvantages of homogeneous hydrogenation are lower availability of homogeneous catalysts and more complicated isolation of products.

Activators and Deactivators of Catalysts

Efficiency of catalysts is affected by the presence of some compounds. Even small amounts of alien admixtures, especially with noble metal catalysts, can increase or decrease the rate of hydrogenation and, in some cases, even inhibit the hydrogenation completely. Moreover some compounds can even influence the selectivity of a catalyst.

Minute quantities of zinc acetate or ferrous sulfate enhance hydrogenation of the carbonyl group in unsaturated aldehydes and cause preferential hydrogenation to unsaturated alcohols [59]. As little as 0.2% of palladium present in a platinum-on-carbon catalyst deactivates platinum for hydrogenolysis of benzyl groups and halogens [28]. Admixtures of stannous chloride (7% of the weight of platinum dioxide) increased the rate of hydrogenation of valeraldehyde ten times, admixture of 6.5% of ferric chloride eight times [25].

On the other hand, some compounds slow down the uptake of hydrogen and may even stop it at a certain stage of hydrogenation. Addition of lead acetate to palladium on calcium carbonate makes the catalyst suitable for selective hydrogenation of triple to double bonds (Lindlar catalyst) [36] (*Procedure 9*, p. 206).

The strongest inhibitors of noble metal catalysts are sulfur and most sulfur compounds. With the exception of modifying a palladium-on-barium sulfate catalyst [35] or platinum oxide [60] for the Rosenmund reduction the presence of sulfur compounds in materials to be hydrogenated over platinum, palladium and rhodium catalysts is highly undesirable. Except for hexacovalent sulfur compounds such as sulfuric acid and sulfonic acids, most sulfur-containing compounds are 'catalytic poisons' and may inhibit hydrogenation very strongly. If such compounds are present as impurities they can be removed by contact with Raney nickel, which combines with almost any form of sulfur to form nickel sulfide. Shaking or stirring of a compound contaminated with sulfur-containing contaminants with Raney nickel makes possible subsequent hydrogenation over noble metal catalysts.

Many nucleophiles act as inhibitors of platinum, palladium and rhodium catalysts. The strongest are mercaptans, sulfides, cyanide and iodide; weaker are ammonia, azides, acetates and alkalis [26].

Acidity or alkalinity of the medium plays a very important role. Hydrogenation of aromatic rings over platinum catalysts requires acid medium. Best results are obtained when acetic acid is used as the solvent. Addition of sulfuric or perchloric acid is of advantage, and in the hydrogenation of pyridine compounds, even necessary. Reduction of weakly basic pyridine and its derivatives to the corresponding strongly basic piperidines generates catalytic poisons.

On the other hand, addition of tertiary amines accelerated hydrogenation of some compounds over Raney nickel [61]. In hydrogenation of halogen derivatives over palladium [62] or Raney nickel [63] the presence of at least one equivalent of sodium or potassium hydroxide was found necessary.

Effects of the Amount of Catalyst, Solvent, Temperature and Pressure

The *ratio of the catalyst to the compound* to be hydrogenated has, within certain limits, a strong effect on the rate of hydrogenation. Platinum group catalysts are used in amounts of 1–3% of the weight of the metal while Raney nickel requires much larger quantities. Doubling the amount of a nickel catalyst in the hydrogenation of cottonseed oil from 0.075% to 0.15% doubled the reaction rate [64]. With 10% as much Raney nickel as ester practically no hydrogenation of the ester took place even at 100°. With 70%, hydrogenation at 50° was more rapid than hydrogenation at 100° with 20–30% catalyst [61]. Some hydrogenations over Raney nickel and copper chromite went well only if 1.5 times as much catalyst as reactant was used [65].

The best *solvents* for hydrogen (pentane, hexane) are not always good solvents for the reactants. Methanol and ethanol, which dissolve only about one third the amount of hydrogen as the above hydrocarbons dissolve, are used most frequently. Other solvents for hydrogenations are ether, thiophen-free benzene, cyclohexane, dioxane and acetic acid, the last one being especially useful in the catalytic hydrogenations of aromatics over platinum metal catalysts. Too volatile solvents are not desirable. In hydrogenations at atmospheric pressures they make exact reading of the volume of consumed hydrogen tedious and inaccurate, and in high-temperature and high-pressure hydrogenations they contribute to the pressure in the autoclave.

Water does not dissolve many organic compounds but it can be used as a solvent, especially in hydrogenations of acids and their salts. It may have some deleterious effects; for example it was found to enhance hydrogenolysis of vinylic halogens [66].

The pH of the solution plays an important rate in the steric outcome of the reaction. Acidic conditions favor *cis* addition and basic conditions favor *trans* addition of hydrogen [22, 67].

Temperature affects the rate of hydrogenation but not as much as is usual with other chemical reactions. The rise in temperature from 50° to 100° and from 25° to 100° caused, respectively, fourfold and 12-fold increases in the rates of hydrogenation of esters over Raney nickel W6 [61]. The increased rate of hydrogenation may reduce the selectivity of the particular catalyst.

Pressure of hydrogen, as expected, increases the rate of hydrogenation considerably but not to the same extent with all compounds. In the hydrogenation of esters over Raney nickel W6 the rate doubled with an increase from 280 to 350 atm [*61*]. In hydrogenation of cottonseed oil in benzene over nickel at 120°, complete hydrogenation was achieved at 30 atm in 3 hours, at 170 atm in 2 hours, and at 325 atm in 1.5 hours [*64*]. High pressure favors *cis* hydrogenation where applicable [*10*] but decreases the stereoselectivity [*8*].

The seemingly peripheral effect of *mixing* must not be underestimated. Since hydrogenation, especially the heterogeneous hydrogenation, is a reaction of three phases, it is important that good contact take place not only between the gas and the liquid but also between hydrogen and the solid, the catalyst. It is therefore essential that the stirring provide for frequent, or better still, permanent contact between the catalyst and the gas. Shaking and fast magnetic stirring is therefore preferred to slow rocking [*68*].

Carrying Out Catalytic Hydrogenation

After a decision has been taken as to what type of hydrogenation, what catalyst and what solvent are to be used, a careful calculation should precede carrying out of the reaction.

In atmospheric or low-pressure hydrogenation the volume of hydrogen needed for a partial or total reduction should be calculated. This is imperative for partial hydrogenations when the reduction has to be interrupted after the required volume of hydrogen has been absorbed. In exact calculations vapor pressure of the solvent used must be considered since it contributes to the total pressure in the apparatus. If oxide-type catalysts are used, the amount of hydrogen needed for the reduction of the oxides to the metals must be included in the calculation.

In high-pressure hydrogenations the calculations are even more important. It is necessary to take into account, in addition to the pressure resulting from heating of the reaction mixture to a certain temperature, an additional pressure increase caused by the temperature rise owing to considerable heat of hydrogenation (approximately 30 kcal/mol per double bond). In hydrogenations of large amounts of compounds in low-boiling solvents, the reaction heat may raise the temperature inside the autoclave above the critical temperature of the components of the mixture. In such a case, when one or more components of the mixture gasify, the pressure inside the container rises considerably and may even exceed the pressure limits of the vessel. A sufficient leeway for such potential or accidental pressure increase should be secured [*68*].

The calculated amounts of the catalyst, reactant and solvent (if needed) are then placed in the hydrogenation vessel. Utmost care must be exercised in loading the hydrogenation container with catalysts which are pyrophoric, especially when highly volatile and flammable solvents like ether, methanol, ethanol, cyclohexane or benzene are used. The solution should be added to the catalyst in the container. If the catalyst must be added to the solution this should be done under a blanket of an inert gas to prevent potential ignition.

The hydrogenation vessel is attached to the source of hydrogen, evacuated, flushed with hydrogen once or twice to displace any remaining air (oxygen may inhibit some catalysts), pressurized with hydrogen, shut off from the hydrogen source, and stirring and heating (if required) are started.

After the hydrogenation is over and the apparatus is cold the excessive pressure is bled off. In high-temperature high-pressure hydrogenations carried out in robust autoclaves it takes considerable time for the assembly to cool down. If correct reading of the final pressure is necessary it is advisable to let the autoclave cool overnight.

Isolation of the products is usually carried out by filtration. Suction filtration is faster and preferable to gravity filtration. When pyrophoric noble metal catalysts and Raney nickel are filtered with suction the suction must be stopped before the catalyst on the filter paper becomes dry. Otherwise it can ignite and cause fire. Where feasible centrifugation and decantation should be used for the separation of the catalyst. Sometimes the filtrate contains colloidal catalyst which has passed through the filter paper. Stirring of such filtrate with activated charcoal followed by another filtration usually solves this problem. Evaporation of the filtrate and crystallization or distillation of the residue completes the isolation.

Isolation of products from homogeneous hydrogenation is more complicated and depends on the catalyst used.

Catalytic Transfer of Hydrogen

A special kind of catalytic hydrogenation is catalytic hydrogen transfer achieved by heating of a compound to be hydrogenated in a solvent with a catalyst and a hydrogen donor – a compound which gives up its hydrogen. The hydrogen donors are hydrazine [*69, 70*], formic acid [*71*], triethylammonium formate [*72, 73*], cyclohexene [*74, 75*], cyclohexadiene [*76, 77*], tetralin [*78*], pyrrolidine [*79*], indoline [*79*], tetrahydroquinoline [*79*], triethysilane [*80*] and others; the catalysts are platinum, palladium or Raney nickel [*81*] (*Procedure 10*, p. 206).

Catalytic hydrogen transfer results usually in *cis* (*syn*) addition of hydrogen and is sometimes more selective than catalytic hydrogenation with hydrogen gas (p. 44).

REDUCTION WITH HYDRIDES AND COMPLEX HYDRIDES

Fifty years after the introduction of catalytic hydrogenation into the methodology of organic chemistry another discovery of comparable importance was published: synthesis [*82*] and applications [*83*] of lithium aluminum hydride and lithium and sodium borohydride.

Treatment of lithium hydride with aluminum chloride gives *lithium aluminum hydride* which, with additional aluminum chloride, affords *aluminum hydride, alane* (*Procedure 11*, p. 206).

[82] $$4\ LiH\ +\ AlCl_3\ \xrightarrow{\ Et_2O\ }\ LiAlH_4\ +\ 3\ LiCl \qquad (3)$$

M.W. 37.95, m.p. 125° (dec.)

$$3\ LiAlH_4\ +\ AlCl_3\ \xrightarrow{\ Et_2O\ }\ 4\ AlH_3\ +\ 3\ LiCl$$

A boron analog – *sodium borohydride* – was prepared by reaction of sodium hydride with trimethyl borate [84] or with sodium fluoroborate and hydrogen [85], and gives, on treatment with boron trifluoride or aluminum chloride, *borane* (*diborane*) [86]. Borane is a strong Lewis acid and forms complexes with many Lewis bases. Some of them, such as complexes with dimethyl sulfide, trimethyl amine and others, are sufficiently stable to have been made commercially available. Some others should be handled with precautions. A spontaneous explosion of a molar solution of borane in tetrahydrofuran stored at less than 15° out of direct sunlight has been reported [87].

[84] $$4\ NaH\ +\ B(OMe)_3\ \longrightarrow\ NaBH_4\ +\ 3\ MeONa \qquad (4)$$

M.W. 37.83

[85] $$4\ NaH\ +\ NaBF_4\ \xrightarrow[360°,\ 50\ atm]{\ H_2\ }\ NaBH_4\ (35\%)\ +\ 4\ NaF$$

[86] $$3\ NaBH_4\ +\ 4\ BF_3\ \longrightarrow\ 4\ BH_3\ +\ 3\ NaBF_4$$

$$2\ BH_3\ \longrightarrow\ B_2H_6$$

M.W. 27.66

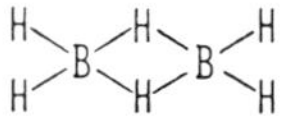

$BH_3 \cdot THF$ $BH_3 \cdot NMe_3$ $BH_3 \cdot SMe_2$

M.W. 85.94   M.W. 72.95   M.W. 75.97

Complex aluminum and boron hydrides can contain other cations. The following compounds are prepared by metathetical reactions of lithium aluminum hydride or sodium borohydride with the appropriate salts of other metals: *sodium aluminum hydride* [88], *magnesium aluminum hydride* [89], *lithium borohydride* [90], *potassium borohydride* [91], *calcium borohydride* [92] and *tetrabutylammonium borohydride* [93].

[88] $NaAlH_4$ [89] $Mg(AlH_4)_2$ (5)

M.W. 54.00   M.W. 86.33

[90] $LiBH_4$ [91] KBH_4

M.W. 21.78   M.W. 53.94
m.p. 275–280°

[92] $Ca(BH_4)_2$ [93] Bu_4NBH_4

M.W. 69.76   M.W. 257.30
m.p. 103–104°

Lithium aluminum hydride and sodium borohydride react with alcohols and form alkoxyaluminohydrides and alkoxyborohydrides: most widely used are *lithium trimethoxy-* [*94*] and *triethoxyaluminum hydride* [*95*], *lithium diethoxyaluminum hydride* [*95*], *lithium tri(tert-butoxy)aluminum hydride* [*96*], *sodium bis(2-methoxyethoxy)aluminum hydride* (Vitride, Red-Al®) [*97, 98*] and *sodium trimethoxyborohydride* [*99*] (*Procedure 12*, p. 207).

(6)

[*94*] $LiAlH(OMe)_3$
M.W. 128.03

[*95*] $LiAlH_2(OEt)_2$
M.W. 126.06

[*96*] $LiAlH(OCMe_3)_3$
M.W. 254.27
m.p. 319°(dec.)

[*97,98*] $NaAlH_2(OCH_2CH_2OMe)_2$
RED-AL®, VITRIDE®
M.W. 202.16

[*99*] $NaBH(OMe)_3$
M.W. 127.91

Replacement of hydrogen by alkyl groups gives compounds like *lithium triethylborohydride* (Super-Hydride®) [*100*], *lithium tris(sec-butyl)borohydride* [*101*] (L-Selectride®) and *potassium tris(sec-butyl)borohydride* (K-Selectride®) [*102*]. Replacement by a cyano group yields *sodium cyanoborohydride* [*103*], a compound stable even at low pH (down to ~ 3), and *tetrabutylammonium cyanoborohydride* [*93*].

(7)

[*100*] $LiBHEt_3$
SUPERHYDRIDE®
M.W. 105.94

[*101*] $LiBH(CHMeEt)_3$
L-SELECTRIDE®
M.W. 190.11

[*102*] $KBH(CHMeEt)_3$
K-SELECTRIDE®
M.W. 222.27

[*103*] $NaBH_3CN$
M.W. 62.84
m.p. >242°(dec.)

[*93*] Bu_4NBH_3CN
M.W. 282.30

Addition of alane and borane to alkenes affords a host of alkylated alanes and boranes with various reducing properties (and sometimes bizarre names): *diisobutylalane* (Dibal-H®) [*104*], *9-borabicyclo[3.3.1]nonane* (9-BBN) (prepared from borane and 1,5-cyclooctadiene) [*105*], *mono-* [*106, 107*] and *diisopinocampheylborane* (B-di-3-pinanylborane) (both prepared from borane and optically active α-pinene) [*108*], *isopinocampheyl-9-borabicyclo[3.3.1]nonane* alias B-3-pinanyl-9-borabicyclo[3.3.1]nonane (3-pinanyl-9-BBN) (prepared from 9-borabicyclo [3.3.1]nonane and α-pinene) [*109*], NB-Enanthrane® prepared from 9-borabicyclo[3.3.1]nonane and nopol benzyl ether) [*110*] and others.*

Lithium aluminum hydride and alanes are frequently used for the preparation of hydrides of other metals. Diethylmagnesium is converted to magnesium hydride [*111*], trialkylchlorosilanes are transformed to trialkylsilanes

---

* The symbol ® in this and the two previous paragraphs indicates trade names of the Aldrich Chemical Company.

(8)

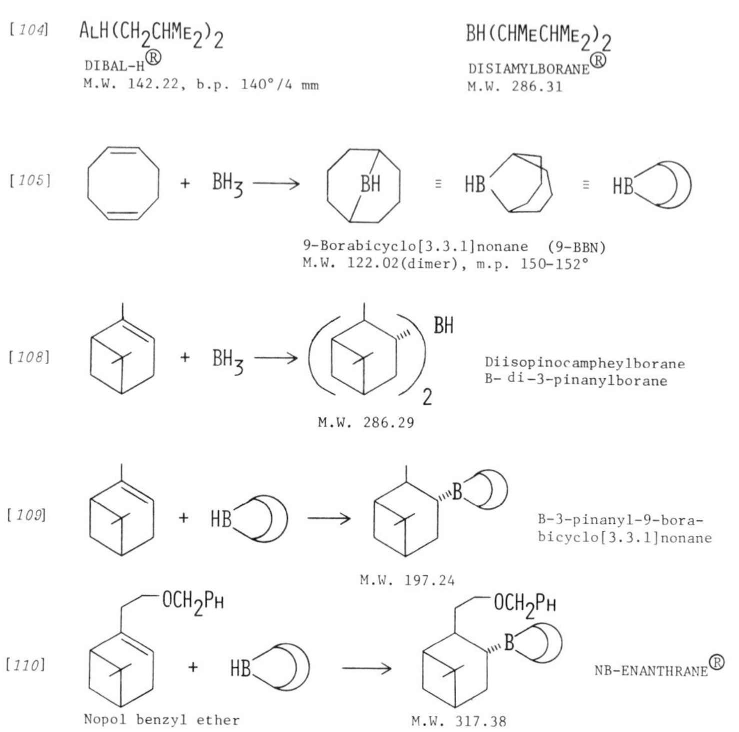

[104] $ALH(CH_2CHME_2)_2$ $BH(CHMECHME_2)_2$
DIBAL-H®   DISIAMYLBORANE®
M.W. 142.22, b.p. 140°/4 mm   M.W. 286.31

[105]

9-Borabicyclo[3.3.1]nonane   (9-BBN)
M.W. 122.02(dimer), m.p. 150-152°

[108]

Diisopinocampheylborane
B- di-3-pinanylborane
M.W. 286.29

[109]

B-3-pinanyl-9-bora-
bicyclo[3.3.1]nonane
M.W. 197.24

[110]

Nopol benzyl ether   M.W. 317.38   NB-ENANTHRANE®

[112], and alkyl and aryltin chlorides to mono-, di- and trisubstituted stannanes of which *dibutyl-* and *diphenylstannanes* and *tributyl-* and *triphenylstan-
nanes* are used most frequently [113, 114]. Their specialty is replacement of
halogens in all types of organic halides.

Reduction of cuprous chloride with sodium borohydride gives *copper hydride* which is a highly selective agent for the preparation of aldehydes from
acyl chlorides [115].

(9)

[112] $4\ ET_3SICL\ +\ LIALH_4 \longrightarrow 4\ ET_3SIH\ +\ LICL\ +\ ALCL_3$
M.W. 116.28
b.p. 109°,d 0.728

[114] $4\ BU_3SNCL\ +\ LIALH_4 \longrightarrow 4\ BU_3SNH\ +\ LICL\ +\ ALCL_3$
M.W. 291.05
b.p. 80°/0.4 mm,d 1.082

[114] $4\ PH_3SNCL\ +\ LIALH_4 \longrightarrow 4\ PH_3SNH\ +\ LICL\ +\ ALCL_3$
M.W. 351.02

[113] $2\ PH_2SNCL_2\ +\ LIALH_4 \longrightarrow 2\ PH_2SNH_2\ +\ LICL\ +\ ALCL_3$
M.W. 274.91

[115] $CUCL\ +\ NABH_4\ +\ 2\ PH_3P \longrightarrow (PH_3P)_2CU \overset{H}{\underset{H}{\diagup}} B \overset{H}{\underset{H}{\diagdown}} +\ NACL$

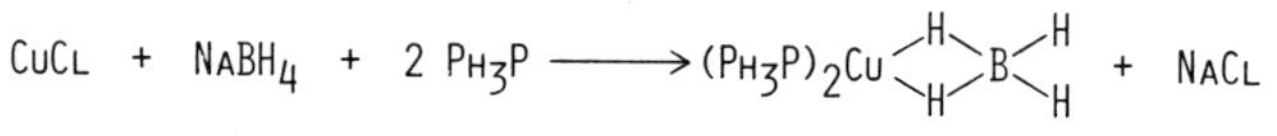

In more recent publications a new, more systematic nomenclature for hydrides and complex hydrides has been adopted. Examples of both nomenclatures are shown below:

Old	*New*
Lithium aluminum hydride	Lithium tetrahydroaluminate
	Lithium tetrahydridoaluminate
Sodium borohydride	Sodium tetrahydroborate
	Sodium tetrahydridoborate
Lithium trialkoxyaluminum hydride	Lithium trialkoxyhydridoaluminate
Sodium bis(2-methoxyethoxy)aluminum hydride	Sodium bis(2-methoxyethoxy)dihydroaluminate
Diisobutylaluminum hydride	Diisobutylalane
Tributyltin hydride	Tributylstannane

Because the majority of publications quoted here utilize the older terminology it is still used predominantly throughout the present book.

Mechanism, Stoichiometry and Stereochemistry of Reductions with Hydrides

The reaction of complex hydrides with carbonyl compounds can be exemplified by the reduction of an aldehyde with *lithium aluminum hydride*. The reduction is assumed to involve a hydride transfer from a nucleophile – tetrahydroaluminate ion – onto the carbonyl carbon as a place of the lowest electron density. The alkoxide ion thus generated complexes the remaining aluminum hydride and forms an alkoxytrihydroaluminate ion. This intermediate reacts with a second molecule of the aldehyde and forms a dialkoxydihydroaluminate ion which reacts with the third molecule of the aldehyde and forms a trialkoxyhydroaluminate ion. Finally the fourth molecule of the aldehyde converts the aluminate to the ultimate stage of tetraalkoxyaluminate ion that on contact with water liberates four molecules of an alcohol, aluminum hydroxide and lithium hydroxide. Four molecules of water are needed to hydrolyze the tetraalkoxyaluminate. The individual intermediates really exist and can also be prepared by a reaction of lithium aluminum hydride

(10)

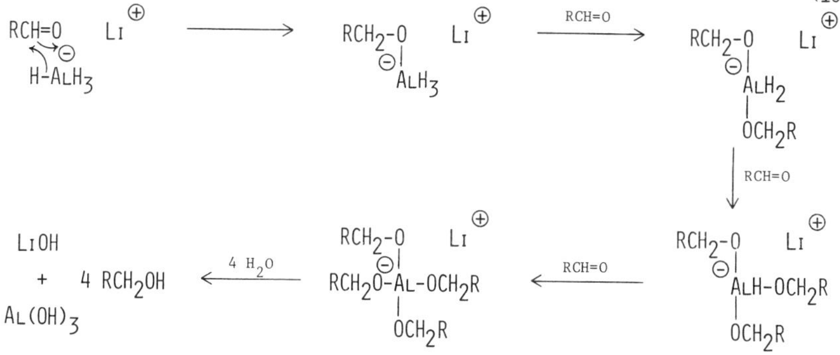

with alcohols. In fact, they themselves are, as long as they contain at least one unreplaced hydrogen atom, reducing agents.

From the equation showing the mechanism it is evident that 1 mol of lithium aluminum hydride can reduce as many as four molecules of a carbonyl compound, aldehyde or ketone. *The stoichiometric equivalent of lithium aluminum hydride* is therefore one fourth of its molecule, i.e. 9.5 g/mol, as much as 2 g or 22.4 liters of hydrogen. Decomposition of 1 mol of lithium aluminum hydride with water generates four molecules of hydrogen, four hydrogens from the hydride and four from water.

The stoichiometry determines the ratios of lithium aluminum hydride to other compounds to be reduced. Esters or tertiary amides treated with one hydride equivalent (one fourth of a molecule) of lithium aluminum hydride are reduced to the stage of aldehydes (or their nitrogen analogs). In order to reduce an ester to the corresponding alcohol, two hydride equivalents, i.e. 0.5 mol of lithium aluminum hydride, is needed since, after the reduction of the carbonyl, hydrogenolysis requires one more hydride equivalent.

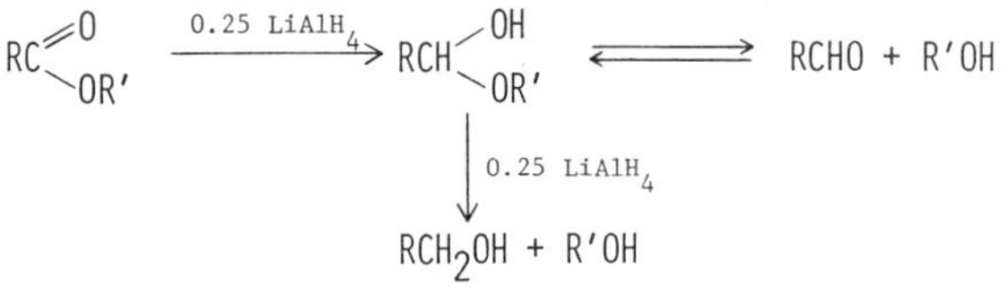

$$
RC\!\!\nearrow^{\!\!O}_{\!\!\searrow OR'} \quad \xrightarrow{\;0.25\ \text{LiAlH}_4\;} \quad RCH\!\!\nearrow^{\!\!OH}_{\!\!\searrow OR'} \quad \rightleftharpoons \quad RCHO + R'OH \tag{11}
$$

$$
\Big\downarrow 0.25\ \text{LiAlH}_4
$$

$$
RCH_2OH + R'OH
$$

Free acids require still an additional hydride equivalent because their acidic hydrogens combine with one hydride ion of lithium aluminum hydride forming acyloxy trihydroaluminate ion. Complete reduction of free carboxylic acids to alcohols requires 0.75 mol of lithium aluminum hydride. The same amount is needed for reduction of monosubstituted amides to secondary amines. Unsubstituted amides require one full mole of lithium aluminum hydride since one half reacts with two acidic hydrogens while the second half achieves the reduction.

$$
4\ RC\!\!\nearrow^{\!\!O}_{\!\!\searrow OH} + \text{LiAlH}_4 \quad \longrightarrow \quad 4\ \Big[RC\!\!\nearrow^{\!\!O}_{\!\!\searrow O}\Big]^{\!\ominus} \ Al^{3\oplus}\ Li^{\oplus} + 4\ H_2 \tag{12}
$$

$$
4\ \Big[RC\!\!\nearrow^{\!\!O}_{\!\!\searrow O}\Big]^{\!\ominus} \ Al^{3\oplus}\ Li^{\oplus} + 2\ \text{LiAlH}_4 \longrightarrow 4\ \Big[RCH_2O\Big]^{\!\ominus} \ Al^{3\oplus}\ Li^{\oplus} + 2\ \text{LiAlO}_2
$$

$$
2\Big\downarrow H_2O
$$

$$
4\ RCH_2OH + \text{LiAlO}_2
$$

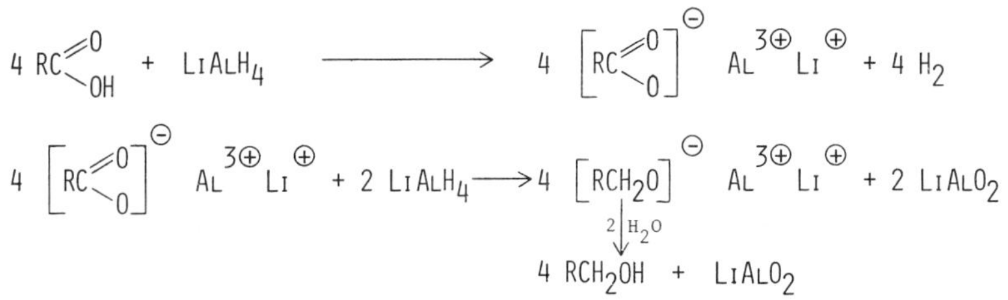

$$
4\ RCO_2H + 3\ \text{LiAlH}_4 + 2\ H_2O = 4\ RCH_2OH + 3\ \text{LiAlO}_2 + 4\ H_2
$$

The stoichiometry of lithium aluminum hydride reductions with other compounds such as nitriles, epoxides, sulfur- and nitrogen-containing compounds is discussed in original papers [*94*, *116*, *117*, *118*] and surveyed in Table 2.

In practice often an excess of lithium aluminum hydride over the stoichiometric amount is used to compensate for the impurities in the assay. In truly precise work, when it is necessary to use an exact amount, either because any overreduction is to be avoided or because kinetic and stoichiometric measurements are to be carried out, it is necessary to determine the contents of the pure hydride in the solution analytically (*Procedure 13*, p. 207).

Theoretical equivalents of various hydrides are listed in Table 2.

Table 2   Stoichiometry of hydrides and complex hydrides*

Starting compound	Product	Stoichiometric hydride equivalent†
R—Cl, Br, I	RH	1
RNO_2	RNH_2	6
$2RNO_2$	RN=NR	8
ROH	RH	2
$R_2\overset{O}{\overset{\diagup\diagdown}{C}}$—$CR'_2$	$R_2C(OH)CHR'_2$	1
RSSR	RSH	2
RSO_2Cl	RSH	6
RCHO, RR'CO	RCH_2OH, RR'CHOH	1
RCO_2H	RCH_2OH	3
ROCl	RCHO	1
	RCH_2OH	2
$(RCO)_2O$	$2\ RCH_2OH$	4
RCO_2R'	RCHO + R'OH	1
	RCH_2OH + R'OH	2
$RCONR_2'$	$RCHO + HNR'_2$	1
	$RCH_2NR'_2$	2
RCONHR'	RCH_2NHR'	3
$RCONH_2$	RCH_2NH_2	4
RCN	RCHO	1
	RCH_2OH	2

*Abstracted from: A. Hajos, *Complex Hydrides and Related Reducing Agents in Organic Synthesis* (listed on p. 100).

† One hydride equivalent is equal to 0.25 equivalent of $LiAlH_4$ or $NaBH_4$, 0.33 equivalent of BH_3 or AlH_3, 0.5 equivalent of $NaAlH_2$ (OCH_2CH_2OMe), and 1 equivalent of $LiAlH(OR)_3$ or R_3SnH.

A similar mechanism and stoichiometry underlie reactions of organic compounds with *lithium* and *sodium borohydrides*. With modified complex hydrides the stoichiometry depends on the number of hydrogen atoms present in the molecule.

The mechanism of reduction by *boranes* and *alanes* differs somewhat from that of complex hydrides. The main difference is in the entirely different chemical nature of the two types. Whereas complex hydride anions are strong nucleophiles which attack the places of lowest electron density, boranes and alanes are electrophiles and combine with that part of the organic molecule which has a free electron pair [*119*]. By a hydride transfer alkoxyboranes or alkoxyalanes are formed until all three hydrogen atoms from the boron or aluminum have been transferred and the borane or alane converted to alkoxyboranes (trialkylborates) or alkoxyalanes.

$$\text{RR'C=O} \xrightarrow{\text{BH}_3 \cdot \text{THF}} \underset{\ominus}{\overset{\oplus}{\text{RR'C=O-B}}}\text{H} \longrightarrow \text{RR'CHO-B-H} \tag{13}$$

$$\downarrow \text{RR'CO}$$

$$\underset{\text{OCHRR'}}{\overset{\text{OCHRR'}}{\text{RR'CHO-B}}} \xleftarrow{\text{RR'CO}} \underset{\text{OCHRR'}}{\overset{\text{H}}{\text{RR'CHO-B}}} \xleftarrow{} \underset{\text{O=CRR'}}{\overset{\text{H}}{\text{RR'CHO-B-H}}}$$

$$\downarrow \text{H}_2\text{O}$$

$$3 \text{ RR'CHOH} + \text{B(OH)}_3$$

Hydrides and complex hydrides tend to approach the molecule of a compound to be reduced from the less hindered side (*steric approach control*). If a relatively uninhibited function is reduced the final stereochemistry is determined by the stability of the product (*product development control*). In addition, torsional strain in the transition state affects the steric outcomes of the reduction.

In many cases also the reduction agent itself influences the result of the reduction, especially if it is bulky and the environment of the function to be reduced is crowded. A more detailed discussion of stereochemistry of reduction with hydrides is found in the section on ketones (p. 114).

Handling of Hydrides and Complex Hydrides

Many of the hydrides and complex hydrides are now commercially available. Some of them require certain precautions in handling. Lithium aluminum hydride is usually packed in polyethylene bags enclosed in metal cans. It should not be stored in ground glass bottles.

Opening of a bottle where some particles of lithium aluminum hydride were squeezed between the neck and the stopper caused a fire [68]. Lithium aluminum hydride must not be crushed in a porcelain mortar with a pestle. Fire and even explosion may result from contact of lithium aluminum hydride with small amounts of water or moisture. Sodium bis(2-methoxy-ethoxy)aluminum hydride (Vitride, Red-Al®) delivered in benzene or toluene solutions also may ignite in contact with water. Borane (diborane) ignites in contact with air and is therefore kept in solutions in tetrahydrofuran or in complexes with amines and sulfides. Powdered lithium borohydride may ignite in moist air. Sodium borohydride and sodium cyanoborohydride, on the other hand, are considered safe.*

Since some of the hydrides and complex hydrides are moisture-sensitive,

*Note added in proof. A severe explosion resulting in serious injury occurred when a screw-cap glass bottle containing some 25 g of sodium borohydride was opened (le Noble, W. J., *Chem. & Eng. News*, 1983, **61** (19), 2).

hypodermic syringe technique and work under inert atmosphere are advisable and sometimes necessary. Because of high toxicity of some of the hydrides (borane) and complex hydrides (sodium cyanoborohydride), and for other safety reasons, all work with these compounds should be done in hoods. Specifics about handling individual hydrides and complex hydrides are to be found in the particular papers or reviews [118] (pp. 223, 257, 258).

Reductions with hydrides and complex hydrides are usually carried out by mixing solutions. Only sodium borohydride and some others are sometimes added portionwise as solids. Since some of the complex hydrides such as lithium aluminum hydride are not always completely pure and soluble without residues, it is of advantage to place the solutions of the hydrides in the reaction flask and add the reactants or their solutions from separatory funnels or by means of hypodermic syringes.

If, however, an excess of the reducing agent is to be avoided and a solution of the hydride is to be added to the compound to be reduced (by means of the so-called 'inverse technique') special care must be taken to prevent clogging of the stopcock of a separatory funnel with undissolved particles.

When sparingly soluble compounds have to be reduced they may be transported into the reaction flask for reduction by extraction in a Soxhlet apparatus surmounting the flask [83, 120].

In many reactions, especially with compounds which are only partly dissolved in the solvent used, efficient stirring is essential. A paddle-type 'trubore' stirrer is preferable to magnetic stirring, particularly when larger amounts of reactants are handled.

Solubilities of the most frequently used hydrides and complex hydrides in most often used solvents are listed in Table 3. In choosing the solvent it is necessary to consider not only the solubility of the reactants but also the boiling points in case the reduction requires heating.

Isolation of the product varies from one reducing agent and one reduced compound to another. The general feature is the decomposition of the re-

Table 3 Solubilities of selected complex hydrides in common solvents (in g per 100 g of solvent) at $25 \pm 5°$*

Solvent	H_2O	MeOH	EtOH	Et_2O	THF	Dioxane	$(CH_2OMe)_2$	Diglyme
Hydride								
LiAlH$_4$				35–40	13	0.1	7 (12)†	5 (8)†
LiBH$_4$		~2.5 (dec.)	~2.5 (dec.)	2.5	28	0.06		
NaBH$_4$	55 (88.5)‡	16.4 (dec.)	20 (dec.)		0.1		0.8	5.5
NaBH$_3$CN	212	very sol.	slightly sol.	0	37.2			17.6
NaBH(OMe)$_3$						1.6 (4.5)†		
LiAlH(OCMe$_3$)$_3$				2	36		4	41

* Abstracted from: A. Hajos, *Complex Hydrides and Related Reducing Agents in Organic Synthesis* (listed on p. 100).

† At 75°.

‡ At 60°.

action mixture after the reaction with ice or water followed by acid or alkali.

Reaction of the products of reduction with *lithium aluminum hydride* with water is very exothermic. This is especially true of reductions in which an excess of lithium aluminum hydride has been used. In such cases it is advisable to decompose the unreacted hydride by addition of ethyl acetate (provided its reduction product – ethanol – does not interfere with the isolation of the products). Then normal decomposition with water is carried out followed by acids [83] or bases [121].

If the reduction has been carried out in ether, the ether layer is separated after the acidification with dilute hydrochloric or sulfuric acid. Sometimes, especially when not very pure lithium aluminum hydride has been used, a gray voluminous emulsion is formed between the organic and aqueous layers. Suction filtration of this emulsion over a fairly large Buchner funnel is often helpful. In other instances, especially in the reductions of amides and nitriles when amines are the products, decomposition with alkalis is in order. With certain amounts of sodium hydroxide of proper concentration a granular by-product – sodium aluminate – may be separated without problems [121].

Isolation of products from the reductions with *sodium borohydride* is in the majority of cases much simpler. Since the reaction is carried out in aqueous or aqueous-alcoholic solution, extraction with ether is usually sufficient. Acidification of the reaction mixture with dilute mineral acids may precede the extraction.

Decomposition of the reaction mixtures with water followed by dilute acids applies also to the reductions with *boranes and alanes*. Modifications are occasionally needed, for example hydrolysis of esters of boric acid and the alcohols formed in the reduction. Heating of the mixture with dilute mineral acid or dilute alkali is sometimes necessary.

The domain of hydrides and complex hydrides is reduction of carbonyl functions (in aldehydes, ketones, acids and acid derivatives). With the exception of boranes, which add across carbon–carbon multiple bonds and afford, after hydrolysis, hydrogenated products, isolated carbon–carbon double bonds resist reduction with hydrides and complex hydrides. However, a conjugated double bond may be reduced by some hydrides, as well as a triple bond to the double bond (p. 44). Reductions of other functions vary with the hydride reagents. Examples of applications of hydrides are shown in *Procedures 14–24* (pp. 207–210).

ELECTROREDUCTION AND REDUCTIONS WITH METALS

Dissolving metal reductions were among the first reductions of organic compounds discovered some 130 years ago. Although overshadowed by more universal catalytic hydrogenation and metal hydride reductions, metals are still used for reductions of polar compounds and selective reductions of specific types of bonds and functions. Almost the same results are obtained by electrolytic reduction.

Mechanism and Stereochemistry

Reduction is defined as acceptance of electrons. Electrons can be supplied by an electrode – cathode – or else by dissolving metals. If a metal goes into solution it forms a cation and gives away electrons. A compound to be reduced, e.g. a ketone, accepts one electron and changes to a radical anion A. Such a radical anion may exist when stabilized by resonance, as in sodium-naphthalene complexes with some ethers [122]. In the absence of protons the radical anion may accept another electron and form a dianion B. Such a process is not easy since it requires an encounter of two negative species, an electron and a radical anion, and the two negative sites are close together. It takes place only with compounds which can stabilize the radical anion and the dianion by resonance.

Rather than accepting another electron, the radical anion A may combine with another radical anion and form a dianion of a dimeric nature C. This intermediate too, is formed more readily in the aromatic series.

In the presence of protons, the initial radical anion A is protonated to a radical D which has two options: either to couple with another radical to form a pinacol E, or to accept another electron to form an alcohol F. The pinacol E and the alcohol F may also result from double protonation of the doubly charged intermediates C and B, respectively.

(14)

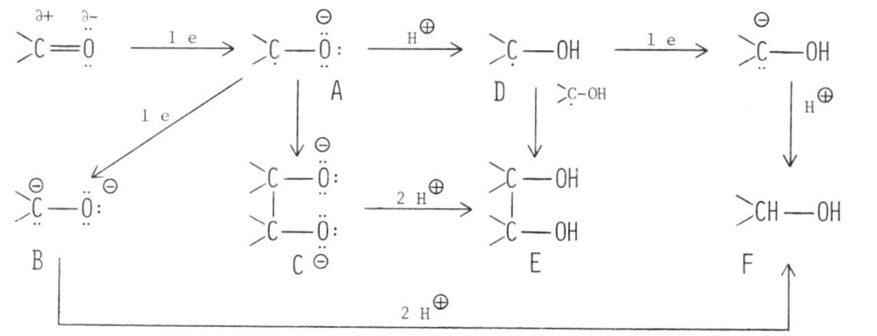

A similar process may take place in the reduction of polar compounds with single bonds. A halogen, hydroxy, sulfhydryl or amino derivative by accepting an electron dissociates into a radical and an anion. In aprotic solvents the two radicals combine. In the case of halogen derivatives the result is Wurtz synthesis. In the presence of protons the anion is protonated and the radical accepts another electron to form an anion that after protonation gives a hydrocarbon or a product in which the substituent has been replaced by hydrogen.

The reductions just described are quite frequent with compounds containing polar multiple bonds such as carbon–oxygen, carbon–nitrogen, nitrogen–oxygen, etc. [123]. Even carbon–carbon bond systems can be reduced in this way, but only if the double bonds are conjugated with a polar group or at least with another double bond or aromatic ring. Dissolving metal reduction

$$R - X \xrightarrow{1\ e} R\cdot + X^{\ominus} \tag{15}$$

$$R - R \xleftarrow{R\cdot} \quad \xrightarrow{1\ e} R^{\ominus}: \xrightarrow{H^{\oplus}} RH$$

of an isolated carbon–carbon double bond is extremely rare and has no practical significance. On the other hand, reduction of conjugated dienes and of aromatic rings is feasible and becomes easier with increasing ability of resonance of the charged intermediates (anthracene and phenanthrene are reduced much more readily than naphthalene and that in turn more readily than benzene). Reductions are especially successful when the source of protons – compounds with acidic hydrogens, alcohols, water, acids – is present in the reducing medium. In such cases dimerization reactions are suppressed and good yields of fully reduced products are obtained (Birch reduction with alkali metals, reduction with zinc and other metals).

Because of the stepwise nature of the reductions by acceptance of electrons the net result of reductions with metals is usually *anti* or *trans* addition of hydrogen. Thus disubstituted acetylenes give predominantly *trans*-alkenes, *cis*-alkenes give racemic mixtures and *threo* forms, and *trans*-alkenes *meso* or *erythro* forms. Reduction of 4-cyclohexanonecarboxylic acid with sodium amalgam gave *trans* 4-cyclohexanolcarboxylic acid (with both substituents equatorial) [124], and reduction of 2-isopropylcyclohexanone with sodium in moist ether gave 90-95% of *trans*-2-isopropylcyclohexanol [125]. In other cases conformational effects play an important role [126] (p. 45).

Electrolytic Reduction

The most important factor in electrolytic reduction (electroreduction) is the nature of the metal used as a *cathode*. Metals of low overvoltage – platinum (0.005–0.09 V), palladium, nickel and iron – give generally similar results of reduction as does catalytic hydrogenation [127]. Cathodes made of metals of high overvoltage such as copper (0.23 V), cadmium (0.48 V), lead (0.64 V), zinc (0.70 V) or mercury (0.78 V) produce similar results to those of dissolving metal reductions.

Another important factor in electroreduction is the *electrolyte*. Most electrolytic reductions are carried out in more or less dilute sulfuric acid but some are done in alkaline electrolytes such alkali hydroxides, alkoxides or solutions of salts like tetramethylammonium chloride in methanol [128] or lithium chloride in alkyl amines [129, 130].

Different results may also be obtained depending on whether the electrolytic cell is divided (by a diaphragm separating the cathode and anode spaces) or undivided [129].

The apparatus for electrolytic reductions (Fig. 5) may be an open vessel for work with non-volatile compounds, or a closed container fitted with a reflux condenser for work with compounds of high vapor pressure. Since the elec-

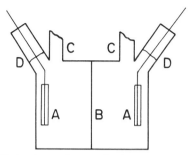

Fig. 5 Electrolytic cell. A, electrodes; B, diaphragm; C, reflux condensers; D, glass seals.

troreduction in exothermic, cooling must sometimes be applied [*131, 132, 133*].

The size of electrodes and the current passed through the electrolytic cell determine the current density, which in most reductions is in the range of 0.05–0.2 A/cm². The current in turn is determined by the imposed voltage and conductivity of the electrolyte. The amount of electricity used for electroreduction of one gram equivalent of a compound is 96,500 coulombs (26.8 ampere-hours). In practice about twice the amount is used (current yield of 50%) (*Procedure 25*, p. 210).

Reductions with Alkali Metals

The reducing powers of metals parallel their relative electrode potentials, i.e. potentials developed when the metal is in contact with a normal solution of its salts. The potential of hydrogen being equal to 0, the potentials or electrochemical series of some elements are as shown in Table 4.

Table 4 **Physical constants and relative electrode potentials of some metals**

	Metal						
	Li	K	Na	Al	Zn	Fe	Sn
Relative electrode potential (volts)	−2.9	−2.9	−2.7	−1.34	−0.76	−0.44	−0.14
Atomic weight	6.94	39.1	22.99	26.98	65.37	55.85	118.69
Melting point (°C)	179	63.7	97.8	660	419.4	1535	231.9
Density	0.53	0.86	0.97	2.70	7.14	7.86	7.28

Metals with large negative potentials, like alkali metals and calcium, are able to reduce almost anything, even carbon-carbon double bonds. Metals with low potentials (iron, tin) can reduce only strongly polar bonds (such as nitro groups but, generally, not carbonyl groups).

Sodium, lithium and potassium dissolve in liquid ammonia (m.p. −77.7°, b.p. $-33.4°$). They form blue solutions which are stable unless traces of metals such as iron or its compounds are present that catalyze the reaction between the metal and ammonia to form alkali amide and hydrogen. Solubilities of lithium, sodium and potassium in liquid ammonia at its boiling point are (in g per 100 g) 10.4, 24.5, and 47.8, which correspond to mole ratios of 0.25, 0.18 and 0.21, respectively [*134*]. In practice usually more dilute solutions are used. The majority of dissolving metal reductions are carried out in the presence of proton donors, most frequently methanol, ethanol and *tert*-butyl alcohol (*Birch reduction*). The function of these donors is to protonate the intermediate anion radicals and thus to cut down on undesirable side reactions such as dimerization of the radical anion and polymerization. Other proton sources such as *N*-ethylaniline, ammonium chloride and others are used less often. Cosolvents such as tetrahydrofuran help increase mutual miscibility of the reaction components.

The reductions are usually carried out in three-necked flasks fitted with efficient cold finger reflux condensers filled with dry-ice and acetone, mechanical or magnetic stirrers, and inlets for ammonia and the metal. Ammonia may be introduced as a liquid from the bottom of an ammonia cylinder, or as a gas which condenses in the apparatus. In the latter case it is of advantage to cool the reaction flasks with dry-ice/acetone baths to accelerate the condensation. When ammonia without traces of water is necessary, it can be first introduced into a flask containing pieces of sodium and then evaporated into the proper reaction flask where it condenses. The opening for the introduction of ammonia is also used for adding into the flask the pieces of metal. After the reduction has been completed ammonia is allowed to evaporate, following the removal of the reflux condenser. If unreacted metal is present it is decomposed by addition of ammonium chloride or, better still, finely powdered sodium benzoate. Evaporation of ammonia is slow. It is best done overnight. It can be sped up by gentle heating of the reaction flask. The final work-up depends on the chemical properties of the products. A similar technique can be used for reductions with calcium (*Procedure 26*, p. 211).

Reductions in liquid ammonia run at atmospheric pressure at a temperature of $-33°$. If higher temperatures are necessary for the reduction, other solvents of alkali metals are used: methylamine (b.p. $-6.3°$), ethylamine (b.p. $16.6°$), and ethylenediamine (b.p. $116-117°$).

Many reductions with sodium are carried out in boiling alcohols: in methanol (b.p. $64°$), ethanol (b.p. $78°$), butanol (b.p. $117-118°$), and isoamyl alcohol (b.p. $132°$). More intensive reductions are achieved at higher temperatures. For example reduction of naphthalene with sodium in ethanol gives 1,4-dihydronaphthalene whereas in boiling isoamyl alcohol tetralin is formed.

Reductions carried out by adding sodium to a compound in boiling alcohols require large excesses of sodium and alcohols. A better procedure is to carry out such reductions by adding a stoichiometric quantity of an alcohol and the compound in toluene or xylene to a mixture of toluene or xylene and a calculated amount of a dispersion of molten sodium [*122*, *135*] (p. 152). Reductions with sodium can also be accomplished in moist ether [*125*] and other solvents (*Procedures 27, 28*, pp. 211, 212).

A form of sodium suitable for reductions in aqueous media is *sodium amalgam*. It is easily prepared in concentrations of 2–4% by dissolving sodium in mercury (*Procedure 29*, p. 212).

Reductions with sodium amalgam are very simple. The amalgam is added in small pieces to an aqueous or aqueous alcoholic solution of the compound to be reduced contained in a heavy-walled bottle. The contents of the stoppered bottle are shaken energetically until the solid pieces of the sodium amalgam change to liquid mercury. After all the sodium amalgam has reacted the mercury is separated and the aqueous phase is worked up. Sodium amalgam generates sodium hydroxide. If the alkalinity of the medium is undesirable, an acid or a buffer such as boric acid [*136*] must be added to keep the pH of the solution in the desirable range.

Reductions with sodium amalgam are fairly mild. Only easily reducible groups and conjugated double bonds are affected. With the availability of sodium borohydride the use of sodium amalgam is dwindling even in the field of saccharides, where sodium amalgam has been widely used for reduction of aldonic acids to aldoses.

Reductions with Magnesium

Applications of magnesium are rather limited. Halogen derivatives, especially bromides, iodides and imidoyl chlorides, react with magnesium to form Grignard reagents that on decomposition with water or dilute acids give compounds in which the halogen has been replaced by hydrogen [*137*, *138*]. The reduction can also be carried out by the simultaneous action of isopropyl alcohol [*139*] and magnesium activated by iodine. Another specialty of magnesium is reduction of ketones to pinacols carried out by magnesium amalgam. The amalgam is prepared *in situ* by adding a solution of mercuric chloride in the ketone to magnesium turnings submerged in benzene [*140*].

Reductions with Aluminum

Reductions with aluminum are carried out almost exclusively with aluminum amalgam. This is prepared by immersing strips of a thin aluminum foil in a 2% aqueous solution of mercuric chloride for 15–60 seconds, decanting the solution, rinsing the strips with absolute ethanol, then with ether, and cutting them with scissors into pieces of approximately 1 cm² [*141*, *142*]. In aqueous and non-polar solvents aluminum amalgam reduces cumulative double bonds [*143*], ketones to pinacols [*144*], halogen compounds [*145*], nitro compounds [*146*, *147*], azo compounds [*148*], azides [*149*], oximes [*150*] and quinones [*151*], and cleaves sulfones [*141*, *152*, *153*] and phenylhydrazones [*154*] (*Procedure 30*, p. 212).

Several reductions have been achieved by dissolving a *nickel–aluminum alloy* containing usually 50% of aluminum in 20–50% aqueous sodium hydroxide in the presence of a reducible compound. Since under such conditions elemental hydrogen is generated by dissolution of aluminum, and since Raney nickel is formed in this process, such reductions have to be considered catalytic hydrogenations rather than dissolving metal reductions. Their outcome certainly points to the former type.

Reductions with Zinc

Zinc has been, and still is, after sodium the most abundantly used metal reductant. It is available mainly in form of zinc dust and in a granular form referred to as mossy zinc.

Zinc dust is frequently covered with a thin layer of zinc oxide which deactivates its surface and causes induction periods in reactions with compounds. This disadvantage can be removed by a proper activation of zinc dust immediately prior to use. Such an activation can be achieved by a 3–4-minute contact with very dilute (0.5–2%) hydrochloric acid followed by washing with water, ethanol, acetone and ether [155]. Similar activation is carried out *in situ* by a small amount of anhydrous zinc chloride [156] or zinc bromide [157] in alcohol, ether or tetrahydrofuran. Another way of activating zinc dust is by its conversion to a zinc–copper couple by stirring it (180 g) with a solution of 1 g of copper sulfate pentahydrate in 35 ml of water [158].

Mossy zinc is activated by conversion to zinc amalgam by brief immersion in a dilute aqueous solution of mercuric chloride and decantation of the solution before the reaction proper (40 g of mossy zinc, 4 g of mercuric chloride, 4 ml of concentrated hydrochloric acid and 40 ml of water [159]). This type of activation is especially used in the Clemmensen reduction which converts carbonyl groups to methylene groups [160] (*Procedure 31*, p. 213).

In the Clemmensen reduction (which does not proceed through a stage of an alcohol) the indispensable strong acid protonates the oxygen and ultimately causes its elimination as water after a transient formation of an unstable organozinc species. It is not impossible that a carbene-like intermediate is formed which may give (and indeed gives) an alkene [161]. When acetophenone is reduced with zinc in deuterium oxide it gives α,α-dideuteroethylbenzene and styrene. If the reduction is carried out with ω,ω,ω-trideuteroacetophenone, α,β,β-trideuterostyrene is obtained [162].

Reductions with zinc are carried out in aqueous [160] as well as anhydrous solvents [163] and at different pHs of the medium. The choice of the reaction conditions is very important since entirely different results may be obtained under different conditions. While reduction of aromatic nitro groups in alkali hydroxides or aqueous ammonia gives hydrazo compounds, reduction in aqueous ammonium chloride gives hydroxylamines, and reduction in acidic medium amines (p. 73). Of organic solvents the most efficient seem to be dimethyl formamide [164] and acetic anhydride [155]. However, alcohols have been used much more frequently. The majority of reductions with zinc are carried out in acids: hydrochloric, sulfuric, formic and especially acetic. Strongly acidic medium is essential for some reductions such as Clemmensen reduction (*Procedures 31-33*, p. 213).

Reductions with zinc are usually performed in hot or refluxing solvents, and some take many hours to complete. After the reaction is over the residual zinc, if any, is filtered off, and the products are isolated in ways depending on their chemical nature. If zinc dust has been used care must be taken in disposing of it since it may be pyrophoric in the dry state.

Zinc is used to a limited extent for reductions of double bonds conjugated with strongly polar groups, for semireduction of alkynes to (*trans*) alkenes and partial reduction of some aromatics. It is especially suited for reductions of carbonyl groups in aldehydes and ketones to methyl or methylene groups, respectively. It is also able to reduce carbonyl compounds to alcohols and pinacols, is excellent for replacement of halogens by hydrogen, and is, with the exception of catalytic hydrogenation, the most versatile reducing agent for aromatic nitro compounds and other nitrogen derivatives.

Reductions with Iron

Iron was one of the first metals employed for the reduction of organic compounds over 130 years ago. It is used in the form of filings. Best results are obtained with 80 mesh grain [*165*]. Although some reductions are carried out in dilute or concentrated acetic acid the majority are performed in water in the presence of small amounts of hydrochloric acid, acetic acid or salts such as ferric chloride, sodium chloride (as little as 1.5-3%) [*165*], ferrous sulfate [*166*] and others. Under these conditions iron is converted to iron oxide, Fe_3O_4. Methanol or ethanol are used to increase the solubility of the organic material in the aqueous medium [*166*] (*Procedure 34*, p. 213).

(16)

$$4 \ RNO_2 \ + \ 9 \ Fe \ + \ 4 \ H_2O \ \longrightarrow \ 4 \ RNH_2 \ + \ Fe_3O_4$$

Iron is occasionally used for the reduction of aldehydes [*167*], replacement of halogens [*168, 169*], deoxygenation of amine oxides [*170*] and desulfurizations [*171, 172*] but its principal domain is the reduction of aromatic nitro compounds (p. 77). The advantage over other reducing agents lies in the possibility of using almost neutral or only a slightly acidic medium and in the selectivity of iron which reduces few functions beside nitro groups.

Reductions with Tin

Once a favorite reducing agent for the conversion of aromatic nitro compounds to amines, tin is used nowadays only exceptionally. The reason is partly unavailability, the necessity for the use of strongly acidic media, and the fact that most reductions achieved by tin can also be accomplished by stannous chloride, which is soluble in aqueous acids and some organic solvents. Reductions with tin require heating with hydrochloric acid, sometimes in the presence of acetic acid or ethanol.

Tin was used for reduction of acyloins to ketones [173], of enediones to diketones [174], of quinones [175] and especially of aromatic nitro compounds [176] (*Procedure 35*, p. 214).

Reductions with Metal Compounds

Compared with tin *stannous chloride* is only one half as efficient since it furnishes only two electrons per mole (tin donates four electrons). This disadvantage is more than offset by the solubility of stannous chloride in water, hydrochloric acid, acetic acid, alcohols and other organic solvents. Thus reductions with stannous chloride can be carried out in homogeneous media (*Procedure 36*, p. 214).

Stannous chloride is used most frequently for the reduction of nitro compounds [177, 178, 179] and of quinones [180, 181]. It is also suitable for conversion of imidoyl chlorides [182] and of nitriles [183] to aldehydes, for transformations of diazonium salts to hydrazines [184], for reduction of oximes [185], and for deoxygenation of sulfoxides to sulfides [186].

Divalent chromium salts show very strong reducing properties. They are prepared by reduction of chromium(III) compounds with zinc [187] or a zinc-copper couple and form dark blue solutions extremely sensitive to air. Most frequently used salts are chromous chloride [188], chromous sulfate [189], and less often chromous acetate. Reductions of organic compounds are carried out in homogeneous solutions in aqueous methanol [190], acetone [191], acetic acid [192], dimethylformamide [193] or tetrahydrofuran [194] (*Procedure 37*, p. 214).

Divalent chromium reduces triple bonds to double bonds (*trans* where applicable) [195], enediones to diones [196], epoxides to alkenes [192] and aromatic nitroso, nitro and azoxy compounds to amines [190], deoxygenates amine oxides [191], and replaces halogens by hydrogen [197, 198].

Titanous chloride (titanium trichloride) is applied in aqueous solutions, sometimes in the presence of solvents increasing the miscibility of organic compounds with the aqueous phase [199, 200]. Its applications are reduction of nitro compounds [201] and cleavage of nitrogen–nitrogen bonds [202] but it is also an excellent reagent for deoxygenation of sulfoxides [203] and amine oxides [199] (*Procedure 38*, p. 214).

Solutions of *low-valence titanium chloride* (*titanium dichloride*) are prepared *in situ* by reduction of solutions of titanium trichloride in tetrahydrofuran or 1,2-dimethoxyethane with lithium aluminum hydride [204, 205], with lithium or potassium [206], with magnesium [207, 208] or with a zinc–copper couple [209, 210]. Such solutions effect hydrogenolysis of halogens [208], deoxygenation of epoxides [204] and reduction of aldehydes and ketones to alkenes [205,

207, 209] (p. 112). Applications of titanium chlorides in reductions have been reviewed [*211*] (*Procedure 39*, p. 215).

Cerium(III) iodide prepared *in situ* from an aqueous solution of ceric sulfate first by reduction with sulfur dioxide followed by sodium iodide reduced α-bromoketones to ketones in 82–96% yields [*212*].

Aqueous solutions of *vanadous chloride* (vanadium dichloride) are prepared by reduction of vanadium pentoxide with amalgamated zinc in hydrochloric acid [*213*]. Reductions are carried out in solution in tetrahydrofuran at room temperature or under reflux. Vanadium dichloride reduces α-halo ketones to ketones [*214*], α-diketones to acyloins [*215*], quinones to hydroquinones [*215*], sulfoxides to sulfides [*216*] and azides to amines [*217*] (*Procedure 40*, p. 215).

Molybdenum trichloride is generated *in situ* from molybdenum pentachloride and zinc dust in aqueous tetrahydrofuran. It is used for conversion of sulfoxides to sulfides [*216*].

Ferrous sulfate is a gentle reducing agent. It is applied in aqueous solutions mainly for reduction of aromatic nitro compounds containing other reducible functions such as aldehyde groups which remain intact [*218*].

Sodium stannite is one of the reagents suitable for replacement of a diazonium group in aromatic compounds by hydrogen [*219*].

A special use of *sodium arsenite* (applied in aqueous alkaline solutions) is partial reduction of trigeminal halides to geminal halides [*220*] and reduction of aromatic nitro compounds to azoxy compounds [*221*].

REDUCTIONS WITH NON-METAL COMPOUNDS

Reductions with Hydrogen Iodide

Hydrogen iodide (M.W. 127.93, m.p. −55°, b.p. −35.4°, density 5.56), one of the oldest reducing agents, is available as a gas but is used mainly in the form of its aqueous solution, hydriodic acid. Azeotropic hydriodic acid boils at 127°, has a density of 1.70, and contains 57% of hydrogen iodide. So-called fuming hydriodic acid is saturated at 0° and has a density of 1.93–1.99.

Hydrogen iodide dissociates at higher temperatures to iodine and hydrogen, which effects hydrogenations. The reaction is reversible. Its equilibrium is shifted in favor of the decomposition by the reaction of hydrogen with organic compounds to be reduced but it can also be affected by removal of iodine; this can be accomplished by allowing iodine to react with phosphorus to form phosphorus triiodide which decomposes in the presence of water to phosphorous acid and hydrogen iodide. In this way, by adding phosphorus to the reaction mixture hydrogen iodide is recycled and the reducing efficiency of hydriodic acid is enhanced [*222*] (*Procedure 41*, p. 215).

Reductions with hydriodic acid are usually accomplished by refluxing organic compounds with the azeotropic (57%) acid. Acetic acid is added to increase miscibility of the acid with the organic compound. If more energetic conditions are needed the heating has to be done in sealed tubes. This technique is necessary when fuming hydriodic acid is employed. At present very few such reactions are carried out.

Milder reductions with hydriodic acid can be accomplished by using more dilute hydriodic acid, or solutions of hydrogen iodide prepared from alkaline iodides and hydrochloric or acetic acid in organic solvents [223].

Selective reduction of α-chloro and α-bromo ketones can be achieved by treatment with sodium iodide and chlorotrimethylsilane in acetonitrile at room temperature [224].

Hydriodic acid is a reagent of choice for reduction of alcohols [225], some phenols [226], some ketones [227, 228], quinones [222], halogen derivatives [223, 229], sulfonyl chlorides [230], diazo ketones [231], azides [232], and even some carbon–carbon double bonds [233]. Under very drastic conditions at high temperatures even polynuclear aromatics and carboxylic acids can be reduced to saturated hydrocarbons but such reactions are hardly ever used nowadays.

In very reactive halogen derivatives such as α-bromo ketones [234] and α-bromothiophene [235] the halogens are replaced by hydrogen with *hydrogen bromide* in acetic acid provided phenol is added to react with the evolved bromine and to affect favorably the equilibrium of the reversible reaction. Hydrogen bromide in acetic acid also reduces azides to amines [232].

Reductions with Sulfur Compounds

Hydrogen sulfide is probably the oldest reducing agent applied in organic chemistry, having been used by Wöhler in 1838. It is a very toxic gas with a foul odor (m.p. −82.9°, b.p. −60.1°, density 1.19) which is sparingly soluble in water (0.5% at 10°, 0.4% at 20°, 0.3% at 30°) and more soluble in anhydrous ethanol (1 g in 94.3 ml at 20°) and ether (1 g in 48.5 ml at 20°). It is usually applied in basic solutions in pyridine [236], in piperidine [237] and most frequently in aqueous ammonia [238, 239, 240] where it acts as *ammonium sulfide or hydrosulfide* (*Procedure 42*, p. 216).

Similar reducing effects are obtained from *alkali sulfides, hydrosulfides* and *polysulfides* [241]. A peculiar reaction believed to be due to sodium polysulfide formed *in situ* by refluxing sulfur in aqueous-ethanolic sodium hydroxide is a conversion of *p*-nitrotoluene to *p*-aminobenzaldehyde [242]. Oxidation of the methyl group by the polysulfide generates hydrogen sulfide which then reduces the nitro group to the amino group.

The reducing properties of organic compounds of sulfur, such as *methyl mercaptan*, show up in partial reduction of trigeminal to geminal dihalides [243]. *Dimethyl sulfide* reduces hydroperoxides to alcohols and ozonides to aldehydes while being converted to dimethyl sulfoxide [244].

The main field of applications of hydrogen and alkali sulfides* is reduction of nitrogen functions in nitro compounds [236, 240], nitroso compounds

---

* Reduction with hydrogen sulfide and its salts is sometimes referred to as the Zinin reduction (Org. Reactions 1973, 20, 455).

[*245*], azo compounds [*241, 246*] and azides [*247*] (all are reduced to amines). Less frequent are reductions of carbonyl groups [*237, 248*] and sulfonyl chlorides (to sulfinic acids) [*249*].

The use of *sulfur dioxide* as a reductant is very limited, mainly to the reduction of quinones to hydroquinones [*250*]. Sulfur dioxide is a gas of pungent odor, soluble in water (17.7% at 0°, 11.9% at 15°, 8.5% at 25°, 6.4% at 35°), in ethanol (25%), in methanol (32%), in ether and in chloroform.

More abundant are reductions with *sodium sulfite* which is applied in aqueous solutions (solubility 24%). Its specialties are reduction of peroxides to alcohols [*251*], of sulfonyl chlorides to sulfinic acids [*252*], of aromatic diazonium compounds to hydrazines [*253*], and partial reduction of geminal polyhalides [*254*] (*Procedure 43*, p. 216).

A very strongly reducing agent is *sodium hydrosulfite* (*hyposulfite, dithionite*) $Na_2S_2O_4$ (M.W. 174.13). Like sulfites it is applied in aqueous solutions. It is widely used in the dyestuff industry for reducing vat dyes. In the laboratory the main applications are reductions of nitroso compounds [*255*], nitro compounds [*256, 257*], azo compounds [*258*] and azides to amines [*259*], and conversion of aromatic arsonic and stibonic acids to arseno compounds [*260*] and stibino compounds [*261*], respectively. Exceptionally, even carbonyl groups have been reduced by boiling solutions of sodium hydrosulfite in aqueous dioxane or dimethylformamide [*262*], and a pyrazine ring was converted to a dihydropyrazine [*263*] (*Procedure 44*, p. 216).

Reductions with Diimide and Hydrazine

Diimide (*diimine, diazene*), N_2H_2 or $HN = NH$, is an ephemeral species which results from decomposition with acids of potassium azodicarboxylate [*264, 265*] from thermal decomposition of anthracene-9,10-diimine [*266, 267*], and of hydrazine [*268, 269*] and its derivatives [*270*]. Although this species has not been isolated, its transient existence has been proven by mass spectroscopy and by its reactions in which it hydrogenates organic compounds with concomitant evolution of nitrogen [*271*].

The reaction of diimide with alkenes is a concerted addition proceeding through a six-membered transition state. It results in predominant *cis* (*syn*) addition of hydrogen and is therefore very useful for stereospecific introduction of hydrogen onto double bonds. If N_2D_2 is used deuterium is added to the double bond. No hydrogen–deuterium exchange takes place as it sometimes does in catalytic hydrogenations [*271, 272*]. Applications of diimide are very rare and are mainly limited to hydrogenation of carbon–carbon multiple bonds [*269, 273, 274*] and azo compounds to hydrazo compounds [*264*].

Hydrazine (M.W. 32.05, b.p. 113.5°, density 1.011) is usually available and used in the form of its hydrate (M.W. 50.06, m.p. − 51.7°, b.p. 120.1°, density 1.032). Its reducing properties are due to its thermal decomposition to hydrogen and nitrogen. Similar decomposition takes place when hydrazine deriva-

(17)

[264] $KO_2CN=NCO_2K$ ———— $\dfrac{2\ AcOH}{23°}$ ————————→

[266] [structure: dibenzo fused ring with NH–NH bridge] ———— $90\text{--}100°$ ————→

                                                                    $[HN=HN]$

[270] $C_7H_7SO_2NHNH_2$ ———— $\dfrac{diglyme}{reflux}$ ————→

[269] $N_2H_4, O_2, CuSO_4$ ———— $\dfrac{EtOH}{50\text{--}60°}$ ————→

$>C=C<$

$H\!\!\diagdown_{N\cdot\cdot N}\diagup\!\!C\diagdown H$

↓

$>CH-CH<$

+

$N\equiv N$

tives such as hydrazones or hydrazides are heated in the presence of alkali. Simple refluxing of hydrazine hydrate with oleic acid caused hydrogenation to stearic acid [275], and heating with aromatic hydrocarbons at 160–280° effected complete hydrogenation of the aromatic rings [276]. It is assumed that such hydrogenations are due to the transient formation of diimide, which is formed from hydrazine in the presence of air. However, the hydrogenation of oleic acid took place even when the air was excluded [275].

Hydrazine is a very effective hydrogen donor in the *catalytic transfer of hydrogen* achieved by refluxing compounds to be reduced with hydrazine in the presence of Raney nickel [277, 278, 279] (p. 13).

Most reactions with hydrazine are carried out with aldehydes and ketones in the presence of alkali. The reduction proper is preceded by formation of hydrazones that decompose in alkaline medium at elevated temperatures to nitrogen and compounds in which the carbonyl oxygen has been replaced by two hydrogens. The same results are obtained by alkaline-thermal decomposition of ready-made hydrazones of the carbonyl compounds. Both reactions are referred to as Wolff–Kizhner reduction [280].

The original reduction carried out in ethanol required heating in sealed tubes. The use of higher boiling solvents (ethylene glycol, di- and triethylene glycol) allows the reaction to be carried out in open vessels [281, 282, 283] (*Huang Minlon reduction*) (*Procedure 45*, p. 216). Conversion of carbonyl groups to methylene groups can also be achieved by decomposition of semicarbazones. Another important reduction in which hydrazine is implicitly involved is the McFadyen–Stevens reduction of acids to aldehydes. The acid is converted to its benzenesulfonyl or *p*-toluenesulfonyl hydrazide and this is decomposed by sodium carbonate at 160° [284, 285].

Reductions with Phosphorus Compounds

Both inorganic and organic phosphorus compounds are used relatively infrequently but their reducing properties are sometimes very specific.

Phosphonium iodide PH_4I (M.W. 161.93, b.p. 62.5°, density 2.86) is prepared somewhat tediously from white phosphorus and iodine [286] and was used, together with fuming hydriodic acid, for reduction of a pyrrole ring to dihydropyrrole [287]. Its applications are very rare.

Hypophosphorous acid H_3PO_2 (M.W. 66.00), usually available as its 50% aqueous solution (density 1.274), is a specific reagent for replacement of aromatic diazonium groups by hydrogen under very mild conditions (at 0–25°) and in fair to good yields [288, 289] (*Procedure 46*, p. 217).

Organic derivatives of phosphorus excel by the strong tendency of trivalent phosphorus to combine with oxygen or sulfur to form phosphine oxides and phosphates or their sulfur analogs.

Triphenylphosphine Ph_3P (M.W. 262.29, m.p. 79–81°, b.p. 377°) reacts spontaneously with all kinds of organic peroxides in ether, petroleum ether or benzene and extracts one atom of oxygen per peroxide bond [290]. It also extrudes sulfur and converts thiiranes to alkenes [291], and performs other reductions [292].

Of *trialkyl phosphites* the most frequently used is triethyl phosphite $(EtO)_3P$ (M.W. 166.16, b.p. 156°, density 0.969) which combines with sulfur in thiir-anes [291, 294] and gives alkenes in respectable yields. In addition, it can extrude sulfur from sulfides [295], convert α-diketones to acyloins [296], convert α-keto acids to α-hydroxy acids [297], and reduce nitroso compounds to hydroxylamines [298] (*Procedure 47*, p. 217).

Strongly reducing properties are also characteristic of *diethyl phosphite* $(EtO)_2PHO$ (M.W. 138.10, b.p. 50–51°/2 mm, density 1.072) which is suitable for partial reduction of geminal dihalides [299], of *sodium O,O-diethylphos-phorotellurate* $(EtO)_2PTeONa$ which replaces halogens by hydrogen in α-halo ketones [300] and transforms epoxides to alkenes [293], and of *ethyl phosphite* $EtOPH_2O$ which converts sulfinates to disulfides [301].

Strong affinity for oxygen and sulfur is also exhibited by phosphines, *tris(dimethylamino)phosphine* and *tris(diethylamino)phosphine*. The former converted a dialdehyde to an epoxide [302], the latter disulfides to sulfides [303].

Reductions with Organic Compounds

Reductive properties of *alcohols* (most frequently ethanol and isopropyl alcohol) derive from their relatively easy dehydrogenation to aldehydes or ketones. They therefore act as hydrogen donors in catalytic transfer of hydro-gen in the presence of catalysts such as Raney nickel [304]. In the presence of cuprous oxide [305] or under irradiation with ultraviolet light [306] ethanol was used for replacement of aromatic diazonium groups by hydrogen (meth-anol is unsuitable for this purpose) [289]. Also an aliphatic diazo group was replaced by two hydrogens in this way [307]. Isopropyl alcohol is utilized for reduction of aldehydes and ketones to alcohols. This can be done either under ultraviolet irradiation [308], or much more often in the presence of anhydrous basic catalysts such as aluminium isopropoxide (Meerwein–Ponn-dorf reduction) [309]. The reduction is a base-catalyzed transfer of hydride ion from isopropoxide ion to the carbonyl function, thus yielding the alkoxide ion of the desired alcohol and acetone. By removal of acetone by distillation the equilibrium of the reversible reaction is shifted in favor of the reduction of the starting carbonyl compound (p. 110). The reduction with alcohols is very selective and particularly suitable for the preparation of unsaturated alcohols from unsaturated carbonyl compounds. Halogens can occasionally be replaced by hydrogen [*310, 311*] (*Procedure 48*, p. 217).

A few examples of reduction of organic compounds by *aldehydes* are recorded. Heating of aqueous formaldehyde with ammonium chloride gives a mixture of all possible methylamines (Eschweiler reaction) [*312*]. Aqueous formaldehyde was also used for replacement of a diazonium group by hydrogen [*289*]. *p*-Tolualdehyde was reduced by treatment with formaldehyde in alkaline medium at 60-70° (crossed Cannizzaro reaction) [*313*]. Salicylaldehyde in alkaline medium reduced aromatic ketones to alcohols [*314*]. Reducing sugars such as glucose were used for the reduction of nitro compounds to azo compounds [*315*] and azoxy compounds [*316*].

Formic acid, anhydrous (M.W. 46.03, m.p. 8.5°, b.p. 100.8°, density 1.22), or a 90% aqueous solution, is an excellent hydrogen donor in catalytic hydrogen transfer carried out by heating in the presence of copper [*71*] or nickel [*71*]. Also its salt with triethylamine is used for the same purpose in the presence of palladium [*72, 73*]. Conjugated double bonds, triple bonds, aromatic rings and nitro compounds are hydrogenated in this way.

Reductions with formic acid alone or in the presence of amines are effected at temperatures of 130-190°. Under these conditions formic acid reduced conjugated double bonds [*317*], pyridine rings in quaternary pyridinium salts to tetrahydro- and hexahydropyridines [*318*], quinolines to tetrahydroquinolines [*319*], and enamines to amines [*320*]. The most important reduction involving formic acid is the Leuckart reaction, achieved by heating of an aldehyde or a ketone with ammonium formate or formamide, or formic acid in the presence of a primary or secondary amine. The final products are primary, secondary or tertiary amines. The Leuckart reduction is an alternative to reductive aminations by hydrogen and gives good yields of amines [*320, 321, 322*] (*Procedure 49*, p. 218).

Organic compounds of sulfur and phosphorus are included in the section on reduction with sulfur and phosphorus compounds (p. 32, 35).

Reducing effects of *organometallic compounds* have been known for a long time. Treatment of ketones and acid derivatives with some Grignard reagents, especially isopropyl- and *tert*-butylmagnesium halides, led sometimes to reduction rather than to addition across the carbonyl function [*323, 324*]. The reduction takes place at the expense of the alkyl group in the Grignard reagent which is converted to an alkene. Although not widely used, reductions with Grignard reagents often give good to excellent yields [*325*].

Similar reductions were effected by heating of triethylaluminum (triethylalane) in ether with aldehydes and ketones. The alcohols are formed by hydrogen transfer from the alkyl groups which are transformed to alkenes.

Similar reductions are achieved by trialkylboranes [*326*]. These reactions, although different in nature from the reductions by hydrides and complex hydrides, were amongst the first applications of boranes and alanes for the reduction of organic compounds.

The common features of *biochemical reductions* are very gentle reaction conditions, unique regioselectivity and high stereospecificity leading frequently to optically active compounds. The reductions are accomplished by enzymes produced by microorganisms (bacteria, yeasts) in aqueous media at temperatures around 30–35°. Most reductions convert carbonyl groups to alcoholic groups [*327*], which is exceedingly useful in the polyfunctional compounds such as steroids [*328*]. Other functions have also been reduced, e.g. carbon–carbon double bonds [*329, 330*] and halogens in halogen compounds [*330*] (*Procedure 50*, p. 218).

THE REDUCTION OF SPECIFIC TYPES OF ORGANIC COMPOUNDS

REDUCTION OF ALKANES AND CYCLOALKANES

In general, reduction of alkanes lacks practical importance. Some cycloalkanes give alkanes in catalytic hydrogenation in quantitative yields: cyclopropane and cyclobutane yielded, respectively, propane and butane over nickel at 120° and 200° [332], and cyclopentane was almost quantitatively cleaved to pentane over 12% platinum on carbon at 300° [333]. The relatively easy fission of the cyclopropane ring must be kept in mind in designing reductions of functions in compounds containing such rings. In 1,2-dimethyl-1,2-diphenylcyclopropane the ring was cleaved between carbons 1 and 2 by hydrogen over platinum, palladium or Raney nickel at 25° and 1 atm [334].

Disubstituted and especially trisubstituted cyclopropanes were hydrogenolyzed by treatment with trialkylsilanes and trifluoroacetic acid (ionic hydrogenation, applicable to compounds forming tertiary carbonium ions) [335]. 1,1,2-Trimethylcyclopropane afforded 2,3-dimethylbutane in 65% yield. The reduction is not stereoselective. 1-Methylbicyclo[3.1.0]hexane gave 83% of *cis*- and 7% of *trans*-dimethylcyclopentane while 1-methylbicyclo[4.1.0]heptane yielded 10.7% of *cis*- and 64.3% of *trans*-1,2-dimethylcyclohexane [335].

REDUCTION OF ALKENES AND CYCLOALKENES

Isolated double bonds in alkenes and cycloalkenes are best reduced by catalytic hydrogenation. Addition of hydrogen is very easy and takes place at room temperature and atmospheric pressure over almost any noble metal catalyst, over Raney nickel, and over nickel catalysts prepared in special ways such as P-1 nickel [13] or complex catalysts, referred to as Nic [49].

The rates of hydrogenation approximately parallel the heats of hydrogenation [336] and are subject to steric effects. Monosubstituted alkenes are reduced faster than di-, tri- and tetrasubstituted ones [14] to such an extent that they can be reduced preferentially, especially over a ruthenium catalyst [337]. Relative half-hydrogenation times of the hydrogenation of alkenes over P-1 nickel boride are illustrative: 1-pentene 1.0, 2-methyl-1-butene 1.6; 2-methyl-2-butene 16; and 2,3-dimethyl-2-butene 45 [13]. The lengths of the chains attached to the double bond carbon and their branching have little effect on the rate of hydrogenation [13]; however, if one of the substituents at the double bond is a benzene ring the hydrogenation is even faster than that of the 1-alkene. The aromatic ring is hydrogenated much more slowly. Relative hydrogenation *rates* of some alkenes and cycloalkenes over nickel on alumina are as follows: styrene 900, 1-hexene 306, cyclopentene 294, cyclo-hexene 150, 1-methyl-3-cyclohexene 134, 1-methyl-1-cyclohexene 5, and ben-zene 1 [14].

Isomerizations may take place over some catalysts more than over others. Regio- as well as stereoisomers may be formed [13, 49]. Such side reactions are undesirable in the reduction of pro-chiral alkenes, and especially in the hydrogenation of some cycloalkenes.

Reduction of cycloalkenes is as easy as that of alkenes, and again catalytic hydrogenation is the method of choice. The rates of hydrogenation are slightly lower than those of 1-alkenes, and are affected by the ring size and by the catalyst used. Double bonds in strained cycloalkenes are hydrogenated even faster than in 1-alkenes [338]. Relative half-hydrogenation times in hydrogen-ations over W-2 Raney nickel and P-1 nickel boride (in parentheses) are, respectively: 1-octene 1.3 (1.0); cyclopentene 2.0 (1.3); cyclohexene 3.5 (2.5); cyclooctene 5.3 (2.0); and norbornene about 90% of that of octene [13]. In a competitive experiment over a nickel boride catalyst 92% of hydrogen was taken up by norbornene and only 8% by cyclopentene [338].

The stereochemistry of the products is strongly affected by the catalyst and slightly affected by the pressure of hydrogen [339]. Hydrogenation of 1,2-dimethylcyclohexene over platinum oxide gave 81.8 and 95.4% of cis-1,2-dimethylcyclohexane at 1 atm and 500 atm, respectively. When hydrogenated over palladium on charcoal or alumina at 1 atm, 1,2-dimethylcyclohexene yielded 73% of trans- and 27% of cis-1,2-dimethylcyclohexane. A increase in the pressure of hydrogen on the yields of cis-1,2-dimethylcyclohexane from 2,3-dimethylcyclohexene and 2-methylmethylenecyclohexene resulted in a slight decrease of the ratio of cis to trans isomer [339].

[339]   (18)

	PtO$_2$	1 atm	81.8%	18.2%
	AcOH	4 atm	83.3%	16.7%
	25°	500 atm	94.5%	5.5%
	Pd(C)	1 atm	27%	73%
	AcOH			
	PtO$_2$	1 atm	76.6%	23.4%
	AcOH	4 atm	70.6%	29.4%
	25°	500 atm	69.7%	30.3%
	Pd(C)	1 atm	27%	73%
	AcOH,25°			
	PtO$_2$	1 atm	70.2%	29.8%
	AcOH	4 atm	65.7%	30.0%
	25°	500 atm	67.3%	32.7%
	Pd(C)	1 atm	39%	61%
	AcOH,25°			

Hydrogenation over Raney nickel was found to be even less stereoselective. 2-, 3- and 4-methylmethylenecyclohexenes gave different mixtures of *cis* and *trans* dimethylcyclohexanes depending not only on the structure of the starting alkene but also on the method of preparation and on the freshness of the catalysts. The composition of the stereoisomers ranged from 27-72% *cis* to 28-73% *trans* [340].

The stereoselectivity of different catalysts in catalytic hydrogenations is discussed in the chapter on catalysts (pp. 4, 50). In addition to catalytic hydrogenation, a few other methods of reduction can be used for saturation of carbon–carbon double bonds. However, their practical applications are no match for catalytic hydrogenation.

The *indirect reduction by boranes* consists of two steps: addition of a borane across the double bond, and decomposition of the alkylborane by an organic acid, usually propionic acid [341].

[341]

$$C_4H_9CH=CH_2 \xrightarrow[BF_3 \cdot Et_2O]{NaBH_4} C_4H_9CH_2CH_2\text{-}BH_2 \longrightarrow C_4H_9CH_2CH_3 \quad 91\%$$

(19)

$$\underset{\substack{| \\ O \overset{\longleftarrow}{=} CC_2H_5}}{\overset{H \quad O}{}} \qquad + \qquad \underset{\substack{| \\ C_2H_5}}{O=C-OBH_2}$$

So-called *ionic hydrogenation* using trialkylsilanes and trifluoroacetic acid is applicable mainly to trisubstituted alkenes and cycloalkenes, compounds which form tertiary carbonium ions [335, 342, 343]. Methylcyclohexene was converted to methylcyclohexane in 67-72% yields on heating for 10 hours at 50° with one equivalent of triethylsilane and two to four equivalents of trifluoroacetic acids [335, 342], and 3-methyl-5α-cholest-2-ene was reduced to 3β-methyl-5α-cholestane in 66% yield with triethylsilane and trifluoroacetic acid in methylene chloride at room temperature [343] (*Procedure 23*, p. 210). The reduction with triethylsilane is not stereoselective. 1,2-Cyclopentene gave 60.5% of *cis*- and 27.5% of *trans*-1,2-cyclopentane, whereas 1,2-dimethylcyclohexene afforded 12.4% of *cis*- and 61.6% of *trans*-1,2-dimethylcyclohexane [335]. Most mono- and disubstituted alkenes do not undergo this reduction. Thus the ionic hydrogenation might be a nice complement to catalytic hydrogenation, which reduces preferentially the non-substituted double bonds. Unfortunately not enough experimental material is yet available to prove the usefulness of the ionic method.

The reduction of an isolated carbon–carbon double bond by other methods is exceptional. Occasionally an isolated double bond has been *reduced electrolytically* [344] and by *dissolving metals*. Sodium under special conditions – using *tert*-butyl alcohol and hexamethylphosphoramide as solvents – reduces alkenes and cycloalkenes in 40-100% yields [345]. 3-Methylenecholestane, for example, afforded 3-methylcholestane in 74% yield on heating for 2 hours at 55-57° with lithium in ethylenediamine [346].

A compound which can accomplish addition of hydrogen across isolated double bonds is *diimide* (*diazene*). A series of cycloalkenes was hydrogenated with more or less stereospecificity using disodium azodicarboxylate as a source of diimide [*347*]. This compound is most probably an intermediate generated in the reaction of alkenes with hydrazine in the presence of air, copper and carbon tetrachloride. 1-Hexene was thus reduced to hexane in 38.7% yield [*268*].

Reductions of isolated double bonds in alcohols, amines, carbonyl compounds, acids, etc., are discussed in the appropriate sections.

REDUCTION OF DIENES AND POLYENES

The behavior of dienes and polyenes – both open chain and cyclic – toward reduction depends especially on the respective positions of the double bonds. Carbon–carbon double bonds separated by at least one carbon atom behave as independent units and can be partly or completely reduced by *catalytic hydrogenation* or by *diimide*. *cis, trans, trans*-1,5,9-Cyclododecatriene treated with hydrazine and air in the presence of copper sulfate (diimide *in situ*) gave *cis*-cyclododecene in 51–76% yield [*269*]. It appears as if the *trans* double bonds were reduced preferentially.

If the double bonds are structurally different to such an extent that their rates of hydrogenation differ considerably they can be reduced selectively. Monosubstituted double bonds are saturated in preference to di- and trisubstituted ones [*13, 348, 349*]. When 2-methyl-1,5-hexadiene was hydrogenated over nickel boride with one mole of hydrogen only the unbranched double bond was reduced, giving 95% yield of 2-methyl-1-hexene [*349*].

The monosubstituted double bond in 4-vinylcyclohexene is hydrogenated over P-1 nickel in preference to the disubstituted double bond in the ring, giving 98% of 4-ethylcyclohexane [*13*]. Similarly in limonene the double bond in the side chain is reduced while the double bond in the ring is left intact if the compound is treated with hydrogen over 5% platinum on carbon at 60° and 3.7 atm (yield 97.6%) [*348*].

Under the conditions which cause migration of the double bonds, e.g. treatment with sodium and especially potassium, the isolated double bonds can become conjugated, and thus they undergo reduction by metals. Some macrocyclic cycloalkadienes were reduced (predominently to *trans* cycloalkenes) by treatment with potassium on alumina in a hydrogen atmosphere [*350*].

In systems of **conjugated double bonds** catalytic hydrogenation usually gives a mixture of all possible products. Conjugated dienes and polyenes can be reduced by *metals: sodium, potassium, or lithium*. The reduction is accomplished by 1,4-addition which results in the formation of a product with only one double bond and products of coupling and polymerization. Isoprene was reduced in 60% yield to 2-methyl-2-butene by sodium in liquid ammonia [*351*]. Reduction of cyclooctatetraene with sodium in liquid ammonia gave a

61% yield of a mixture of 1,3,6- and 1,3,5-cyclooctatriene, the latter being a result of isomerization of the initially formed 1,3,6-triene [*352*].

Side reactions can be partly cut down when the reduction by alkali metals is carried out in the presence of proton donors such as triphenylmethane, fluorene, or secondary amines (*N*-alkyanilines). The products of the initial 1,4-addition are protonated as formed before coupling and polymerization start [*353*, *354*, *355*]. 1,3-Butadiene was reduced with lithium in ether in the presence of *N*-ethylaniline to *cis*-2-butene in 92.5% yield [*354*], and similarly *cis*,*cis*-1,3-cyclooctadiene in the presence of *N*-methylaniline to. *cis*-cyclooctene in 98% yield [*355*]. With sodium in liquid ammonia the yields of reduction of conjugated dienes were much lower [*353*].

Double bonds conjugated with aromatic rings and with carbonyl, carboxyl, nitrile and other functions are readily reduced by catalytic hydrogenation and by metals. These reductions are discussed in the appropriate sections: aromatics, unsaturated aldehydes and ketones, unsaturated acids, their derivatives, etc.

REDUCTION OF ALKYNES AND CYCLOALKYNES

Partial reduction of acetylenes (alkynes) gives alkenes; total reduction gives alkanes. Total reduction is no problem since it can be easily accomplished by catalytic hydrogenation, exceptionally also with diazene. Partial reduction is more desirable, especially with internal alkynes, which, depending on the reduction method, can be converted either to *cis* alkenes or *trans* alkenes. It thus provides means for obtaining the particular olefinic stereoisomers. **Catalytic hydrogenation gives exclusively or at least predominantly *cis* alkenes, reduction with metals and metal salts predominantly *trans* alkenes.** Hydrides afford both stereoisomers depending on the nature of the hydride and reaction conditions. While reduction with hydrides and metals or metal salts stops with negligible exceptions at the stage of alkenes, catalytic hydrogenation gives alkenes only under strictly controlled conditions.

The rates of hydrogenation of the double and triple bond do not differ appreciably. With some catalysts the double bond is hydrogenated even faster. This explains why it is very difficult to prevent total hydrogenation by stopping the reaction after the consumption of just one mole of hydrogen.

Partial hydrogenation of the triple bond was achieved with catalysts over which the rate of hydrogenation decreases after the absorption of the first mole of hydrogen: *nickel and palladium*.

Hydrogenation using Raney nickel is carried out under mild conditions and gives *cis* alkenes from internal alkynes in yields ranging from 50 to 100% [*356*, *357*, *358*, *359*, *360*]. Half hydrogenation of alkynes was also achieved over nickel prepared by reduction of nickel acetate with sodium borohydride (P-2 nickel, nickel boride) [*349*, *361*, *362*] or by reduction with sodium hydride [*49*], or by reduction of nickel bromide with potassium-graphite [*363*]. Other catalysts are palladium on charcoal [*364*], on barium sulfate [*365*, *366*], on barium carbonate [367], or on calcium carbonate [366], especially if the noble metal is partly deactivated by quinoline [368, 369], lead acetate [36] or potassium hydroxide [370]. The most popular catalyst is palladium on calcium carbonate deactivated by lead acetate (*Lindlar's catalyst*) [36]. However, some of the above catalysts are claimed to be better. Palladium on carbon has also been used for semihydrogenation of acetylenes using trimethylammonium formate *in lieu* of hydrogen [72]. With terminal alkynes where no stereo-isomeric products are formed, reduction with *sodium* in liquid ammonia gives alkenes of higher purity than catalytic hydrogenation [360]. Reduction of alkynes by *lithium aluminum hydride* in tetrahydrofuran or its mixture with diglyme gave high yields of *trans* isomers [371]. When toluene was used as the solvent, 2-pentyne yielded 97.6% of pentane at 125°. This rather exceptional total reduction of alkyne to alkane was probably due to the presence of hydrogen resulting from the decomposition of the hydride at 125° in the autoclave. Lithium aluminum hydride was proven earlier to be a catalyst for hydrogenation of alkenes and alkynes with elemental hydrogen [372].

Lithium methyldiisobutylaluminum hydride prepared *in situ* from diisobu-tylalane and methyllithium reduced 3-hexyne to pure *trans*-3-hexene in 88% yield [373]. On the other hand, *diisobutylalane* gave 90% yield of *cis*-3-hexene [373, 374]. Deuterodiborane prepared *in situ* from lithium aluminum deuter-ide and boron trifluoride etherate was used for converting cyclodecyne to *cis*-1,2-dideuterocyclodecene [375] (43% yield, 81% purity). *Borane* and disi-amylborane (di-*sec*-amylborane) gave, respectively, 68–80% and 90% of high purity *cis*-3-hexene [376].

Ref	Conditions			
[49]	Nic$_w$, MeOH,25°,1 atm,80%(isol.)	97%		3%
[349]	NiB$_2$	98–99%		
[361]	P–2 Ni,EtOH,97%(80% isol.)	99–99.5%		
[49]	Nic,EtOH,25°,1 atm,98%	100%		
[371]	LiAlH$_4$,THF,diglyme,117–150°,91–97%	1.6–3.9%	96.1–98.4%	0
[374]	iso-Bu$_2$AlH,<45°,91%	95%	3%	
[373]	LiAlHMeisoBu$_2$,88%		100%	
[376]	NaBH$_4$·BF$_3$,diglyme, 25°, 68–80%	100%		
[376]	sec-Am$_2$BH,diglyme,90%	100%		
[345]	Na,HMPA,tert-BuOH		14%	70%

Another hydride, *magnesium hydride* prepared *in situ* from lithium alu-minum hydride and diethylmagnesium, reduced terminal alkynes to 1-alkenes in 78–98% yields in the presence of cuprous iodide or cuprous *tert*-butoxide, and 2-hexyne to pure *cis*-2-hexene in 80–81% yields [111]. Reduction of alkynes by lithium aluminum hydride in the presence of transition metals gave alkenes with small amounts of alkanes. Internal acetylenes were reduced predominantly but not exclusively to *cis* alkenes [377, 378].

Alkynes can be reduced *electrolytically*. Internal alkynes gave 65–80% yields of *cis* alkenes when electrolysed in 10% sulfuric acid in ethanol at spongy nickel cathode [*127*], or predominantly *trans* alkenes if the electrolysis was carried out in a methylamine solution of lithium chloride. The yields of the alkenes and the ratios of *trans* to *cis* alkenes varied depending on whether the electrolysis was carried out in divided or undivided cells (yields 24–80%, composition of product 89–99% of *trans* alkene) [*379*].

A universal method for conversion of internal alkynes to alkenes is reduction with *alkali metals* in liquid ammonia. Open-chain acetylenes were reduced in high yields to *trans* alkenes with sodium [*360*] or lithium [*380*]. *trans* Cycloalkenes were obtained by sodium reduction of cycloalkynes in 14-membered rings [*126, 381*]. Cycloalkynes with smaller rings do not show such clean-cut stereochemistry. Although cyclononyne gave, on reduction with sodium in liquid ammonia, 63.5% of pure *trans*-cyclononene [*382*] the stereochemistry of reductions of cyclodecyne, tetramethylcyclodecyne and cyclododecyne is very complex and the outcome is strongly affected by the size of the ring and by reaction conditions [*126, 381*]. The ratios of *cis* to *trans* cycloalkenes depend strongly on the metal used. Lithium gave 91% of *trans*- and 9% of *cis*-cyclodecene while sodium and potassium afforded practically pure *cis*-cyclodecene [*381*]. Cyclododecyne was reduced by both metals predominantly to *trans*-cyclododecene. Sodium in liquid ammonia yielded a mixture containing 81% of *trans*- and 19% of *cis*-cyclododecene [*381*] whereas lithium gave 95.5% of *trans*- and 4.5% of *cis*-cyclododecyne. The reductions are also influenced by the presence of a proton donor such as ethanol which favors direct *cis* reduction. In the absence of ethanol in liquid ammonia the formation of *cis* isomer is probably a result of reduction of a cumulene produced by isomerization of the acetylene by sodium amide [*126, 381*]. The high content of the *cis* isomer in the reduction of cyclodecyne is hard to understand, especially since cyclononyne gave pure *trans*-cyclononene [*382*].

Simple acetylenes, i.e. those in which the triple bond is not conjugated with an aromatic ring, carbonyl or carboxyl functions, are not reduced by sodium in amyl alcohol, or by zinc [*360*]. Conjugation, however, changes the reducibility of the triple bond dramatically (*vide infra*).

In **acetylenes containing double bonds** the triple bond was selectively reduced by controlled treatment with hydrogen over special catalysts such as palladium deactivated with quinoline [*368*] or lead acetate [*36*], or with triethylammonium formate in the presence of palladium [*72*]. 1-Ethynylcyclohexene was hydrogenated to 1-vinylcyclohexene over a special nickel catalyst (Nic) in 84% isolated yield [*49*].

Stereoselective reductions can be further accomplished by reagents which usually do not reduce double bonds (unless conjugated with carbonyl or carboxyl functions): *lithium aluminum hydride* [*383*] or *zinc* [*384*].

Diacetylenes having an internal and a terminal triple bond can be reduced selectively at the internal triple bond if they are first converted to sodium acetylides at the terminal bond by sodamide prepared *in situ* from sodium in liquid ammonia. Undeca-1,7-diyne was in this way converted to *trans*-7-undecene-1-yne in 75% yield [*385*].

Complete reduction of alkynes to alkanes is easily accomplished by catalytic hydrogenation, especially using palladium [*386, 387*], platinum oxide and active nickel catalysts [*359*].

If the triple bond is conjugated with an aromatic ring or with carbonyl groups complete reduction can be achieved with reagents which are capable of reducing conjugated double bonds (p. 42).

Reduction of triple bonds in compounds containing aromatic rings and functional groups is dealt with in the appropriate sections.

REDUCTION OF CARBOCYCLIC AROMATICS

Reduction of carbocyclic aromatics is generally more difficult than that of alkenes, dienes and alkynes. Any reduction destroys resonance-stabilized systems and consequently has to overcome resonance energy. The heat of hydrogenation of the *cis*-disubstituted carbon–carbon double bond in *cis*-2-butene is 28.6 kcal. Since the resonance energy of benzene is 36 kcal, uptake of two hydrogen atoms is an endothermic reaction. In partial hydrogenation of naphthalene, whose resonance energy is 61 kcal, destruction of one aromatic ring means loss of resonance energy of only 25 kcal (61 − 36). Reduction of the 9,10-double bonds in anthracene and phenanthrene is accompanied by losses of resonance energy of only 12 and 20 kcal, respectively. In accordance with the energetics the reduction of naphthalene is easier than that of benzene, and reduction of anthracene and phenanthrene in turn easier than that of naphthalene. The trend is noticeable in the increasing ease of reduction of the four above aromatics, be it by catalytic hydrogenation, by reduction with metals, or by reduction with other compounds.

The *catalytic hydrogenation* of benzene and its homologs is more difficult than that of unsaturated aliphatic hydrocarbons, and more difficult than that of carbonyl functions, nitriles, halogen compounds, nitro compounds and others. It is therefore possible to hydrogenate the above functions preferentially. Rates of hydrogenation of benzene homologs decrease with increasing numbers of substituents [*14*]. If the relative rate of hydrogenation of benzene over nickel on alumina equals 1, the rates of hydrogenation of its homologs are: for toluene 0.4, xylene 0.22, trimethylbenzene 0.1, tetramethylbenzene 0.04, pentamethylbenzene 0.005, and hexamethylbenzene 0.001 [*14*].

Hydrogenation of benzene and its homologs at room temperature and atmospheric or slightly elevated pressure (2–3 atm) requires very active catalysts: platinum oxide, noble metals, and possibly very active Raney nickel. Excellent results were obtained by hydrogenation using *Adams' catalyst (platinum dioxide)* [*388, 389*]. Benzene and its homologs were completely hydrogenated in 2–26 hours using 0.2 g of the catalyst per 0.2 mol of the aromatic compound, acetic acid as the solvent, temperatures of 25–30°, and hydrogen pressures of 2–3 atm [*388*]. (*Procedure 1*, p. 201.) Under high pressure (215 atm) such hydrogenations were completed within 12–30 minutes [8]. Even better results were obtained using *rhodium–platinum oxides* prepared by fusion of rhodium chloride and chloroplatinic acid (or ammonium chloroplatinate) in 3 : 1 ratio with sodium nitrate in the manner used for the preparation of platinum dioxide (p. 5). Hydrogenations of aromatics over this catalyst took place at 24–26° and at atmospheric pressure at times considerably shorter than with platinum dioxide alone [38].

(21)

$$C_6H_5CH_3 \quad
\begin{array}{l}
\text{[388]} \quad H_2/PtO_2, \text{ AcOH, } 25\text{–}30°, \text{ 2–3 atm, 195 min} \\
\text{[8]} \quad H_2/PtO_2, \text{ AcOH, } 25°, \text{ 215 atm, 12 min} \longrightarrow \\
\text{[38]} \quad H_2/RhO_2 \cdot PtO_2, \text{ AcOH, } 24\text{–}26°, \text{ 1 atm, 110 min}
\end{array}
\quad C_6H_{11}CH_3$$

Hydrogenation over *rhodium* (5% on alumina) in acetic acid at room temperature and 3–4 atm of hydrogen was successfully used for reduction of benzene rings in compounds containing functions which would be hydrogenolyzed over platinum or palladium catalyst [390]. Palladium at low temperature usually does not reduce benzene rings and is therefore suited for hydrogenolysis of benzyl derivatives (pp. 150, 151).

Specially prepared highly active *Raney nickel* (W-6) can also be used for hydrogenation of some benzene homologs under the same mild conditions that are used with noble metals [45]. In diphenyl selective hydrogenation of one benzene ring is possible.

Hydrogenation of benzene and its homologs over other nickel catalysts requires higher temperatures (170–180°) and/or higher pressures [43, 48, 391].

Copper chromite (Adkins' catalyst) does not catalyze hydrogenation of benzene rings.

Stereochemistry of catalytic hydrogenation of benzene homologs is somewhat controversional. The old simplistic view that platinum, palladium and osmium favor *cis* products whereas nickel favors *trans* products, and that *cis* products rearrange to *trans* products when heated with nickel at higher temperatures (175–180°) [391] has had to be modified as a result of later observations. Treatment for 1 hour of *cis*-1,2-, *cis*-1,3-, and *cis*-1,4-dimethylcyclohexane over platinum oxide at 85° and 84 atm gave only up to 3%, and over Raney nickel at 200° and 84 atm only up to 4% of the corresponding *trans* isomers [9]. Hydrogenation using platinum oxide over a 35–85° temperature range up to 8.4 atm pressure gave up to 15.6% of *trans*-1,2-, up to 19% of *trans*-1,3-, and up to 33.4% of *trans*-1,4-dimethylcyclohexane from *o*-, *m*-, and *p*-xylene, respectively. The corresponding contents of the *trans* isomers after treatment of the three xylenes over Raney nickel at 198–200° at 8.6 atm were: 11%, 21%, and 5% respectively (less than over platinum!) [9]. Other authors found that increase in hydrogen pressure from 0.5 to 300 atm increased the amounts of *cis* isomers from 88% to 93% for *cis*-1,2-, from 77% to 85% for *cis*-1,3-, and from 70% to 82% for *cis*-1,4-dimethylcyclohexane over platinum oxide at 25° [*10*]. Above all, the ratios of *cis* to *trans* product depend also on the groups linked to the benzene ring [*9, 10*].

Whereas catalytic hydrogenation always converts benzene and its homologs to cyclohexane derivatives, *electrolysis* using platinum electrodes and lithium chloride in anhydrous methylamine yields partially hydrogenated products. When the electrolytic cell was undivided, the products were 1,4-cyclohexadiene and its homologs. When the cell was divided by an asbestos diaphragm, cyclohexene and its homologs were obtained in high purities and yields ranging from 44% to 85% [*129, 132*] (*Procedure 25*, p. 210).

Benzene and its homologs can be converted to the corresponding cyclohexadienes and cyclohexenes, and even cyclohexanes, by treatment with *dissolving metals: lithium, sodium, potassium or calcium* in liquid ammonia or amines. Conversions are not complete, and the ratio of cyclohexadienes to cyclohexenes depends on the metal used, on the solvent, and on the presence of hydrogen donors (alcohols) added to the ammonia or amine [*392, 393, 394*].

Addition of benzene in methanol to an excess (25–100%) of sodium in liquid ammonia afforded 84–88% yields of 1,4-cyclohexadiene [*392*]. *o*-Xylene under the same conditions gave 70%–92% yield of 1,2-dimethyl-1,4-cyclohexadiene [*392, 395*]. Lithium in neat ammonia at 60° gave 91% of 1,4-cyclohexadiene and 9% of cyclohexene (conversion 58.4%), while calcium under the same conditions yielded 21% of cyclohexadienes and 79% of cyclohexene in conversions of 13–60% [*393*]. If benzene dissolved in ether was added to the compound $Ca(NH_3)_6$, preformed by dissolving calcium in liquid ammonia and evaporating ammonia, a 75% conversion to pure cyclohexene was achieved [*394*].

Similar results were achieved when benzene was reduced with alkali metals in anhydrous methylamine at temperatures of 26–100°. Best yields of cyclohexene (up to 77.4%) were obtained with lithium at 85° [*396*]. Ethylamine [*397*] and especially ethylenediamine are even better solvents [*398*]. Benzene was reduced to cyclohexene and a small amount of cyclohexane [*397, 398*]; ethylbenzene treated with lithium in ethylamine at −78° gave 75% of 1-ethylcyclohexene whereas at 17° a mixture of 45% of 1-ethylcyclohexene and 55% of ethylcyclohexane was obtained [*397*]. Xylenes (*m*- and *p*-) yielded nonconjugated 2,5-dihydro derivatives, 1,3-dimethyl-3,6-cyclohexadiene and 1,4-dimethyl-1,4-cyclohexadiene, respectively, on reduction with sodium in liquid ammonia in the presence of ethanol (in poor yields) [*399*]. Reduction of diphenyl with sodium or calcium in liquid ammonia at −70° afforded mainly 1-phenylcyclohexene [*400*] whereas with sodium in ammonia at 120–125° mainly phenylcyclohexane [*393*] was formed.

Dissolving metal reductions of the benzene rings are especially important with functional derivatives of benzene such as phenols, phenol ethers and carboxylic acids (pp. 80, 82, 93 and 140).

It is of interest to mention reduction of benzene and its homologs to the corresponding hexahydro derivatives by heating with anhydrous *hydrazine*.

Best yields were obtained with mesitylene (92%). However, this reaction is not very attractive since it requires heating in autoclaves at 250° [276].

Aromatic hydrocarbons with side chains containing double bonds can be easily reduced by catalytic hydrogenation regardless of whether the bonds are isolated or conjugated. Double bonds are saturated before the aromatic ring is reduced. Hydrogenation of styrene to ethylbenzene is one of the fastest catalytic hydrogenations [14].

Saturation of double bonds can also be achieved by *catalytic transfer of hydrogen* [74, 76]. Styrene was quantitatively reduced to ethylbenzene by 1,4-cyclohexadiene in the presence of iodine under ultraviolet irradiation [76], and *trans*-stilbene to 1,2-diphenylethane in 90% yield by refluxing for 46 hours with cyclohexene in the presence of palladium on charcoal [74].

Double bonds conjugated with benzene rings are reduced *electrolytically* [344] (p. 23). Where applicable, stereochemistry can be influenced by using either catalytic hydrogenation or *dissolving metal* reduction [401] (p. 24). Indene was converted to indane by sodium in liquid ammonia in 85% yield [402] and acenaphthylene to acenaphthene in 85% yield by reduction with *lithium aluminum hydride* in carbitol at 100° [403]. Since the benzene ring is not inert toward alkali metals, nuclear reduction may accompany reduction of the double bond. Styrene treated with *lithium in methylamine* afforded 25% of 1-ethylcyclohexene and 18% of ethylcyclohexane [404].

Aluminum amalgam in wet tetrahydrofuran reduced cumulene 1,1,4,4-tetraphenylbutatriene to 1,1,4,4-buta-1,2-diene, which on hydrogenation over Raney nickel was converted to 1,1,4,4-tetraphenylbutane [143].

(22)

[143] $Ph_2C=C=C=CPh_2$ $\xrightarrow[\substack{THF,H_2O \\ 25°}]{AlHg}$ $Ph_2C=C=CH-CHPh_2$ 70%

$\downarrow H_2$ Raney Ni, 25°

$Ph_2CHCH_2CH_2CHPh_2$

Triple bonds in side chains of aromatics can be reduced to double bonds or completely saturated. The outcome of such reductions depends on the structure of the acetylene and on the method of reduction. If the triple bond is not conjugated with the benzene ring it can be handled in the same way as in aliphatic acetylenes. In addition, *electrochemical reduction* in a solution of lithium chloride in methylamine has been used for partial reduction to alkenes (*trans* isomers, where applicable) in 40–51% yields (with 2,5-dihydroaromatic alkenes as by-products) [379]. Aromatic acetylenes with triple bonds conjugated with benzene rings can be *hydrogenated over Raney nickel* to *cis* olefins [356], or to alkyl aromatics over rhenium sulfide catalyst [54]. Electroreduction in methylamine containing lithium chloride gives 80% yields of alkyl aromatics [379].

Reduction of diphenylacetylene with *sodium* in methanol or with *zinc* yielded stilbene, whilst reduction with sodium in ethanol gave 1,2-diphenylethane. Unfortunately the yields are not reported. Reduction to 1,2-diphen-ylethane was accomplished with *hydrogen* [387] and with *diimide* in 80% yield [273]. An excellent reagent for partial *trans* reduction of acetylenes, chromous sulfate, converted phenylacetylene to styrene in 89% yield [195].

(23)

$$PhC \equiv CPh \longrightarrow PhCH = CHPh + PhCH_2CH_2Ph$$

[359]	H_2, Raney Ni, MeOH, 4 atm	87% (cis)	
[387]	H_2, Pd, AcOH, 25°		99.3%
[379]	electro, LiCl, MeNH$_2$	2% (trans)	78%
[273]	N_2H_4, O_2, Cu		80%

Condensed aromatic hydrocarbons are reduced much more easily than those of the benzene series, both catalytically and by dissolving metals.

Depending on the catalysts, *catalytic hydrogenation* converts **naphthalene** to tetrahydronaphthalene (tetralin), or *cis-* or *trans-*decalin (decahydro-naphthalene) [8, 405, 406, 407]. Tetrahydronaphthalene was converted to *cis-*decalin by hydrogenation over platinum oxide [8].

(24)

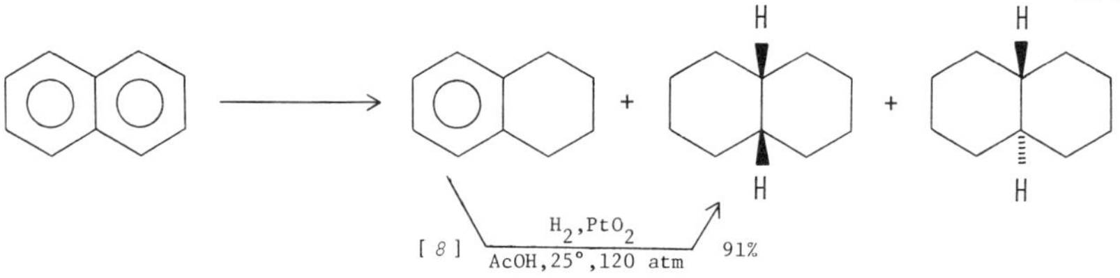

[405]	H_2/CuCr$_2$O$_4$,200°,150-200 atm	80%	
[8]	H_2/PtO$_2$,AcOH,25°,130 atm	71%	21%
[406]	H_2/Pt	75%	25%
[406]	H_2/Ni, 160-162°	22%	78%

Dissolving metals reduce naphthalene to a host of products depending on the reaction conditions. Heating naphthalene with sodium to 140–145° [408] or treating naphthalene with sodium in liquid ammonia at temperatures below −60° gives 1,4-dihydronaphthalene, which slowly rearranges to 1,2-dihydronaphthalene [400]. Since the latter has the double bond conjugated with the remaining benzene ring it is reduced to 1,2,3,4-tetrahydronaphthal-ene [400]. This is also the final product when the reduction is carried out with enough sodium in liquid ammonia at −33° [400]. In the presence of alcohols, sodium in liquid ammonia reduced naphthalene [409, 410] and 1,4-dihydro-naphthalene to 1,4,5,8-tetrahydronaphthalene [410].

Lithium in ethylamine converted naphthalene to a mixture of 80% of 1,2,3,4,5,6,7,8-octahydronaphthalene ($\triangle^{9,10}$-octalin) and 20% of 1,2,3,4,5,6,7,10-octahydronaphthalene [411]. Potassium in methylamine re-duced tetralin to $\triangle^{9,10}$-octalin in 95% yield [396] while sodium in ammonia and methanol yielded 98% of 1,2,3,4,5,8,-hexahydronaphthalene [412]. Lith-

(25)

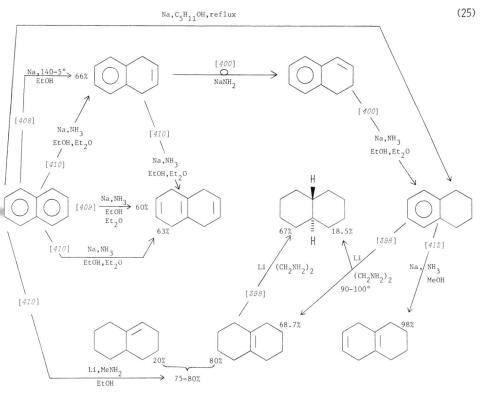

ium in ethylenediamine reduced tetralin to 68.7% of $\triangle^{9,10}$ -octalin and 18.5% of *trans*-decalin [398].

Reduction of naphthols, naphthyl ethers, naphthylamines and naphthoic acids are discussed on pp. 80, 82, 93 and 140, respectively.

Acenaphthylene heated with *lithium aluminum hydride* in carbitol at 120° gave 97% yield of acenaphthene [403]. **Anthracene** is reduced very easily to the 9,10-dihydro compound by *catalytic hydrogenation* [78] and by *sodium* [413]. Further reduction is achieved by catalytic hydrogenation using different catalysts and reaction conditions [78, 407] and by *lithium* in ethylenediamine [414].

(26)

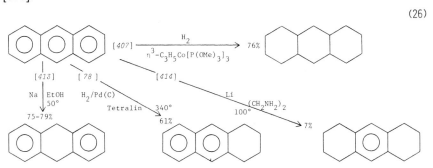

(27)

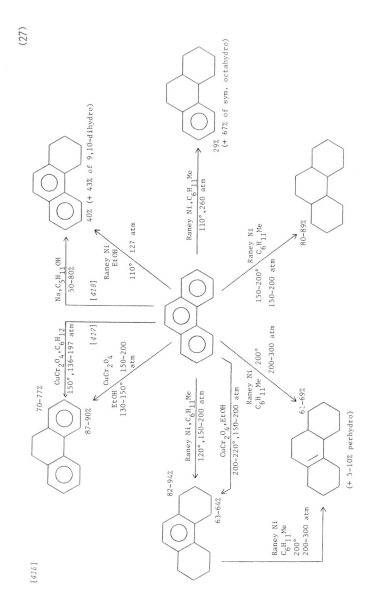

Phenanthrene is converted to partially and totally reduced phenanthrene by *catalytic hydrogenation* [*415, 416, 417*]. Partial reduction to 1,2,3,4-tetra-hydrophenanthrene can also be achieved by *sodium* [*418*].

REDUCTION OF HETEROCYCLIC AROMATICS

Reduction of **furan** and its homologs is readily achieved by *catalytic hydrogenation* using palladium [*419*], nickel [420] or Raney nickel [*421*] as catalysts. Too energetic conditions are to be avoided since hydrogenated furan rings are easily hydrogenolyzed [*422, 423*].

(28)

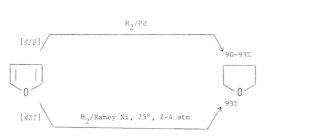

Benzofuran was reduced over Raney nickel at 90° and 85 atm to 2,3-dihydrobenzofuran, at 190° and 119 atm to octahydrobenzofuran, and further hydrogenation over copper chromite at 200–300° caused hydrogenolysis to 2-ethylcyclohexanol [*422*].

3,4-Dihydrobenzopyran (**chromane**) and its methyl homologs were reduced by *lithium* in ethylamine and dimethylamine to 3,4,5,6,7,8-hexahydrobenzopyrans in 84.5% yields [*424*].

Catalytic hydrogenation of **thiophene** poses a problem since noble metal catalysts are poisoned, and Raney nickel causes desulfurization. Best catalysts proved to be cobalt polysulfide [*425*], dicobalt octacarbonyl [*426*], rhenium heptasulfide [*53*] and rhenium heptaselenide [*54*]. The last two require high temperatures (230–260°, 250°) and high pressures (140, 322 atm) and give 70% and 100% of tetrahydrothiophene (thiophane, thiolene), respectively.

Similar procedures to those used for the catalytic hydrogenation of furan and thiophene can usually be applied to the functional derivatives of these heterocycles.

Pyrrole and its homologs can be reduced completely or partially. Complete reduction is best achieved by *catalytic hydrogenation*. Pyrrole itself was hydrogenated with difficulty, and not in high yields, over platinum [*427*], nickel [*43*], Raney nickel or copper chromite [*428*]. Over Raney nickel at 180° and 200–300 atm only 47% yield of pyrrolidine was obtained with 48% of the pyrrole unreacted. On the other hand, pyrrole homologs and derivatives are readily hydrogenated and give better yields, especially if the substituents are attached to nitrogen [*428, 429, 430*]. 1-Methyl- and 1-butylpyrrole hydrogenated over platinum oxide gave high yields of the corresponding *N*-alkyl pyrrolidines [*427*]. 2,5-Dimethylpyrrole in acetic acid over rhodium at 60°

and 3 atm afforded 70% of *cis*-2,5-dimethylpyrrolidine [*431*], and 2-butylpyr-
role in acetic acid over platinum oxide at room temperature and 3 atm gave
94% of 2-butylpyrrolidine [429]. 1-Phenylpyrrole was hydrogenated over
Raney nickel at 135° and 200–300 atm to 63% of 1-phenylpyrrolidine and
30% of 1-cyclohexylpyrrolidine [*428*] and 2-phenylpyrrole gave, over Raney
nickel at 165° and 250–300 atm, 15% of 2-phenylpyrrolidine and 27% of 2-
cyclohexylpyrrolidine whereas, over copper chromite at 200°, it gave a 55%
yield of 2-phenylpyrrolidine [*430*]. Some pyrrole derivatives may be hydrogenated over platinum oxide at room temperature [*432*] (p. 113), others over
Raney nickel at 50–180° and 200–300 atm in yields of up to 98% [*428*]. Copper
chromite as the catalyst is inferior to Raney nickel since it requires temperatures of 200–250° [*428*].

(29)

(RECOVERED)

[*430*]	H$_2$/Raney Ni			
	165°, 250–300 atm	40%	15%	27%
	200°, 250–300 atm			69%
[*430*]	H$_2$/CuCr$_2$O$_4$			
	200°, 250–300 atm	20%	55%	

Partial and total reduction of *N*-methylpyrrole was achieved with *zinc*
[*433*], and of a pyrrole derivative to a dihydropyrrole derivative with *phosphonium iodide* [*287*] (p. 34, 35). The pyrrole ring is not reduced by sodium.

Pyrrole derivatives having double bonds in the side chains are first reduced
at the double bonds and then in the pyrrole ring. 2-(2-Butenyl)pyrrole gave
88% yield of 2-butylpyrrole over platinum oxide in ether; further hydrogenation in acetic acid gave a 94% yield of 2-butylpyrrolidine [*429*].

Reduction of the pyrrole ring in **benzopyrrole (indole)** is discussed on p. 56
[*430*].

Pyridine and its homologs can be reduced completely to hexahydro derivatives, or partially to dihydro- and tetrahydropyridines. *Catalytic hydrogenation* is faster than with the corresponding benzene derivatives and gives only
completely hydrogenated products. Partial reduction can be achieved by
different methods (pp. 55, 56).

Hydrogenation of the pyridine ring takes place under very mild conditions
using palladium [*434*], platinum oxide [*434, 435*] or rhodium [*431, 434, 436*].
With these metals the reaction must be carried out in acidic media, best in
acetic acid, since the hydrogenated products are strong bases which deactivate
the catalysts. Hydrogenation over Raney nickel [*437*] and copper chromite
[*50*] requires high temperatures and high pressures. Use of alcohols as solvents
for these hydrogenations should be avoided because alkylation on nitrogen
could occur. A catalyst which does not require acidic medium and which can
be used in the presence of alcohols is ruthenium dioxide [*438*]. A thorough

treatment of catalytic hydrogenation of pyridine is published in *Advances in Catalysis* [*439*].

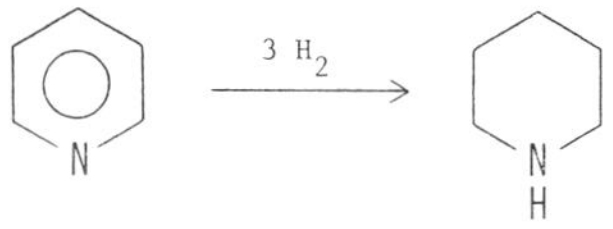

(30)

[*50*]	$CuCr_2O_4$, 220°, 100–150 atm	50%
[*437*]	Raney Ni, 200°, 130–300 atm	83%
[*438*]	RuO_2, 95°, 70–100 atm	100%

Lithium aluminum hydride dissolves in pyridine and forms lithium tetrakis-(*N*-dihydropyridyl)aluminate which itself is a reducing agent for purely aromatic ketones [*440, 441*].

The total and partial reduction of pyridine and its homologs and derivatives was achieved by *electroreduction* and by reduction with *sodium* in alcohols. Both methods give mixtures of tetrahydropyridines and hexahydropyridines. The heteropolar bonds are more receptive to electrons than the carbon–carbon bonds. Consequently the reduction initially gives 1,2-dihydro- and 1,4-dihydropyridines. These rearrange according to the scheme shown below. The 1,2-dihydro intermediate is further reduced to 1,2,3,6-tetrahydro and 1,2,5,6-tetrahydro products (identical if the starting pyridine compound is symmetrical). The 1,4-dihydro intermediate rearranges to the 3,4-dihydro compound which, on further reduction, gives a 1,2,3,4-tetrahydro derivative; this undergoes an internal enamine–imine shift and is ultimately reduced to the hexahydropyridine product. Both tetrahydro derivatives contain carbon–carbon double bonds insulated from the nitrogen and do not undergo further reduction to piperidines.

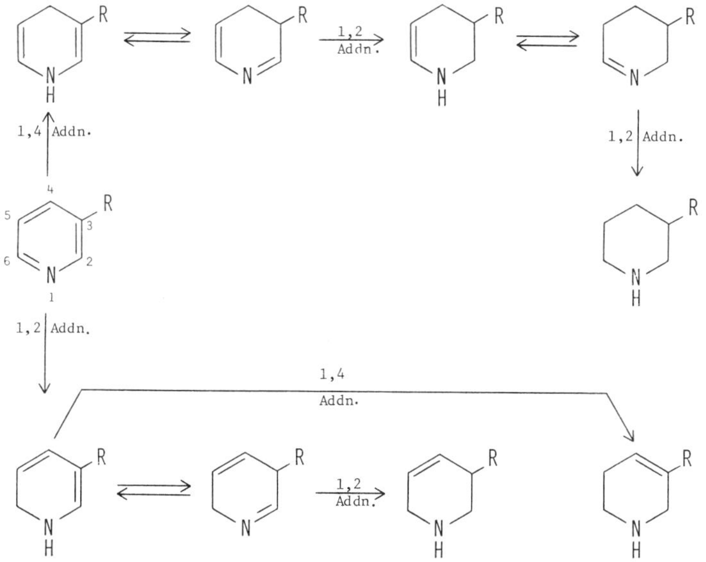

(31)

Reductions of this type were applied to pyridine [442], alkyl pyridines and to many pyridine derivatives: alcohols [443], aldehydes [443], ketones [443] and acids [444]. In compounds containing both pyridine and benzene rings sodium exclusively reduces the pyridine ring [445]. Pyridine was not reduced by zinc and other similar metals.

The **pyridine ring** is easily reduced in the form of its **quaternary salts** to give hexahydro derivatives by catalytic hydrogenation [446], and to tetrahydro and hexahydro derivatives by reduction with *alane (aluminum hydride)* [447], *sodium aluminum hydride* [448], *sodium bis(2-methoxyethoxy)aluminum hydride* [448], *sodium borohydride* [447], *potassium borohydride* [449], *sodium* in ethanol [444, 450], and *formic acid* [318]. Reductions with hydrides give predominantly 1,2,5,6-tetrahydro derivatives while *electroreduction* and reduction with formic acid give more hexahydro derivatives [451, 452].

(32)

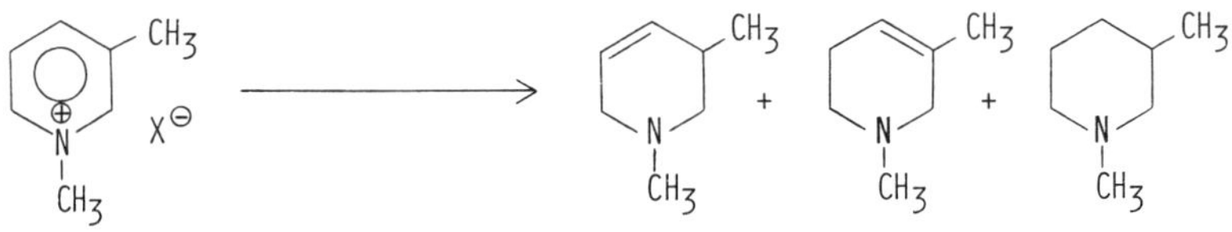

		Yield	%	%	%
[447,451]	X=I; NaBH$_4$;	79.1%	5.0	91.5	3.5
[447]	X=I; LiAlH$_4$;	98%	15.5	84.5	0
[447]	X=I; AlH$_3$;	88%	16.0	84.0	0
[449]	X=I; KBH$_4$;	72%	7.5	82.5	10
[318]	X=Br; HCO$_2$H	68%		45	55
[449,452]	X=MeOSO$_3$ electro redn.		11	17	72
[448]	X=I; NaAlH$_4$;	44%	20	35	5
[448]	X=I; NaAlH$_2$(OCH$_2$CH$_2$OCH$_3$)$_2$;	51%	27	32	20

Double bonds in **alkenyl pyridines** may be *hydrogenated* under mild conditions (Raney nickel at room temperature) to give alkyl pyridines [450]. If the double bond is conjugated with the pyridine ring *sodium* in alcohol will reduce both the double bond and the pyridine ring in good yields [450].

In pyridylpyrrole derivatives the pyridine ring was hydrogenated preferentially, giving piperidylpyrroles, which were further hydrogenated to piperidylpyrrolidines over platinum oxide in acetic acid at room temperature and 2 atm [453].

The *N*-methylpyrrole ring in nicotyrine—1-methyl-2-(3′-pyridyl)pyrrole— was reduced to nicotine both by catalytic hydrogenation [454] and by *zinc* [455] in 40% and 12% yield, respectively.

The double bond in **indole** and its homologs and derivatives is reduced easily and selectively by *catalytic hydrogenation* over platinum oxide in ethanol and fluoroboric acid [456], by *sodium borohydride* [457], by *sodium cyanoborohydride* [457], by *borane* [458, 459], by *sodium in ammonia* [460], by *lithium* [461] and by *zinc* [462]. Reduction with *sodium borohydride* in acetic acid can result in alkylation on nitrogen giving *N*-ethylindoline [457].

(33)

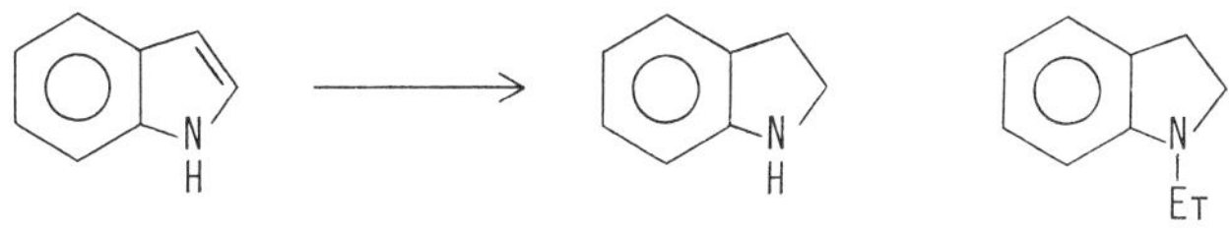

[457]	NaBH$_3$CN	88%	
[457]	NaBH$_4$,AcOH		86%
[458]	BH$_3$·NEt$_3$	80%	
[459]	BH$_3$·C$_5$H$_5$N,AcOH	86%	
[462]	Zn,H$_3$PO$_4$,70–80°	64%	

Catalytic hydrogenation may selectively reduce the double bond, or reduce the aromatic ring as well depending on the reaction conditions used [430, 463, 464]. Only exceptionally has the benzene ring been hydrogenated in preference to the pyrrole ring [463, 465].

(34)

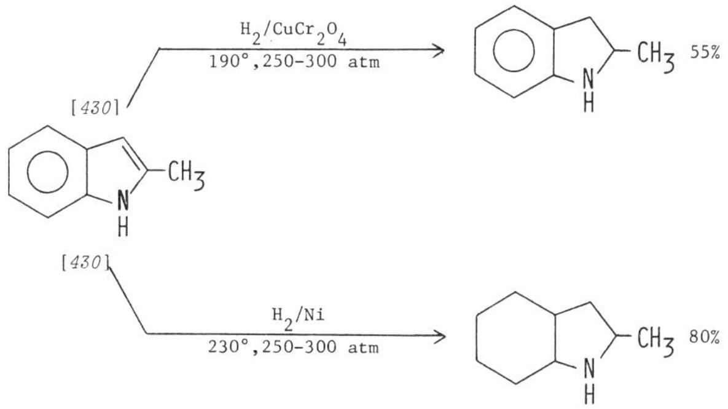

In indole's isomer **pyrrocoline** (1-azabicyclo[4,3,0]nonatetraene) *catalytic hydrogenation* over palladium in acidic medium reduced the pyrrole ring [466], in neutral medium the pyridine ring [467].

(35)

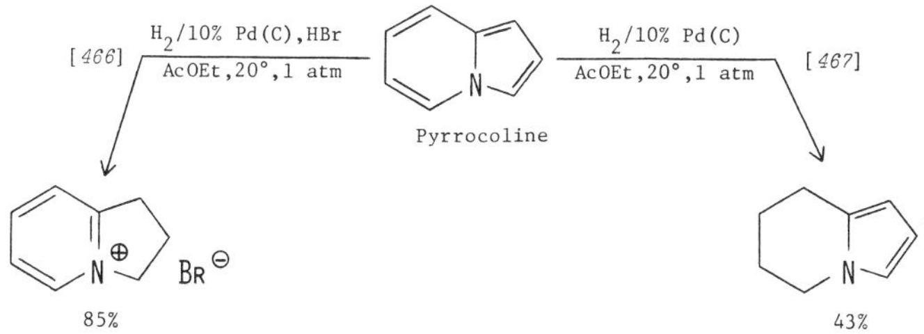

Carbazoles, too, can be reduced partially to dihydrocarbazoles by *sodium* in liquid ammonia [460], or to tetrahydrocarbazoles by sodium in liquid ammonia and ethanol [460] or by sodium borohydride [457]. Carbazole was converted by *catalytic hydrogenation* over Raney nickel or copper chromite to 1,2,3,4-tetrahydrocarbazole, 1,2,3,4,10,11-hexahydrocarbazole, and do-decahydrocarbazole in good yields [430].

In **quinoline** and its homologs and derivatives it is usually the pyridine ring which is reduced first. *Sodium* in liquid ammonia converted quinoline to 1,2-dihydroquinoline [468]. 1,2,3,4-Tetrahydroquinoline was obtained by *catalytic hydrogenation* (yield 88%) [43, 469] and by reduction with *borane* [459] and sodium cyanoborohydride [470], 5,6,7,8-tetrahydroquinoline by hydrogenation over platinum oxide or 5% palladium or rhodium on carbon in trifluoroacetic acid (yields 69–84%) [471]. Vigorous hydrogenation gave *cis*- and *trans*-decahydroquinoline [43, 469].

(36)

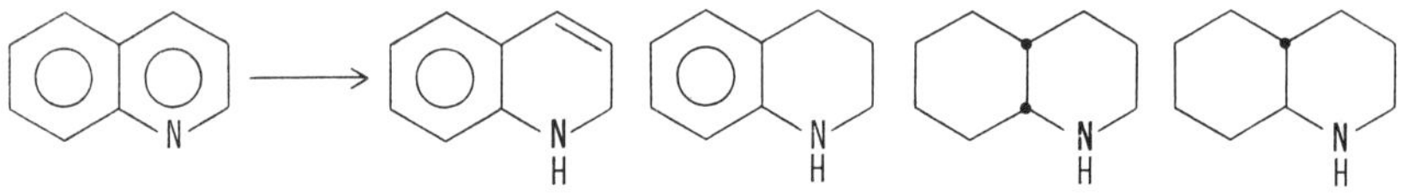

[468]	Na,NH$_3$	75.5%			
[468]	H$_2$,Raney Ni,EtOH		56%		
[43]	H$_2$,Ni,150°,160 atm		96%		
[469]	H$_2$,Pt,40°,2-3 atm		100%		
[50]	H$_2$,CuCr$_2$O$_4$,190°,150-200 atm		100%		
[469]	H$_2$,Pt,H$_2$PtCl$_4$,AcOH,H$_2$O,40°,2-3 atm			20%	80%
[469]	H$_2$,Pt,H$_2$PtCl$_4$,AcOH,HCl,40°,2-3 atm			65%	35%
[459]	BH$_3$·C$_5$H$_5$N,AcOH		71%		
[470]	NaBH$_3$CN,AcOH		71%		

Quinoline homologs and derivatives, including those with double bonds in the side chains, were reduced selectively by *catalytic hydrogenation* over platinum oxide (side chain double bonds), and to dihydro- and tetrahydroquinolines by *sodium* in butanol, by *zinc* and formic acid, and by *triethylammonium formate* [319, 472]. Catalytic hydrogenation of quinoline and its derivatives has been thoroughly reviewed [439].

Isoquinoline was converted to 1,2,3,4-tetrahydroisoquinoline in 89% yield by reduction with *sodium* in liquid ammonia and ethanol [473], and to a mixture of 70–80% *cis*- and 10% *trans*-decahydroisoquinoline by *catalytic hydrogenation* over platinum oxide in acetic and sulfuric acid [474]. Without sulfuric acid the hydrogenation stopped at the tetrahydro stage. Catalytic hydrogenation of isoquinoline and its derivatives is the topic of a review in *Advances in Catalysis* [439].

Dehydroquinolizinium salts containing quaternary nitrogen at the bridgehead of bicyclic systems were easily hydrogenated over platinum to quinolizidines [475].

(37)

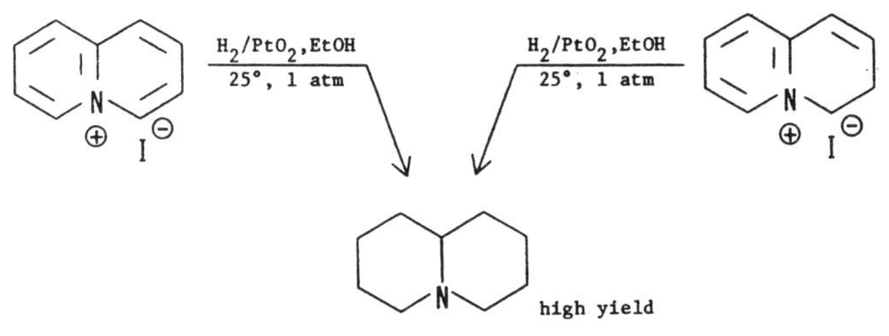

Quinolizidine

Reaction conditions used for reduction of **acridine** [430, 476], partly hydrogenated **phenanthridine** [477] and **benzo[f]quinoline** [477] are shown in Schemes 38–40. Hydrogenation over platinum oxide in trifluoroacetic acid at 3.5 atm reduced only the carbocyclic rings in acridine and **benzo[h]quinoline,** leaving the pyridine rings intact [471].

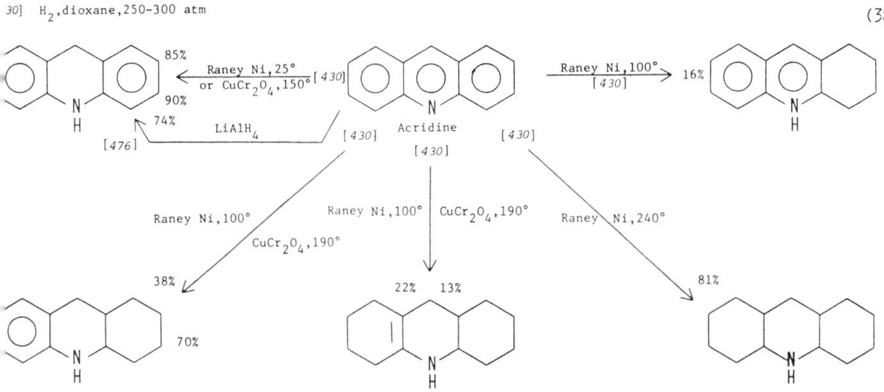

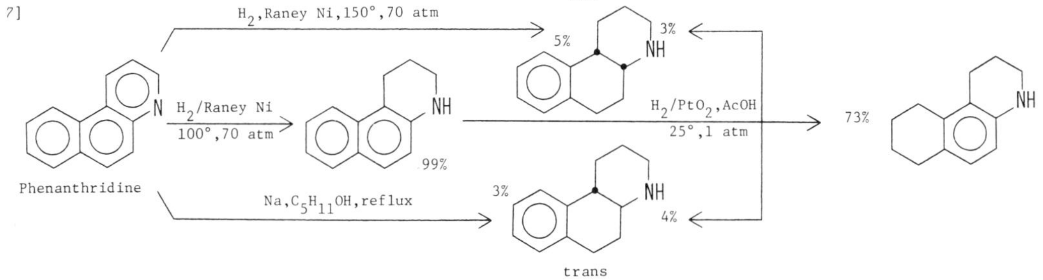

Aromatic heterocycles containing two nitrogen atoms are best reduced by catalytic hydrogenation. Other reduction methods may cleave some of the rings. Even catalytic hydrogenation causes occasional hydrogenolysis.

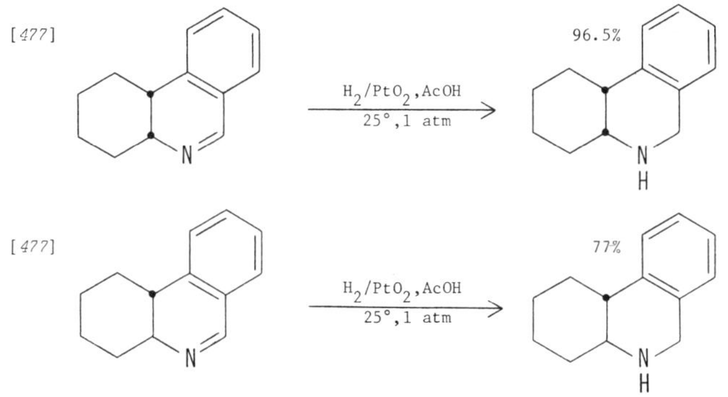

Pyrazole was *hydrogenated* over palladium on barium sulfate in acetic acid to 4,5-dihydropyrazole ($\triangle^2$-pyrazoline), and 1-phenylpyrazole at 70–80° to 1-phenylpyrazolidine [478]. In benzopyrazole (*indazole*) and its homologs and derivatives the six-membered ring is hydrogenated preferentially to give 4,5,6,7-tetrahydroindazoles over platinum, palladium or rhodium in acetic, hydrochloric, sulfuric and perchloric acid solutions in 45–96% yields [479].

Imidazole was converted by *hydrogenation* over platinum oxide in acetic anhydride to 1,3-diacetylimidazolidine in 80% yield, and *benzimidazole* similarly to 1,3-diacetylbenzimidazoline in 86% yield [480]. While benzimidazole is very resistant to hydrogenation over platinum at 100° and over nickel at 200° and under high pressure, 2-alkyl- or 2-aryl-substituted imidazoles are reduced in the benzene ring rather easily. 2-Methylbenzimidazole was hydrogenated over platinum oxide in acetic acid at 80–90° to 2-methyl-4,5,6,7-tetrahydrobenzimidazole in 87% yield [481]. Lithium aluminum hydride reduced benzimidazole to 2,3-dihydrobenzimidazole [476].

Six-membered rings with two nitrogen atoms behave differently depending on the position of the nitrogen atoms. 1,2-Diazines (pyridazines) are very stable to catalytic hydrogenation [482]. 1,3-Diazine (pyrimidine) and its 2-methyl, 4-methyl and 5-methyl homologs were easily *hydrogenated* in aqueous hydrochloric acid over 10% palladium on charcoal to 1,4,5,6-tetrahydropyrimidines in yields of 97–98% [483].

1,4-Diazines (pyrazines) afforded hexahydro derivatives by *catalytic hydrogenation* over palladium on charcoal at room temperature and 3–4 atm [484, 485] (yield of 2-butylpiperazine was 62% [484], yield of 2,5-diphenylpiperazine 80% [485]). *Electrolytic reduction* gave unstable 1,2-, 1,4- and 1,6-dihydropyrazines [485].

In condensed systems containing rings with two nitrogen atoms it is almost exclusively the heterocyclic ring which is reduced preferentially. 4-Phenylcinnoline was *hydrogenated* in acetic acid over palladium almost quantitatively to 4-phenyl-1,4-dihydrocinnoline [486], and this over platinum to 4-phenyl-1,2,3,5-tetrahydrocinnoline [487]. 4-Methylcinnoline gave either 1,4-dihydrocinnoline or 3-methylindole on *electrolytic reduction* under different conditions [488].

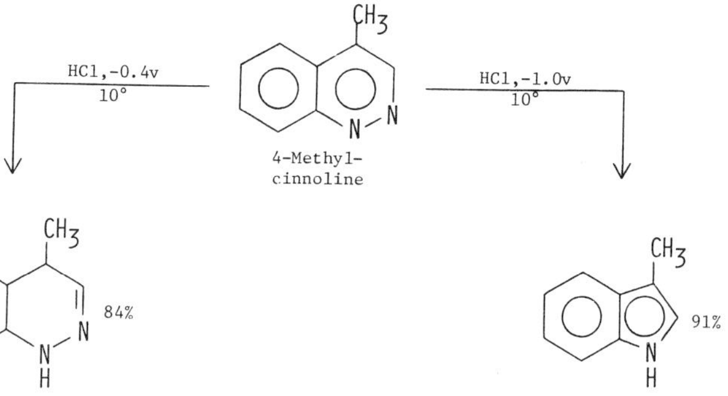

(41)

4-Methyl-
cinnoline

Quinazoline was *reduced by hydrogen* over platinum oxide to 3,4-dihydro-quinazoline [*489*], and by *sodium borohydride* in trifluoroacetic acid to 1,2-dihydroquinazoline [*490*].

(42)

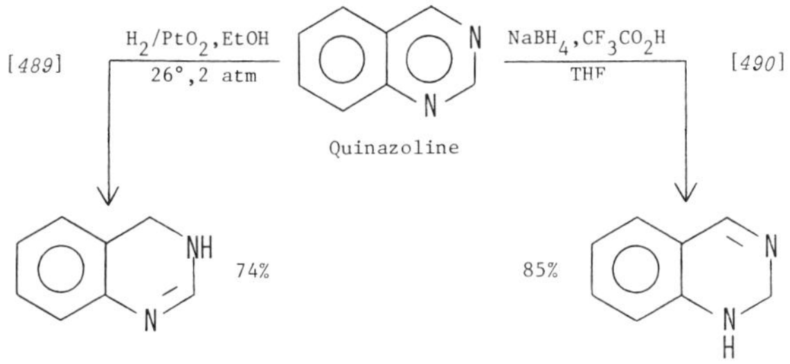

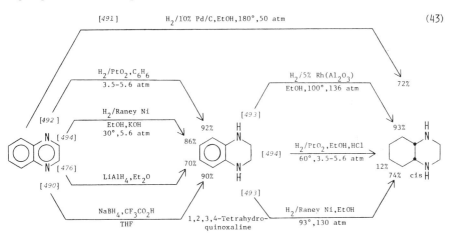

Reduction of **quinoxalines** was carried out by *catalytic hydrogenation* [*491, 492, 493, 494*], with *sodium borohydride* [*490*] or with *lithium aluminum hydride* [*476*] to different stages of saturation.

Phthalazine and its homologs and derivatives are easily hydrogenolyzed. *Electroreduction* in alkaline medium gave 1,2-dihydrophthalazines [*495*], while in acidic media [*495*] and on *catalytic hydrogenation* [*496*], the ring was cleaved to yield *o*-xylene-α,α′-diamine.

(44)

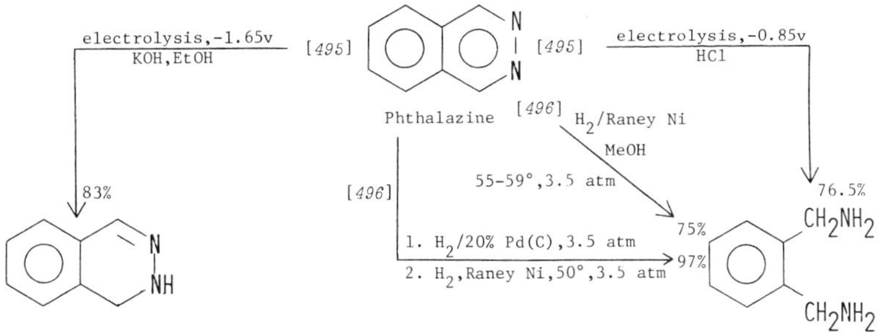

Phenazine (9,10-diazaanthracene) was partially reduced by *lithium aluminum hydride* to 9,10-dihydrophenazine (yield 90%) [476] and totally reduced by *catalytic hydrogenation* over 10% palladium on carbon in ethanol at 180° and 50 atm to tetradecahydrophenazine (80% yield) [491]. Catalytic hydrogenation of **1,10-phenanthroline** afforded 1,2,3,4-tetrahydro- and *sym*-octahydrophenanthroline [497].

(45)

[497]

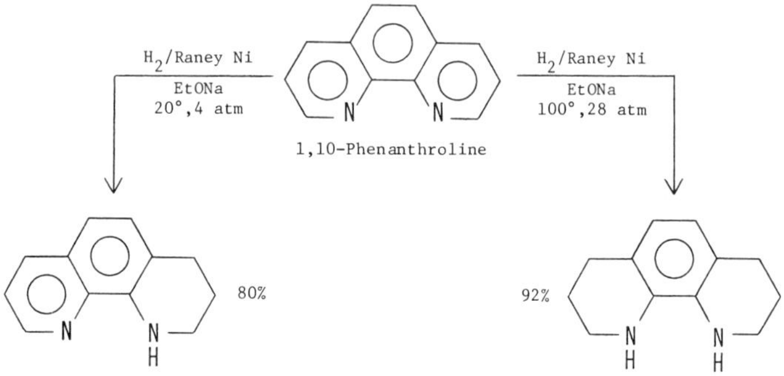

1,10-Phenanthroline

80%   92%

Because of easily controlled reaction conditions catalytic hydrogenation was and is used frequently for partial reductions of complex natural products and their derivatives [498, 499]. A random example is reduction of yobyrine [498].

[498]

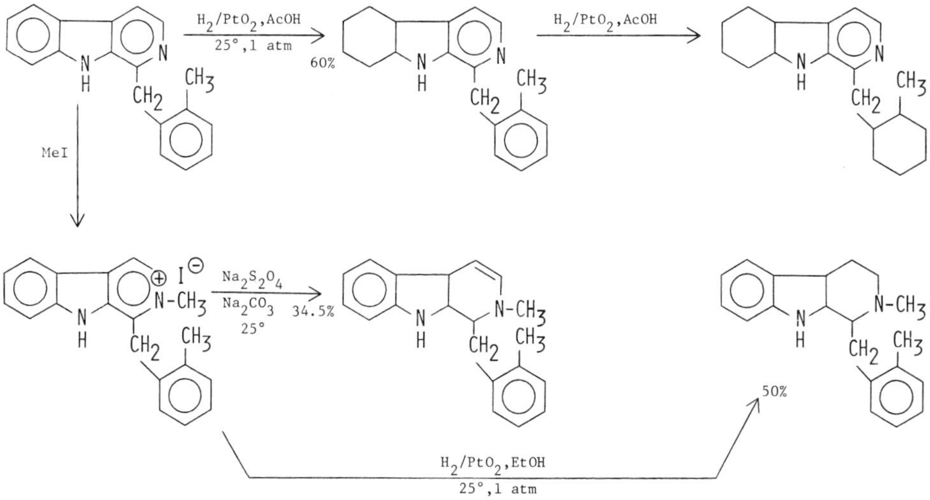

REDUCTION OF HALOGEN DERIVATIVES OF HYDROCARBONS AND BASIC HETEROCYCLES

For the replacement of halogen by hydrogen – hydrogenolysis of the carbon–halogen bond – many reduction methods are available. Since breaking of carbon–halogen bonds is involved, the ease of hydrogenolysis decreases in the series iodide, bromide, chloride and fluoride and increases in the series aliphatic < aromatic < vinylic < allylic < benzylic halides. In catalytic hydrogenation the rate of reduction increases from primary to tertiary halides whereas in reductions with complex hydrides, where the reagent is a strong nucleophile, the trend is just the opposite.

Reduction of Haloalkanes and Halocycloalkanes

Alkyl fluorides are for practical purposes resistant to reduction. In a few instances, where fluorine in fluoroalkanes was replaced by hydrogen, very energetic conditions were required and usually gave poor yields [500, 501]. For this reason it is not difficult to replace other halogens present in the molecule without affecting fluorine.

Alkyl chlorides are with a few exceptions not reduced by mild catalytic hydrogenation over platinum [502], rhodium [40] and nickel [63], even in the presence of alkali. Metal hydrides and complex hydrides are used more successfully: various *lithium aluminum hydrides* [506, 507], *lithium copper hydrides* [501], *sodium borohydride* [504, 505], and especially different *tin hydrides (stannanes)* [503, 508, 509, 510] are the reagents of choice for selective replacement of halogen in the presence of other functional groups. In some cases the reduction is stereoselective. Both *cis-* and *trans*-9-chlorodecalin, on reductions with triphenylstannane or dibutylstannane, gave predominantly *trans*-decalin [509].

(47)

$$C_8H_{17}Cl \longrightarrow C_8H_{18}$$

[506]	LiAlH$_4$,THF,25°,24 hrs	73%
[506]	NaAlH$_4$,THF,25°,24 hrs	38%
[506]	LiAlH(OMe)$_3$,THF,25°,24 hrs	19%
[506]	LiBHEt$_3$,THF,25°,24 hrs	73%
[507]	2 LiAlH(OMe)$_3$·CuI,THF,25°,15 hrs	96%
[505]	NaBH$_4$,DMSO,45°–50°,4 hrs	42%
[506]	NaBH$_4$,diglyme,25°,24 hrs	2.5%
[506]	LiAlH(OMe)$_3$,LiBH$_4$, or BH$_3$·SMe$_2$	0
[504]	*NaBH$_4$,DMSO,85°,48 hrs	67.2%
[503]	*Bu$_3$SnH,hν,7 hrs	70%

*2-Chlorooctane

Alkyl bromides and especially **alkyl iodides** are reduced faster than chlorides. *Catalytic hydrogenation* was accomplished in good yields using Raney nickel in the presence of potassium hydroxide [63] (*Procedure 5*, p. 205). More frequently, bromides and iodides are reduced by hydrides [508] and complex hydrides in good to excellent yields [501, 504]. Most powerful are *lithium triethylborohydride* and *lithium aluminum hydride* [506]. *Sodium borohydride* reacts much more slowly. Since the complex hydrides are believed to react by an S_N2 mechanism [505, 511], it is not surprising that secondary bromides and iodides react more slowly than the primary ones [506]. The reagent prepared from trimethoxylithium aluminum deuteride and cuprous iodide was found to be stereospecific giving, surprisingly, *endo-* and *exo-*2-deuteronorbornane from *endo-* or *exo-*2-bromonorbornane, respectively, with complete retention of configuration [*507*]. Reductions with *stannanes* follow practically a free radical mechanism and are stereoselective [*509*]. The order of reducing power of different stannanes was found to be: Ph_2SnH_2, Bu_3SnH > Ph_3SnH, Bu_2SnH_2 > Bu_3SnH [*508*] (*Procedure 24*, p. 210). Replacement of bromine or iodine can also be accomplished by reduction with metals, e.g. *magnesium* [*137*] and *zinc* [*157*], and with metal salts such as *chromous sulfate* [*193*].

(48)

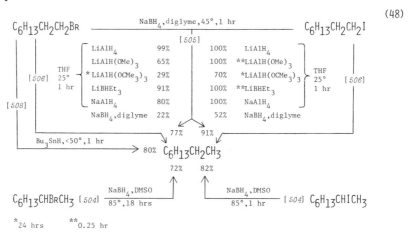

Geminal dihalides undergo partial or total reduction. The latter can be achieved by *catalytic hydrogenation* over platinum oxide [*512*], palladium [*512*] or Raney nickel [*63, 512*]. Both partial and total reduction can be accomplished with *lithium aluminum hydride* [*513*], with *sodium bis(2-methoxyethoxy)aluminum hydride* [*514*], with *tributylstannane* [*503, 514*], *electrolytically* [*515*], with *sodium* in alcohol [*516*] and with *chromous sulfate* [*193, 197*]. For partial reduction only, *sodium arsenite* [*220*] or *sodium sulfite* [*254*] are used.

(49)

[*514*]	$NaAlH_2(OCH_2CH_2OMe)_2$	syn 17%	anti 48%
[*514*]	Bu_3SnH	syn 59%	anti 23%
[*515*]	Hg cathode, LiCl		
	MeOH 90%, H_2O 10%	67–76%	13–14%
	AcOH	64–72%	16–18%
	DMF	65–73%	15–17%

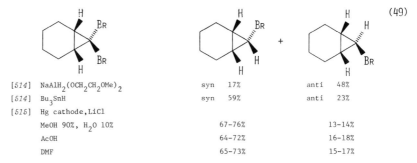

Trigeminal trihalides are completely reduced by *catalytic hydrogenation* over palladium [*62*] and Raney nickel [*63*], and partially reduced to dihalides or monohalides by *electrolysis* using mercury cathode [*518*], by *aluminum*

amalgam [145], by *sodium arsenite* [220, 517] or by *sodium sulfite* [254] in good to excellent yields (*Procedure 43*, p. 216). The trichloromethyl group activated by an adjacent double bond or carbonyl group is partly or completely reduced by *sodium amalgam* and *zinc* [519, 520] (p. 142).

[518]   (50)

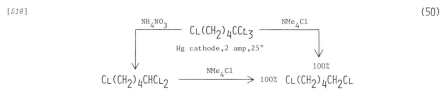

Selective replacement of halogens is possible if their reactivities are different enough. Examples are the replacement of two chlorine atoms in 1,2-dichloro-hexafluorocyclobutane by hydrogens with *lithium aluminum hydride* at 0° [521], and the reduction of 1-chloro-1-fluorocyclopropane with *sodium* in liquid ammonia at −78° which gave 40-84% yield of fluorocyclopropane with retention of configuration [522]. Both chlorine and iodine were replaced by hydrogen in 1-chloro-2-iodohexafluorocyclobutane using lithium aluminum hydride [523] (62% yield), bromine by hydrogen in 1-bromo-1-chloro-2,2,2-trifluoroethane with *zinc* in methanol [524] (yield 59%), and one bromine atom by hydrogen in 1,1-dibromo-1-chlorotrifluoroethane with *sodium sulfite* [254] (80-90% yield).

In **vicinal dihalides** the halogens may be replaced by hydrogen using *hydrogenation* over Raney nickel [63] or *lithium* in *tert*-butanol [525]. Alkenes are frequently formed using *alkylstannanes* [508, 526]. Such dehalogenations are carried out usually with magnesium and zinc but are not discussed in this book. 1,3-Dihalides are either reduced or converted to cyclopropanes depending on the halogens and the reducing reagents used [527].

(51)

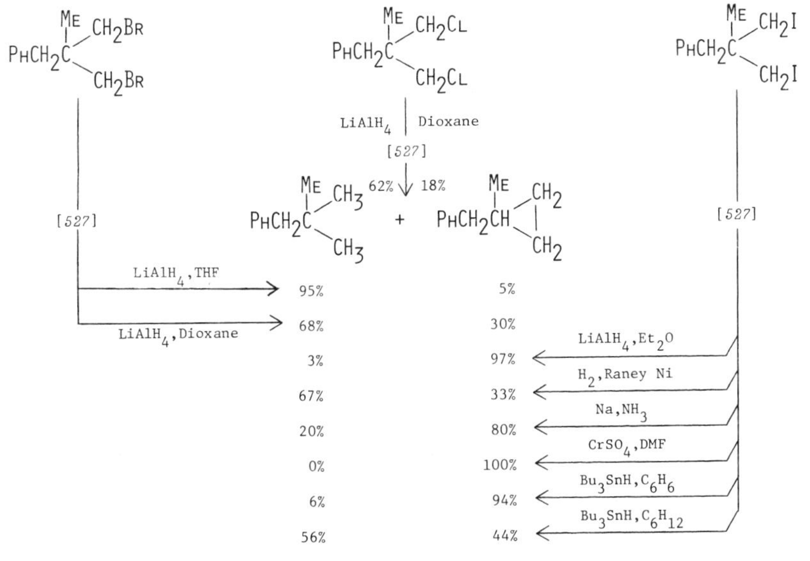

| | Me⎯CH₃ ⎯ + ⎯ Me⎯CH₂ | |
| | PhCH₂C⎯CH₃   PhCH₂CH⎯CH₂ | |

PhCH₂C(Me)(CH₂Br)CH₂Br   PhCH₂C(Me)(CH₂Cl)CH₂Cl   PhCH₂C(Me)(CH₂I)CH₂I

LiAlH₄ | Dioxane
[527]

Me⎯CH₃ 62% ↓ 18% Me⎯CH₂
PhCH₂C⎯CH₃   +   PhCH₂CH⎯CH₂

[527]   [527]

Conditions		
LiAlH₄,THF	95%	5%
LiAlH₄,Dioxane	68%	30%
LiAlH₄,Et₂O	3%	97%
H₂,Raney Ni	67%	33%
Na,NH₃	20%	80%
CrSO₄,DMF	0%	100%
Bu₃SnH,C₆H₆	6%	94%
Bu₃SnH,C₆H₁₂	56%	44%

Reduction of Haloalkenes, Halocycloalkenes and Haloalkynes

The ease of hydrogenolytic replacement of halogens in halogenated unsaturated hydrocarbons depends on the mutual position of the halogen and the multiple bond. In *catalytic hydrogenation* **allylic or propargylic halogen** is replaced very easily with the concomitant saturation of the multiple bond. **Vinylic** and **acetylenic halogens** are hydrogenolyzed less readily than the allylic ones but more easily than the halogens in saturated halides. The multiple bond is again hydrogenated indiscriminately [63].

The conspicuously high speed of hydrogenolysis of allylic and vinylic halogens as compared with those removed further from the multiple bonds or halogens in alkanes implies that the multiple bond participates in a multicenter transition state [63]. Such a mechanism would account even for the surprisingly easy hydrogenolysis of allylic and vinylic fluorine [66, 528, 529, 530].

Hydrogenations of all types of unsaturated halides are carried out over Raney nickel in the presence of potassium hydroxide at room temperature and atmospheric pressure with yields of 32–92%. Even perhalogenated alkenes such as perchlorocyclopentadiene are reduced to the saturated parent compounds [63].

In the absence of potassium hydroxide hydrogenation of vinylic and allylic chlorides to the saturated chlorides (with only partial or no hydrogenolysis of chlorine) is accomplished over 5% platinum, 5% palladium and, best of all, 5% rhodium on alumina [40].

In reductions of allyl and vinyl halides with other reducing agents the double bond is usually conserved. Allylic halides are readily reduced to alkenes in high yields by treatment with *chromous sulfate* in dimethyl formamide [193]. Allylic halogen may be replaced in preference to the vinylic halogen. In 2-fluoro-1,1,3-trichloro-1-propene *lithium aluminum hydride* reduced only the allylic chlorine and neither of the vinylic halogens [531]. If, however, the vinylic halogen is considerably more reactive, it can be replaced preferentially. In 2-iodo-4,4,4-trichloro-1,1,1-trifluoro-2-butene reduction with *zinc* replaced only the vinylic iodine and none of the allylic chlorines, giving 1,1,1-trifluoro-4,4,4-trichloro-2-butene in 63% yield [532]. In reductions with *sodium* or *lithium* in *tert*-butyl alcohol both the allylic and vinylic chlorines were replaced indiscriminately [533]. Surprisingly, *chromous acetate* hydrogenolyzed chlorine once removed from a double bond in preference to both allylic and vinylic chlorines [198]. By reduction with zinc dust in deuterium oxide in the presence of sodium iodide and cupric chloride all six chlorines in perchloro-1,3-butadiene are replaced by deuterium in 80–85% yield [534].

Strongly nucleophilic *lithium aluminum hydride* and *sodium borohydride* in diglyme replace vinylic halogens including fluorine [521, 535], sometimes under surprisingly gentle conditions [521]. Because the carbon atom linked to fluorine is more electrophilic than that bonded to chlorine, fluorine in 1-

chloroperfluorocyclopentene was replaced by hydrogen preferentially to chlorine using sodium borohydride (yield 88%) [535].

(52)

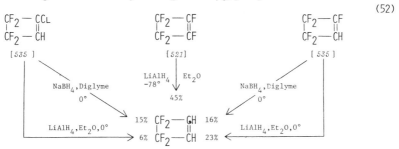

Reduction of Haloaromatics

In the aromatic series both **benzylic halogens** and **aromatic halogens** can be replaced by hydrogen by means of *catalytic hydrogenation*, benzylic halogens much faster than the aromatic ones [502]. Palladium on calcium carbonate [62] and, better still, Raney nickel [63, 536] in the presence of potassium hydroxide are the catalysts of choice for hydrogenations at room temperature and atmospheric pressure with good to excellent yields. Also *hydrogen transfer* from triethylammonium formate over 5% palladium on charcoal proved successful in refluxing aromatic halogen by hydrogen at 50-100° in good yields [317a]. In polyhaloaromatics partial replacement of halogens can be achieved by using amounts of potassium hydroxide equivalent to the number of halogens to be replaced [536].

Benzylic halides are reduced very easily using complex hydrides. In α-chloroethylbenzene *lithium aluminium deuteride* replaced the benzylic chlorine by deuterium with inversion of configuration (optical purity 79%) [537]. *Borane* replaced chlorine and bromine in chloro- and bromodiphenylmethane, chlorine in chlorotriphenylmethane and bromine in benzyl bromide by hydrogen in 90-96% yields. Benzyl chloride, however, was not reduced [538]. Benzylic chlorine and bromine in a *sym*-triazine derivative were hydrogenolyzed by sodium iodide in acetic acid in 55% and 89% yields, respectively [539].

The difference in the reactivity of benzylic versus aromatic halogens makes it possible to reduce the former ones preferentially. *Lithium aluminum hydride* replaced only the benzylic bromine by hydrogen in 2-bromomethyl-3-chloronaphthalene (yield 75%) [540]. Sodium borohydride in diglyme reduces, as a rule, benzylic halides but not aromatic halides (except for some iodo derivatives) [505, 541]. Lithium aluminum hydride hydrogenolyzes benzyl halides and aryl bromides and iodides. Aryl chlorides and especially fluorides are quite resistant [540, 542]. However, in polyfluorinated aromatics, because of the very low electron density of the ring, even fluorine was replaced by hydrogen using lithium aluminum hydride [543].

Bromobenzene was reduced also with *sodium bis(2-methoxy-*

ethoxy)aluminum hydride [544] and with *triethylsilane* in the presence of palladium [112]. *Triphenylstannane* can be used for selective reductions of aryl halides as it can replace aromatic bromine but not chlorine by hydrogen [545].

(53)

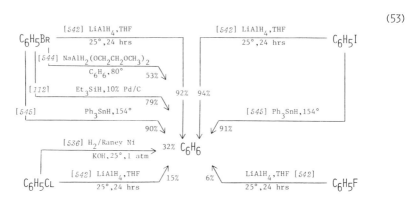

Occasionally metals have been used for the hydrogenolysis of aromatic halides. *Sodium* in liquid ammonia replaced by hydrogen even fluorine bonded to the benzene ring [546], *zinc* with potassium hydrogenolyzed bromobenzene [157] and zinc in acetic acid replaced both α-bromine atoms but not the β-bromine in 2,3,5-tribromothiophene [547]. *Magnesium* can be used indirectly for conversion of an aromatic bromide or iodide to a Grignard reagent, which on treatment with water gives the parent compound.

An interesting case is the hydrogenolysis of aryl chlorides, bromides and iodides in respective yields of 72%, 72% and 82% on irradiation in *isopropyl alcohol* [310].

Halogens in aromatic heterocycles are replaced by hydrogen in *catalytic hydrogenations* without the reduction of the aromatic rings [548]. In pyridine and quinoline and their derivatives halogens in α and γ positions are replaced especially easily. (This is quite understandable since such halogen derivatives are actually disguised imidoyl chlorides or their vinyl analogs, and those compounds undergo easy hydrogenolysis.) Catalytic hydrogenation over palladium in acetic acid and alkaline acetate [549, 550, 551] or over Raney nickel in alkaline medium [552, 553] hydrogenolyzed α- and γ-chlorine atoms in yields up to 94%. All three chloropyridines [63] and bromopyridines [536, 554] were converted almost quantitatively to pyridine over Raney nickel at room temperature and atmospheric pressure.

Halogens in α and γ positions to the nitrogen atoms are distinctly more reactive than those in the β positions also toward nucleophilic reagents. In β-chlorotetrafluoropyridine *lithium aluminum hydride* replaced by hydrogen the fluorine in the γ position in preference to chlorine in the β position [555]. In pyrimidine derivatives chlorine α to nitrogen was hydrogenolyzed in 89% yield by refluxing with *zinc dust* [556] and in 78% yield by heating at 100–110° with azeotropic *hydriodic acid* [557]. The reactive chlorine in 3-chloro-4,5-

benzopyrazole was replaced with hydrogen by refluxing for 24 hours with hydriodic acid and phosphorus (yield 82–86%) [558].

Replacement of halogen in compounds with functional groups is discussed in the appropriate chapters.

REDUCTION OF NITRO, NITROSO, DIAZO AND AZIDO DERIVATIVES OF HYDROCARBONS AND BASIC HETEROCYCLES

Nitrogen-containing substituents of the above type undergo exceedingly easy reduction by many reagents. Nitro and nitroso groups behave differently depending whether they are bonded to tertiary or aromatic carbon, or else to carbon carrying hydrogens. In the latter case tautomerism generates forms which are reduced in ways different from those of the simple nitro or nitroso groups.

(54)

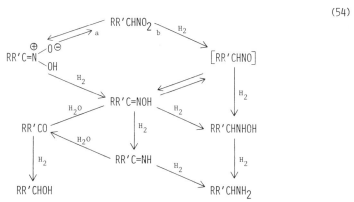

Aliphatic nitro compounds with the nitro group on a tertiary carbon were reduced to amines with *aluminum amalgam* [146] or *iron* [559]. 2-Nitro-2-methylpropane afforded *tert*-butylamine in 65–75% yield [146]. Even some secondary nitroalkanes were *hydrogenated* to amines. *trans*-1,4-Dinitrocyclohexane was converted to *trans*-1,4-diaminocyclohexane with retention of configuration. This may be considered as an evidence that the intermediate nitroso compound is reduced directly and not after tautomerization to the isonitroso compound [560] (see Scheme 54).

(55)

[560]

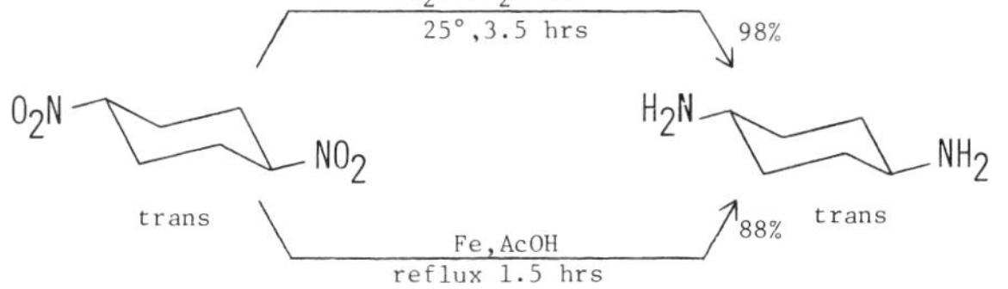

$$O_2N \diagdown\diagup NO_2 \quad \xrightarrow[\substack{H_2/PtO_2,AcOH \\ 25°,3.5 \text{ hrs}}]{98\%} \quad H_2N \diagdown\diagup NH_2$$

trans   trans

Fe,AcOH
reflux 1.5 hrs   88%

Generally primary nitro compounds are reduced to aldoximes and secondary to ketoximes by metal salts. On reduction with *stannous chloride* 1,5-

dinitropentane gave 55–60% yield of the dioxime of glutaric dialdehyde [177]; 3β,5α-dichloro-6β-nitrocholestane with *chromous chloride* gave 60% yield of oxime of 3β-chloro-5α-hydroxycholestan-6-one [194]; and nitrocyclohexane was reduced to cyclohexanone oxime *catalytically* [561] or with *sodium thiosulfate* [562].

(56)

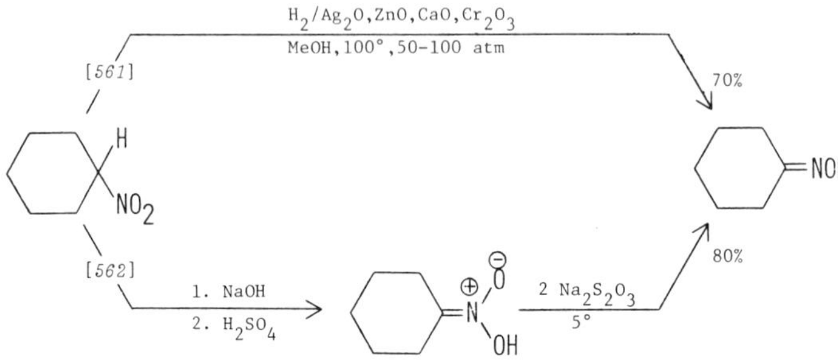

Although primary and secondary nitro compounds may be converted, respectively, to aldehydes and ketones by consecutive treatment with alkalis and sulfuric acid (Nef's reaction) the same products can be obtained by reduction with *titanium trichloride* (yields 45–90%) [563] or *chromous chloride* (yields 32–77%) [190]. The reaction seems to proceed through a nitroso rather than an aci-nitro intermediate [563] (Scheme 54, route b).

Reduction of the nitro group is one of the easiest reductions to accomplish. However, several instances have been recorded of preferential reduction of carbon–carbon double bonds in unsaturated nitro compounds. In 4-nitrocyclohexene with a bulky substituent in position 5 the double bond was saturated by *hydrogenation* over platinum oxide without change of the nitro group [564], and even a pyridine ring was hydrogenated in preference to a remote nitro group under special conditions (platinum oxide, acetic acid, room temperature, 3.5 atm; yield 90%) [565]. Diazo and azido groups, however, are reduced preferentially to nitro groups (pp. 75, 76).

The outcome of the reduction of primary or secondary **nitro compounds with conjugated double bonds** depends on the mode of addition of hydrogen, which in turn depends on the reagents used. 1,4-Addition results in the reduction of the carbon–carbon double bond and leads to a saturated nitro compound. Dimeric compounds formed by coupling of the half-reduced intermediates sometimes accompany the main products [566]. 1-Nitrooctane was prepared in 85% yield by reducing 1-nitrooctene, and β-nitrostyrene gave 48% of β-nitroethylbenzene and 43% of 2,4-dinitro-1,3-diphenylbutane on reduction with *sodium borohydride* [566]. Saturated nitro compounds were also obtained by specially controlled reduction with *lithium aluminum hydride* [567] and by treatment of the unsaturated nitro derivatives with *formic acid* and triethylamine [317]. Ring-substituted β-nitrostyrenes were reduced to β-nitroethylbenzenes by homogeneous *catalytic hydrogenation* using tris(triphenylphosphine)rhodium chloride in 60–90% yields [56], and to βaminoethylbenzenes by hydrogenation over 10% palladium on carbon in aqueous hydrochloric acid at 85° and 35 atm (yield 82%) [*568*].

If the addition of hydrogen takes place in a 1,2-mode the products are oximes, hydroxylamines, amines, and carbonyl compounds resulting from the hydrolysis of the oximes [*567*]. Oximes and carbonyl compounds also result from reductions of α,β-unsaturated compounds with iron [*569*, *570*] and an oxime was prepared by catalytic hydrogenation of a β-nitrostyrene derivative over palladium in pyridine (yield 89%) [*571*].

(57)

		$PhCH_2CH(Me)NO_2$	$PhCH_2C(Me)=NOH$		$PhCH_2CH(Me)NHOH$		$PhCH_2CH(Me)NH_2$
[*567*]	$LiAlH_4$ (1.5 mol)				23%		44%
	$LiAlH_4$ (1.25 mol)						37%
	$LiAlH_4$ (0.75 mol)		16%		34%		8%
	$LiAlH_4$ (0.3 mol)	38–56%					
[*569*]	Fe, $FeCl_3$, H_2O, HCl (trace), MeOH, 85–95°		60–65%	$\xrightarrow[HCl]{H_2O}$		$PhCH_2COMe$	
[*569*]	Fe, $FeCl_3$, H_2O, HCl, 85–95°		13%			56%	

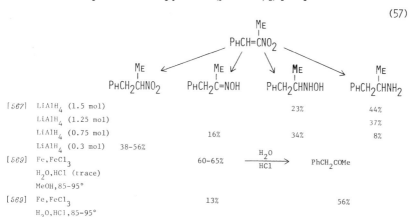

Aromatic nitro compounds were among the first organic compounds ever reduced. The nitro group is readily converted to a series of functions of various degrees of reduction: very exceptionally to a nitroso group, more often to a hydroxylamino group and most frequently to the amino group. In addition azoxy, azo and hydrazo compounds are formed by combination of two molecules of the reduction intermediates (Scheme 58).

(58)

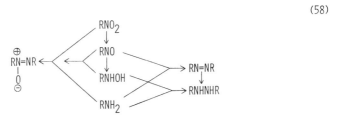

With the exception of the nitroso stage, all the intermediate stages of the reduction of nitro compounds can be obtained by controlled *catalytic hydrogenation* [*572*], and all reduction intermediates were prepared by reduction with appropriate *hydrides or complex hydrides*. However, the outcome of many hydride reductions is difficult to predict. Therefore more specific reagents are preferred for partial reductions of nitro compounds.

By controlling the amount of hydrogen and the pH of the reaction, hydroxylamino, azoxy, azo, hydrazo and amino compounds were obtained in good yields by catalytic hydrogenation over 2% palladium on carbon at room temperature and atmospheric pressure [572].

Nitroso compounds are usually not obtained directly but rather by reoxidation of hydroxylamino compounds or amines. **Hydroxylamino compounds** are prepared by *electrolytic reduction* using a lead anode and a copper cathode [573], by *zinc* in an aqueous solution of ammonium chloride [574] or by *aluminum amalgam* [147], generally in good yields.

Azoxybenzene was synthesized in 85% yield by reduction of nitrobenzene with sodium arsenite [221]. Nitrotoluenes and 2,5-dichloronitrobenzene were converted to the corresponding **azoxy compounds** by heating to 60-90° with hexoses (yields up to 74%) [316]. Some ring-substituted nitrobenzenes were converted to azoxy compounds, some other to azo compounds by *sodium bis(2-methoxyethoxy)aluminum hydride* [575].

Reduction of nitro compounds with *zinc* in alkali hydroxides gives **azo** and **hydrazo compounds.** With an excess of zinc hydrazo compounds are obtained which are easily reoxidized to azo compounds (*Procedure 32*, p. 213).

Direct preparation of **azo compounds** in good yields is accomplished by treatment of nitro compounds with *lithium aluminum hydride* [576], with *magnesium aluminum hydride* [577], with *sodium bis(2-methoxyethoxy)aluminum hydride* [575], with *silicon* in alcoholic alkali [331] or with *zinc* in strongly alkaline medium [578]. Hydrazobenzene was obtained by controlled hydrogenation of nitrobenzene in alkaline medium (yield 80%) [572] and by reduction with *sodium bis(2-methoxyethoxy)alumium hydride* (yield 37%) [544].

Reduction of azo and hydrazo compounds is discussed on pp. 95, 96.

(59)

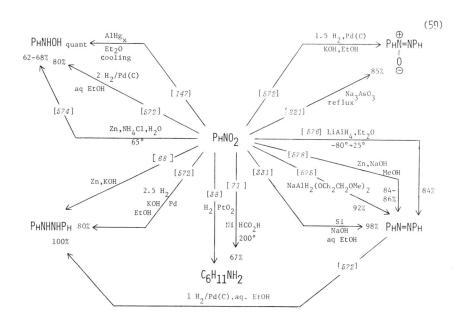

Complete reduction of nitro compounds to amines is accomplished by *catalytic hydrogenation*. The difference in the rate of hydrogenation of the nitro group and of the aromatic ring is so large that the hydrogenation may be easily regulated not to reduce the aromatic ring. Catalysts used for the conversion of aromatic nitro compounds to amines are: platinum oxide [*38*, *579*], rhodium–platinum oxide [*38*], palladium [*38, 580*], Raney nickel [*45*], copper chromite [*50*], rhenium sulfide [*53*] and others. Instead of hydrogen *hydrazine* [*277, 278*], *formic acid* [*71*] or *triethylamine formate* [*73*] have been used for catalytic reduction. Hydrides and complex hydrides are not used for complete reduction to amines since they usually reduce nitro compounds to azo and other derivatives, or else do not reduce the nitro group at all (like sodium borohydride [*581*], diborane, lithium hydride, lithium triethylboro-hydride and sodium trimethoxyborohydride).

The most popular reducing agent for conversion of aromatic nitro compounds to amines is *iron* [*166*]. It is cheap and gives good to excellent yields [*165, 582*]. The reductions are usually carried out in aqueous or aqueous alcoholic media and require only catalytic amounts of acids (acetic, hydrochloric) or salts such as sodium chloride, ferrous sulfate or, better still, ferric chloride [*165*]. Thus the reductions are run essentially in neutral media. The rates of the reductions and sometimes even the yields can be increased by using iron in the form of small particles [*165*]. Iron is also suitable for reduction of complex nitro derivatives since it does not attack many functional groups [*583*].

(60)

$PhNO_2$ $\xrightarrow{\hspace{6cm}}$ $PhNH_2$

[*572*]	H$_2$/2% Pd(C),aq. EtOH,25°,1 atm	>90%
[*38*]	H$_2$/PtO$_2$,EtOH,25°,1 atm	quant.
[*38*]	H$_2$/PtO$_2$-RhO$_2$,EtOH,25°,1 atm	quant.
[*53*]	H$_2$/Re$_2$S$_7$,EtOH,25°,63 atm	100%
[*50*]	H$_2$/CuCr$_2$O$_4$,175°,250-300 atm	100%
[*277*]	N$_2$H$_4$·H$_2$O,Raney nickel,PhMe,EtOH,reflux	95%
[*71*]	HCO$_2$H,Cu,200°	quant.
[*585*]	Zn,HCl	55%
[*586*]	SnCl$_2$,HCl	+
[*165*]	Fe,H$_2$O,FeCl$_3$,100°	45.6-100%

Zinc reduces nitro groups in different ways depending mainly on the pH. Reduction to amino groups can be achieved in aqueous alcohol in the presence of calcium chloride [*584*]. In strong hydrochloric acid the aromatic ring may be chlorinated in positions *para* or *ortho* to the amino group [*585*]. The same is true of reduction with tin [*586*]. Other common reducing agents are *titanium trichloride* [*201*], *stannous chloride* [*587*], *ferrous sulfate* [*218*], *hydrogen sulfide* [*236*] or its salts [*239, 241*], and *sodium hydrosulfite* (hyposulfite, dithionite) [*256, 257*] (*Procedure 44*, p. 216).

Most of the reagents used for the preparation of aromatic amines from nitro compounds can be used for complete reduction of **polynitro compounds** to polyamines [*166, 582*]. However, not all of them are equally suited for partial or selective reductions of just one or one particular nitro group.

Partial reduction of dinitro compounds was accomplished by *catalytic hydrogen transfer* using palladium on carbon and cyclohexene [*588*] or triethylammonium formate [*73*]. Iron was found superior to stannous chloride because it does not require strongly acidic medium which dissolves the initially formed nitro amine and makes it more susceptible to complete reduction [*165, 589*] (*Procedure 34*, p. 213). However, *stannous chloride* is frequently used for selective reductions in which one particular nitro group is to be reduced preferentially or exclusively [*587*]. Other reagents for the same purpose are *titanium trichloride* [*590*], *hydrogen sulfide* in pyridine [*236*] and *sodium* or *ammonium sulfide* [*236, 238, 591*]. Surprisingly enough it is usually the sterically less accessible *ortho* nitro group which is reduced.

Triethylammonium formate in the presence of palladium reduced the less hindered *para* nitro group in 2,4-dinitrotoluene but the more hindered *ortho* nitro group in 2,4-dinitrophenol, 2,4-dinitroanisole, 2,4-dinitroaniline and 2,4-dinitroacetanilide in yields of 24–92% [*73*].

(61)

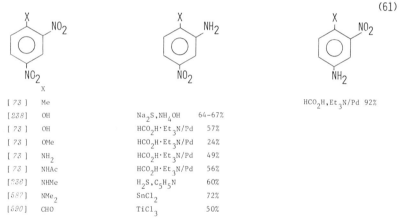

	X		
[*73*]	Me		
[*238*]	OH	Na_2S,NH_4OH	64–67%
[*73*]	OH	$HCO_2H \cdot Et_3N/Pd$	57%
[*73*]	OMe	$HCO_2H \cdot Et_3N/Pd$	24%
[*73*]	NH_2	$HCO_2H \cdot Et_3N/Pd$	49%
[*73*]	NHAc	$HCO_2H \cdot Et_3N/Pd$	56%
[*238*]	NHMe	H_2S,C_5H_5N	60%
[*587*]	NMe_2	$SnCl_2$	72%
[*590*]	CHO	$TiCl_3$	50%

$HCO_2H,Et_3N/Pd$ 92%

Reduction of aromatic nitro group takes preference to the reduction of the aromatic ring. Under certain conditions, however, even the benzene ring was reduced. Hydrogenation of nitrobenzene over platinum oxide or rhodium–platinum oxide in ethanol yielded aniline while in acetic acid cyclohexylamine was produced [*38*]. Heating of nitrobenzene with formic acid in the presence of copper at 200° gave a 100% yield of aniline, whereas similar treatment in the presence of nickel afforded 67% of cyclohexylamine [*71*].

Halogenated nitro compounds may suffer replacement of halogen during some reductions: thus 9-bromo-10-nitrophenanthrene reduced with *zinc* and acid, with *stannous chloride*, with *ammonium sulfide*, or with *hydrazine* in the presence of palladium gave 9-aminophenanthrene (87.5% yield). Reduction to 9-bromo-10-aminophenanthrene was accomplished with iron in dilute acetic acid (yield 35%) [*278*].

2-Iodo-7-nitrofluorene gave 2-iodo-7-aminofluorene in 85% yield by reduction with *hydrazine* in the presence of Raney nickel [*277*], and 2-iodo-6-nitronaphthalene and 3-iodo-6-nitronaphthalene afforded the corresponding iodoaminonaphthalenes in almost quantitative yields on treatment with aqueous-alcoholic solutions of sodium hydrosulfite (hyposulfite, dithionite) [*257*].

A very reliable reagent for reduction^5 of aromatic nitro groups in the presence of sensitive functions is *iron*, which does not affect carbon–carbon double bonds [*583*].

Chlorine was replaced by hydrogen in 2,4,6-trinitrochlorobenzene by heating with sodium iodide and acetic acid at 100° without the reduction of any nitro group [*223*].

Primary and secondary **nitroso compounds** tautomerize to isonitroso compounds – oximes of aldehydes and ketones, respectively. Their reductions are dealt with in the sections on derivatives of carbonyl compounds (pp. 106, 132).

Tertiary and aromatic nitroso compounds are not readily accessible; consequently not many reductions have been tried. Nitrosobenzene was converted to azobenzene by *lithium aluminum hydride* (yield 69%) [*592*], and *o*-nitrosobiphenyl to carbazole, probably via a hydroxylamino intermediate, by treatment with *triphenylphosphine or triethyl phosphite* (yields 69% and 76%, respectively) [*298*]. Nitrosothymol was transformed to aminothymol with *ammonium sulfide* (yield 73–80%) [*245*], and α-nitroso-β-naphthol to α-amino-β-naphthol with *sodium hydrosulfite* (yield 66–74%) [*255*].

Non-functionalized **aliphatic diazo compounds** are fairly rare, and so are their reductions. Good examples of the reduction of diazo compounds to either amines or hydrazones are found with α-diazo ketones and α-diazo esters (pp. 124, 125, 160).

Aromatic diazonium compounds which are prepared readily by diazotization of primary amines can be converted either to their parent compounds by replacement of the diazonium group with hydrogen, or to hydrazines by reduction of the diazonium group. Both reactions are carried out at room temperature or below using reagents soluble in aqueous solutions, and usually give high yields.

The former reaction is accomplished by *sodium borohydride* (yields 48–77%) [*593*], *sodium stannite* (yield 73%) [*219*], *hypophosphorous acid* (yields 52–87%) [*288, 594*] (*Procedure 46*, p. 217), *ethanol* (yields 79–93%) [*305, 306*], *hexamethylphosphoric triamide* (yields 71–94%) [*595*] and others [*289*]. Reducing agents suitable for the replacement of the diazonium groups by hydrogen usually do not attack other elements or groups, so they can be used for compounds containing halogens [*219, 288, 595*], nitro groups [*305*] and other functions (p. 142). Replacement of the aromatic amino group via diazotization was accomplished even in a compound containing an aliphatic amino group since at pH lower than 3 only the aromatic amino group was diazotized [*594*].

(62)

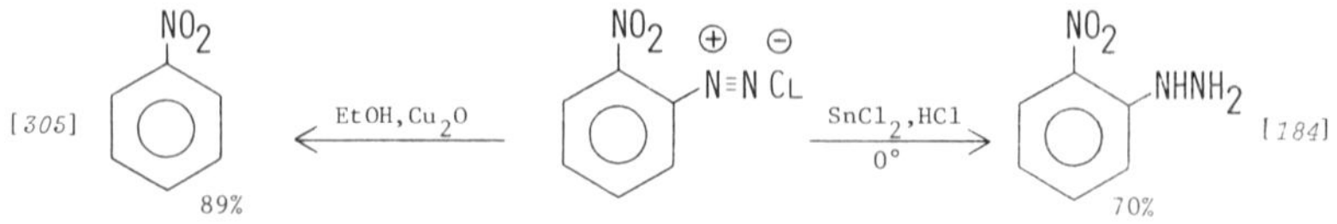

Conversion of diazonium salts to hydrazines was achieved by reduction with *zinc* in acidic medium (yields 85–90%) [596], with *stannous chloride* (yield 70%) [184], and *sodium sulfite* (yields 70–80%) [253, 597]. Nitro groups present in the diazonium salts survive the reduction unharmed [184, 253].

Azido groups are reduced to amino groups most readily. *Catalytic hydrogenation* is carried out over platinum (yield 81%) [598]. Since the volume of nitrogen eliminated during the reduction of the azido group equals that of the hydrogen used, it is impossible to follow the progress of the reduction by measurement of volume or pressure. The easy hydrogenation of the azido group made it possible to reduce an azido compound to an amino compound selectively over 10% palladium on charcoal without hydrogenolyzing a benzyloxycarbonyl group and even in the presence of divalent sulfur in the same molecule [599].

Lithium aluminum hydride reduced β-azidoethylbenzene to β-aminoethylbenzene in 89% yield [600]. The azido group was also reduced with *aluminum amalgam* (yields 71–86%) [149], with *titanium trichloride* (yields 54–83%) [601], with *vanadous chloride* (yields 70–95%) [217] (*Procedure 40*, p. 215), with *hydrogen sulfide* (yield 90%) [247], with *sodium hydrosulfite* (yield 90%) [259], with *hydrogen bromide* in acetic acid (yields 84–97%) [232], and with *1,3-propanedithiol* (yields 84–100%) [602]. Unsaturated azides were reduced to unsaturated amines with *aluminum amalgam* [149] and with *1,3-propanedithiol* [602].

Neither aromatic halogens [232, 602] nor nitro groups were affected during the reductions of the azido group [232, 247, 602]. α-Iodo azides gave, on reduction, aziridines or alkenes depending on the substituents and on the reagents used [603].

[603]   (63)

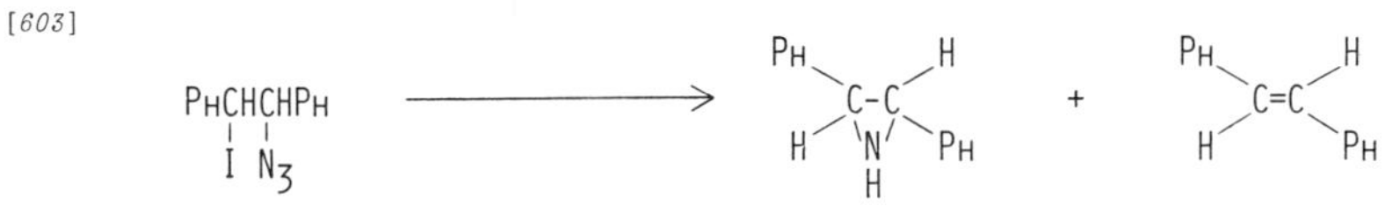

LiAlH$_4$		80%
LiAlH$_4$ + SnCl$_2$		80%
NaBH$_4$,diglyme		96%
LiAlHCl$_3$	83%	
B$_2$H$_6$,NaOH	87%	

REDUCTION OF ALCOHOLS AND PHENOLS AND THEIR SUBSTITUTION DERIVATIVES

Reduction of saturated alcohols to the parent hydrocarbons is not easy and requires rather energetic conditions. Catalytic hydrogenation over molybdenum or tungsten sulfides at 310–350° and 76–122 atm converted tertiary, secondary and primary alcohols to the corresponding alkanes in yields up to 96% [604].

Many hydroxy compounds would not survive such harsh treatment; therefore other methods must be used. Some alcohols were hydrogenolyzed with chloroalanes generated *in situ* from lithium aluminum hydride and aluminum chloride, but the reaction gave alkenes as by-products [605]. Tertiary alcohols were converted to hydrocarbon on treatment at room temperature with triethyl- or triphenylsilane and trifluoroacetic acid in methylene chloride (yields 41–92%). Rearrangements due to carbonium ion formation occur [343].

An elegant method for the replacement of hydroxy groups in alcohols by hydrogen is the addition of alcohol to dialkylcarbodiimide followed by catalytic hydrogenation of the intermediate *O*-alkoxy-*N*,*N'*-dialkylisoureas over palladium at 40–80° and 1–60 atm (yields 52–99%) [606].

[606] (64)

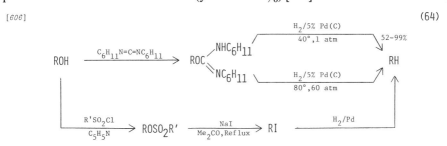

Another procedure for the hydrogenolysis of a hydroxyl is the conversion of the hydroxy compound to its arenesulfonyl ester, exchange of the sulfonyloxy group by iodine using sodium iodide, and replacement of the iodine by hydrogen via catalytic hydrogenation or other reductions (p. 63). Hydrogenolysis of hydroxyl is somewhat easier in polyhydric alcohols. Diols gave alcohols on hydrogenation over copper chromite at 200–250° and 175 atm, after hydrogenolysis of one hydroxyl group [420]. Vicinal diols were converted to alkenes stereospecifically by a special treatment with trialkyl phosphites [607]. Similar reaction can also be achieved by heating of vicinal diols with low-valent titanium obtained by reduction of titanium trichloride with potassium [206].

Results of the **reduction of unsaturated alcohols** depend on the respective positions of the hydroxyl and the double bond. Since the hydroxyl group is fairly resistant to hydrogenolysis by catalytic hydrogenation almost any catalyst working under mild conditions may be used for saturation of the double bond with conservation of the hydroxyl [608]. In addition, sodium in liquid ammonia and lithium in ethylamine reduced double bonds without affecting the hydroxyl in non-allylic alcohols [608].

With few exceptions [609], unsaturated alcohols with hydroxyls in allylic positions undergo hydrogenolysis and give alkenes. Such reductions were achieved by *chloroalanes* made from lithium aluminum hydride and alu-

(65)

[608]

CH₂CH₂OH → H₂/Raney Ni, 150 atm, 85% → CH₂CH₂OH

Na/NH₃, EtOH 95%

Li/EtNH₂ 60%

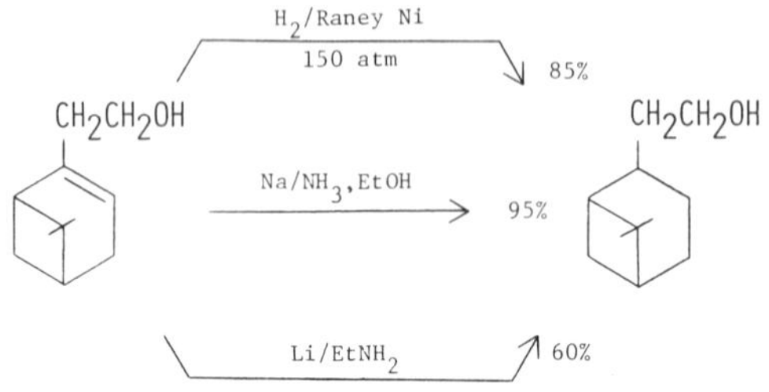

minum chloride [610] and by *sodium* in liquid ammonia [608, 611]. Rearrangement of the double bond usually takes place [611], and dehydration to a diene may be another side reaction.

(66)

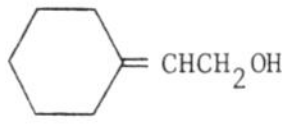

		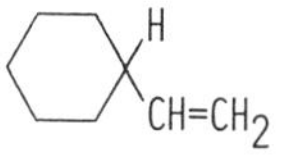	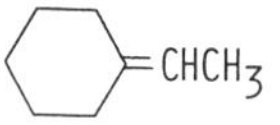	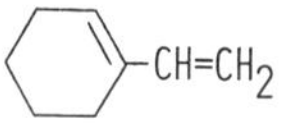	
⬡=CHCH₂OH	LiAlH₄/AlCl₃ Et₂O, 25°	10%	22%	32%	[610]
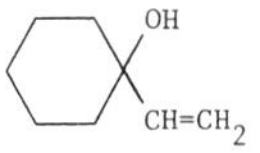 ⬡–OH	LiAlH₄/AlCl₃	13%	30%	20%	[610]
⬡ CH=CH₂	Na/NH₃, EtOH, Et₂O		50%		[611]
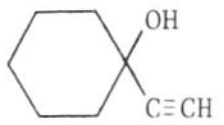 ⬡ C≡CH OH	Na/NH₃, EtOH, Et₂O		+		[611]

Replacement of an allylic hydroxyl without saturation or a shift of the double bond was achieved by treatment of some allylic-type alcohols with *triphenyliodophosphorane* (Ph₃PHI), *triphenyldiiodophosphorane* (Ph₃PI₂) or their mixture with triphenyl phosphine (yields 24–60%) [612]. Still another way is the treatment of an allylic alcohol with a *pyridine–sulfur trioxide* complex followed by reduction of the intermediate with *lithium aluminum hydride* in tetrahydrofuran (yields 6–98%) [613]. In this method saturation of the double bond has taken place in some instances [613].

Acetylenic alcohols, usually of propargylic type, are frequently intermediates in the synthesis, and selective reduction of the triple bond to a double bond is desirable. This can be accomplished by carefully controlled catalytic hydrogenation over deactivated palladium [36, 364, 365, 366, 368, 370], by reduction with lithium aluminum hydride [383, 384], zinc [384] and chromous sulfate [195]. Such partial reductions were carried out frequently in alcohols in which the triple bonds were conjugated with one or more double bonds [36, 368, 384] and even aromatic rings [195].

Hydroxy compounds in the aromatic series behave differently depending on whether the hydroxy group is phenolic, benzylic or more remote from the ring.

Phenolic hydroxyl is difficult to hydrogenolyze. It can be replaced by hydrogen if the phenol is first added to dialkylcarbodiimide to form the *O*-aryl-*N,N'*-dialkylisourea, which is then hydrogenated over 5% palladium on carbon (yields 82–99%) [614]. Alternatively, phenols are converted by treatment with cyanogen bromide to arylcyanates; these are then treated with diethylamine to give O-aryl-N,N-diethylisoureas which on hydrogenation over 5% palladium on carbon in ethanol or acetic acid at 20 or 45° afford hydrocarbons or hydroxyl-free derivatives in 85–94% yields [615]. Good to high yields are also obtained when phenols are transformed to aryl ethers by reaction with 2-chlorobenzoxazole or 1-phenyl-5-chlorotetrazole and the aryl ethers are hydrogenated over 5% palladium on carbon in benzene, ethanol or tetrahydrofuran at 35° [616].

Distillation of phenols with zinc dust [617] or with dry lithium aluminum hydride [618] also results in hydrogenolysis of phenolic hydroxyls but because the reaction requires very high temperatures hardly any other function in the molecule can survive, and the method has only limited use.

In some phenols, such as α-naphthol and 9-hydroxyphenanthrene, the hydroxyl group has been replaced by hydrogen by refluxing with 57% hydriodic acid and acetic acid (yields 52% and 96%, respectively) [226].

In contrast to phenolic hydroxyl, **benzylic hydroxyl** is replaced by hydrogen very easily. In catalytic hydrogenation of aromatic aldehydes, ketones, acids and esters it is sometimes difficult to prevent the easy hydrogenolysis of the benzylic alcohols which result from the reduction of the above functions. A catalyst suitable for preventing hydrogenolysis of benzylic hydroxyl is platinized charcoal [28]. Other catalysts, especially palladium on charcoal [619], palladium hydride [619], nickel [43], Raney nickel [619] and copper chromite [620], promote hydrogenolysis. In the case of chiral alcohols such as 2-phenyl-2-butanol hydrogenolysis took place with inversion over platinum and palladium, and with retention over Raney nickel (optical purities 59–66%) [619].

Benzylic alcohols were also converted to hydrocarbons by *sodium borohydride* [621], by *chloroalane* [622], by *borane* [623], by *zinc* [624], and by *hydriodic acid* [225, 625], generally in good to excellent yields. Hydrogenolysis of benzylic alcohols may be accompanied by dehydration (where feasible) [622].

A mixture of titanium trichloride and 0.33 equivalent of lithium aluminum hydride in dimethoxyethane causes coupling of the benzyl residues: benzyl alcohol thus affords bibenzyl in 78% yield [204].

(67)

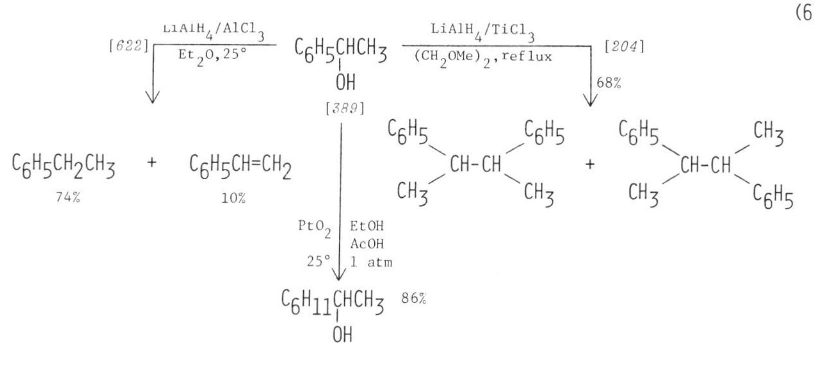

Hydrogenolyses of benzyl-type ethers, esters and amines are discussed on pp. 82, 93, 150 and 151, respectively.

Hydroxylic groups in positions α to heterocyclic aromatics undergo hydrogenolysis in catalytic hydrogenation. In the case of furfuryl alcohol, hydrogenation also reduces the aromatic nucleus and easily cleaves the furan ring giving, in addition to α-methylfuran and tetrahydrofurfuryl alcohol, a mixture of pentanediols and pentanols [38, 420].

Vinylogs of benzylic alcohols, e.g. cinnamyl alcohol, undergo easy saturation of the double bond by catalytic hydrogenation over platinum, rhodium–platinum and palladium oxides [39] or by reduction with lithium aluminum hydride [609]. In the presence of acids, catalytic hydrogenolysis of the allylic hydroxyl takes place, especially over platinum oxide in acetic acid and hydrochloric acid [39].

(68)

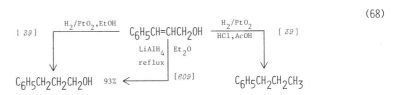

Aromatic hydroxy compounds, both phenols and alcohols, can be **reduced in the aromatic rings.** Catalytic hydrogenation over platinum oxide [8], rhodium–platinum oxide [38], nickel [43] or Raney nickel [45] gave alicyclic alcohols: phenol yielded cyclohexanol by hydrogenation over platinum oxide at room temperature and 210 atm (yield 47%) [8], by heating with hydrogen and nickel at 150° and 200–250 atm (yield 88–100%) [43] or by hydrogenation over Urushibara nickel at 70–110° at 66 atm (yield 79%) [48]. 2-Phenylethanol gave 2-cyclohexylethanol at 175° and 210–260 atm (yield 66–75%) [43], and hydroquinone gave cis-1,4-cyclohexanediol by hydrogenation over nickel at 150° and 170 atm [43], or over Raney nickel at room temperature at 2–3 atm [8]. Hydrogenation of the aromatic ring in benzylic alcohols is usually preceded by hydrogenolysis of the alcohol group prior to the saturation of the ring. However, under special conditions, using platinum oxide as the catalyst, ethanol with a trace of acetic acid as the solvent and room temperature and atmospheric pressure, 1-phenylethanol was converted in 86% yield to 1-cyclohexylethanol [389]. β-Naphthol was under different conditions hydrogenated in the substituted ring, in the other ring, or in both [8, 45]. The results of partial hydrogenation of naphthols depend on the position of the hydroxyl (α or β) and on the catalyst used (Raney nickel [626, 627, 628] or copper chromite [626]).

Partial reduction of naphthols was accomplished by sodium in liquid ammonia. In the absence of alcohols small amounts of 5,8-dihydro-α-naphthol or 5,8-dihydro-β-naphthol were isolated. In the presence of tert-amyl alcohol, α-naphthol gave 65–85% yield of 5,8-dihydro-α-naphthol while β-naphthol afforded 55–65% of β-tetralone [399] (Procedure 26, p. 211). Lithium in liquid ammonia and ethanol reduced α-naphthol to 5,8-dihydro-α-naphthol in 97–99% yield (629).

(69)

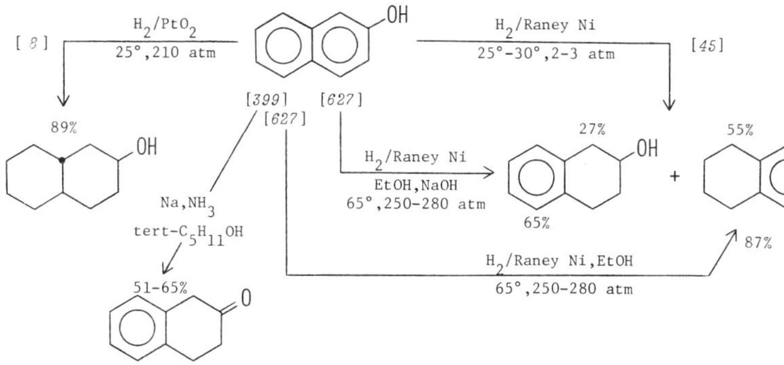

Halogenated saturated alcohols are reduced to **alcohols** by catalytic hydrogenation over Raney nickel [63]. In **halogenated phenols** and naphthols the halogens are hydrogenolyzed over the same catalyst in high yields [536]. A rather exceptional replacement of all three atoms of fluorine by hydrogen in *o*-hydroxybenzotrifluoride is due to the electron-releasing effect of the hydroxy group, which enhances nucleophilic displacement by the hydride anion [630]. Bromohydrins treated with titanium dichloride prepared from lithium aluminum hydride and titanium trichloride give alkenes non-stereospecifically: both *erythro-* and *threo-*5-bromo-6-decanols gave 80/20 and 70/30 mixtures of *trans-* and *cis*-5-decene in 91% and 82%, respectively [204].

In **bromohydrins,** hydrogenation over Raney nickel may lead to epoxides rather than to alcohols since regular Raney nickel contains enough alkali to cause dehydrobromination. Pure hydrogenolysis of bromine can be achieved over Raney nickel which has been washed free of alkali by acetic acid [631].

Nitro alcohols were reduced to **amino alcohols** by catalytic hydrogenation over platinum [632] and with iron [559], and nitrosophenols [255] and nitrophenols [256] to aminophenols with sodium hydrosulfite, sodium sulfide [238] or tin [176]. Bromine atoms in 2,6-di-bromo-4-nitrophenol were not affected [176].

REDUCTION OF ETHERS AND THEIR DERIVATIVES

Open-chain aliphatic ethers are completely resistant to hydrogenolysis. Cyclic ethers (for epoxides, see p. 83) may undergo reductive cleavage under strenuous conditions. The tetrahydrofuran ring was cleaved in vigorous hydrogenations over Raney nickel [420] and copper chromite [420] to give, ultimately, alcohols.

Tetrahydrofuran itself is not entirely inert to some hydrides although it is a favorite solvent for reductions with these reagents. A mixture of lithium aluminum hydride and aluminum chloride produced butyl alcohol on prolonged refluxing in yields corresponding to the amount of alane generated [633].

Under reasonable conditions tetrahydropyran ring is not affected. Catalytic hydrogenation of 1,2-dihydropyran over Raney nickel at room temperature and 2.7 atm saturated the double bond to give quantitative yield of tetrahydropyran [634].

Vinyl ethers were reductively cleaved by lithium, sodium or potassium in liquid ammonia especially in the absence of alcohols (except *tert*-butyl alcohol) [635]. A mixture of 1-methoxy-1,3- and 1-methoxy-1,4-cyclohexadiene gave in this way first methoxycyclohexene and, on further reduction, cyclohexene [635]. Reductive cleavage of α-alkoxytetrahydrofurans and pyrans will be discussed in the chapter on acetals (p. 104).

Allylic ethers were reduced by treatment with lithium in ethylamine to alkenes [636]. **Benzyl ethers** are hydrogenolyzed easily, even more readily than benzyl alcohols [637]. 3,5-Bis(benzyloxy)benzyl alcohol gave 3,5-dihydroxybenzyl alcohol on hydrogenation over palladium on carbon at room temperature and atmospheric pressure in quantitative yield [638]. Hydrogenolysis of benzylic ethers can also be achieved by refluxing the ether with cyclohexene (as a source of hydrogen) in the presence of 10% palladium on carbon in the presence of aluminum chloride [639].

Cleavage of a benzyloxy bond was also accomplished by treatment with sodium in liquid ammonia [640] or by refluxing with sodium and butyl alcohol [641]. On the other hand, aluminum amalgams reduced a nitro group but did not cleave benzyl ether bond in dibenzylether of 2-nitrohydroquinone [642].

Ether linkage in **aryl alkyl ethers** is usually stable to reduction, and hydrogenolysis of the central methoxy group in trimethyl ether of pyrogallol by refluxing with sodium in ethanol belongs to exceptions [643]. Catalytic hydrogenation does not affect the ether bonds. In 4-[2,3-dimethoxyphenyl]-5-nitrocyclohexene only the double bond (not even the nitro group) was reduced by hydrogenation over platinum oxide [564]. In methyl β-naphthyl ether hydrogenation over Raney nickel at 130° and 24 atm reduced the substituted ring to give a tetrahydro derivative but did not break the ether bond [628].

Partial reduction of benzene and naphthalene rings in aryl alkyl ethers was accomplished by dissolving metal reductions. Anisole treated with lithium [644] or with sodium [635] and ethanol in ammonia gave first 1-methoxy-1,4-cyclohexadiene which rearranged to the conjugated 1-methoxy-1,3-cyclohexadiene. Further reduction afforded 1-methoxycylohexene and cyclohexene [635]. Similar reductions were accomplished in methoxyalkylbenzenes [399]. 2-Ethoxynaphthalene was reduced by sodium in ethanol to 3,4-

[635]   (70)

dihydro-2-ethoxynaphthalene which on hydrolysis afforded β-tetralone in 40–50% yield [*645*].

REDUCTION OF EPOXIDES, PEROXIDES AND OZONIDES

Epoxides (oxiranes) can be reduced in different ways. So-called deoxygenation converts epoxides to alkenes. Such a reaction is very useful since it is the reversal of epoxidation of alkenes, and since both these reactions combined represent temporary protection of a double bond.

Reduction to alkenes was accomplished by refluxing the epoxides with *zinc dust in acetic acid* [*646*] (yields 69–70%) or with *zinc–copper couple in ethanol* [*647*], (yields 8–95%), by heating at 65° with *titanium dichloride* prepared *in situ* from titanium trichloride and 0.25 mol of lithium aluminum hydride [*204*] (yields 11–79%), by treatment with *chromous chloride*–ethylenediamine complex in dimethyl formamide at 90° [*648*], by treatment with a mixture of *butyllithium and ferric chloride* in tetrahydrofuran at room temperature (yields 57–92%) [*649*], or by treatment with *alkali O,O-diethyl phosphorotellurate* prepared *in situ* from diethyl sodium phosphite and tellurium (yields 39–91%) [*293*]. The latter procedure is faster with terminal than with internal epoxides and with *Z* than with *E* compounds, and is stereospecific. The other methods give either mixtures of isomers or predominantly *trans* alkenes. The method was applied to epoxides of sesquiterpenes [*647*] and steroids [*646*]. Deoxygenation of epoxide by chromous chloride was carried without affecting double bonds and/or carbonyls present in the same molecule [*192, 650*]. On the other hand, catalytic hydrogenation over 5% palladium on carbon was used to reduce a double bond in 95% yield without deoxygenating the epoxide [*651*].

More frequent than the deoxygenation of epoxides to alkenes is **reduction of epoxides to alcohols.** Its regiospecificity and stereospecificity depend on the reducing agents.

Hydrogenolysis of epoxides to alcohols by *catalytic hydrogenation* over platinum requires acid catalysis. 1-Methylcyclohexene oxide was reduced to a mixture of *cis*- and *trans*-2-methylcyclohexanol [*652*]. Steroidal epoxides usually gave axial alcohols stereospecifically: 4,5-epoxycoprostan-3α-ol afforded cholestan-3α,4β-diol [*652*].

Reagents of choice for reduction of epoxides to alcohols are hydrides and complex hydrides. A general rule of regioselectivity is that the nucleophilic complex hydrides such as *lithium aluminum hydride* approach the oxide from the less hindered side [*511, 653*], thus giving more substituted alcohols. In contrast, hydrides of electrophilic nature such as *alanes* (prepared *in situ* from lithium aluminum hydride and aluminum halides) [*653, 654, 655*] or *boranes*, especially in the presence of boron trifluoride, open the ring in the opposite direction and give predominantly less substituted alcohols [*656, 657, 658*]. As far as stereoselectivity is concerned, lithium aluminum hydride yields *trans* products [*511*] whereas electrophilic hydrides predominantly *cis* products

[*656, 658*]. Both these rules have exceptions, as shown by a study of reduction of deuterated styrene-2,2-d_2 oxide with deuteroalanes [*655*].

(71)

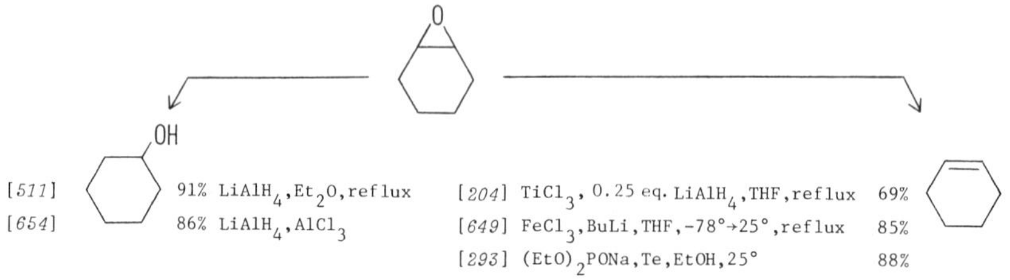

O
OH
[511] 91% LiAlH4,Et2O,reflux
[654] 86% LiAlH4,AlCl3
[204] TiCl3, 0.25 eq. LiAlH4,THF,reflux 69%
[649] FeCl3,BuLi,THF,-78°→25°,reflux 85%
[293] (EtO)2PONa,Te,EtOH,25° 88%

Sterically hindered epoxides are sometimes easier to reduce by dissolving metals: 2-methyl-2,3-epoxybutane gave, after treatment with *lithium* in ethylenediamine at 50° for 1 hour, 74% of 3-methyl-2-butanol and 8% of 2-methyl-2-butanol [*659*].

(72)

CH3
O
CH3
OH
CH3
OH
CH3
OH

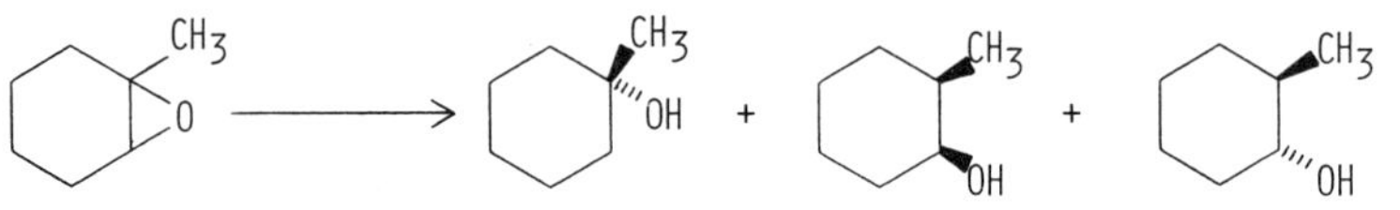

[*652*]	H_2/PtO_2, AcOEt, $HClO_4$		42%	58%
[*656*]	$NaBH_3CN/BF_3 \cdot Et_2O$, THF, 25°	3%	84%	trace
[*658*]	$BH_3/NaBH_4$, THF, 0°	26%	74%	
[*658*]	$BH_3/LiBH_4$, THF, 0°	26%	74%	
[*658*]	AlH_3/2 $AlCl_3$, THF, 0°	10%	19%	12%
[*659*]	Li, $(CH_2NH_2)_2$, 50°	93%		

Unsaturated epoxides are reduced preferentially at the double bonds by *catalytic hydrogenation*. The rate of hydrogenolysis of the epoxides is much lower than that of the addition of hydrogen across the carbon–carbon double bond. In α,β-unsaturated epoxides *borane* attacks the conjugated double bond at β-carbon in a *cis* direction with respect to the epoxide ring and gives allylic alcohols [*660*]. Similar complex reduction of epoxides occurs in α-keto epoxides (p. 126).

[*660*] (73)

CH3
H
C=C
H
CH-CH2
O
1. BH3·THF
THF,-10°-25°
2. H2O,NaOH
66%
H
H
C=C
CH3CH2
CH2OH

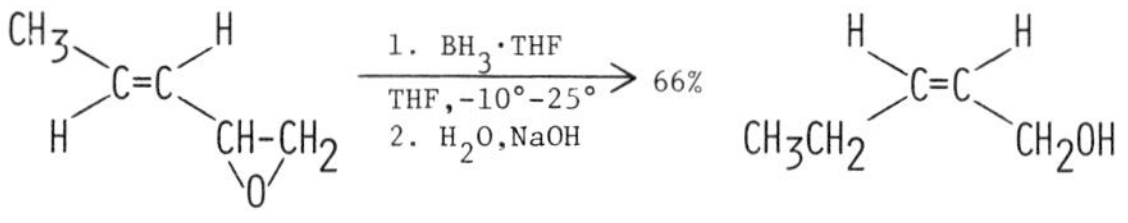

Oxygen–oxygen bond in peroxides and hydroperoxides is cleaved very easily by *catalytic hydrogenation* over platinum oxide [*661, 662*], over palladium [*75, 662, 663, 664*] or over Raney nickel [*665*]. Hydroperoxides yield alcohols [*661*]; unsaturated hydroperoxides yield unsaturated and saturated alcohols [*75*]. Unsaturated cyclic peroxides (endoperoxides) gave saturated peroxides [*662*], unsaturated diols [*662, 663*] or saturated diols [*664*]. Similar results were obtained by reduction with *zinc* [*666*].

(74)

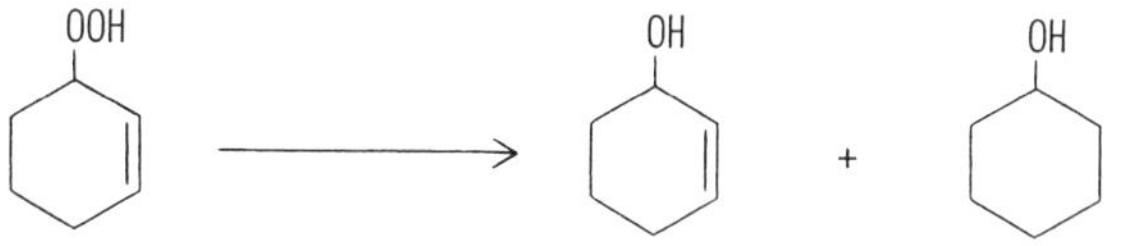

[75]	H$_2$/Pd,25°,1 atm	63%	7%
[75]	H$_2$/Pd/Pb(OAc)$_2$/Al$_2$O$_3$	78%	4%
[75]	H$_2$/Pd/Pb(OAc)$_2$,Quinoline,25°,1 atm	84%	1%
[290]	Ph$_3$P,Et$_2$O	71%	

A very convenient way of reducing hydroperoxides to alcohols in high yields is treatment with 20% aqueous solution of *sodium sulfite* at room or elevated temperatures [251], or with *tertiary phosphines* in organic solvents [290, 667].

(75)

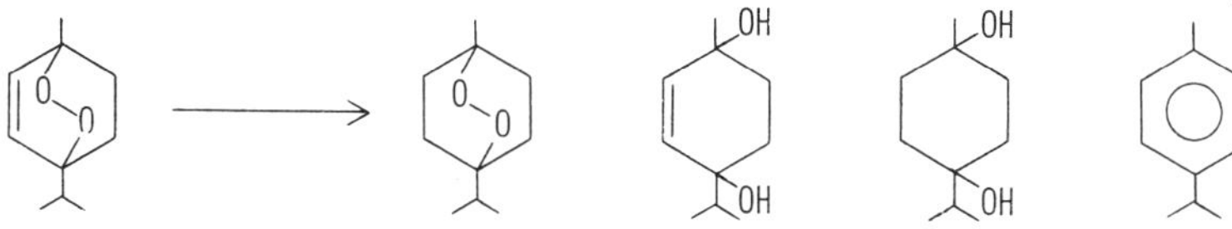

[663]	H$_2$/Pd black		53%
[662]	H$_2$/Pd(C)	23%	47%
[664]	H$_2$/Pd,MeOH	~100%	
[668]	(EtO)$_3$P,160–170°	30%	

Triethyl- and triphenylphosphine have been used for 'deoxygenation' not only of hydroperoxides to alcohols but also of dialkyl peroxides to ethers, of diacyl peroxides to acid anhydrides, of peroxy acids and their esters to acids or esters, respectively, and of endoperoxides to oxides [290] in good to excellent yields. The deoxygenation of ascaridole to 1-methyl-4-isopropyl-1,4-oxido-2-cyclohexene [290] was later challenged (the product is claimed to be *p*-cymene instead [668]).

Apart from the conversion of peroxides to useful products, it is sometimes necessary to reduce peroxides, and especially hydroperoxides formed by auto-oxidation. Such compounds are formed especially in hydrocarbons containing branched chains, double bonds or aromatic rings, and in ethers such as diethyl ether, diisopropyl ether, tetrahydrofuran, dioxane, etc. Since most peroxidic compounds decompose violently at higher temperatures and could cause explosion and fire it is necessary to remove them from liquids they contaminate. Water-immiscible liquids can be stripped of peroxides by shaking with an aqueous solution of sodium sulfite or ferrous sulfate. A simple and efficient way of removing peroxides is treatment of the contaminated compounds with 0.4 nm molecular sieves [669].

Reduction of ozonides is very useful, especially when aldehydes are the desired products. Ozonides are easily *hydrogenolyzed over palladium* [670], or reduced by *zinc* in acetic acid [671], usually in good yields. Ozonolysis of methyl oleate followed by hydrogenation over 10% palladium on charcoal gave 58% yield of nonanal and 60% yield of methyl 9-oxononanoate [670] while reduction of the ozonide with zinc dust in acetic acid gave the respective yields of 81% and 85%. Ozonolysis of 6-chloro-2-methyl-2-heptene in acetic acid followed by treatment with zinc dust in ether gave 55% yield of 4-chloropentanol [671].

An elegant one-pot reduction of ozonides consisting of the treatment of a crude product of ozonization of an alkene in methanol with *dimethyl sulfide* (36% molar excess) gives 62–97% yields of very pure aldehydes [244].

REDUCTION OF SULFUR COMPOUNDS (EXCEPT SULFUR DERIVATIVES OF ALDEHYDES, KETONES AND ACIDS)

Reduction of sulfur compounds will be discussed in the sequence divalent, tetravalent and hexavalent sulfur derivatives.

Sulfur analogs of hydroxy compounds, **thiols (mercaptans and thiophenols),** behave differently from their oxygen counterparts. While hydrogenolysis of alcohols and especially phenols to parent hydrocarbons is far from easy, conversion of thiols to the parent hydrocarbons is achieved readily by several reagents. Benzyl mercaptan was desulfurized to toluene by treatment with *triethyl phosphite* [672] (*Procedure 47*, p. 217). Refluxing with *iron* in ethanol and acetic acid desulfurized the mercapto group in 2-mercapto-6-methylbenzothiazole and gave 82% yield of 6-methylbenzothiazole. The sulfidic sulfur in the heterocyclic ring remained intact [171]. A general method for the desulfurization of thiols is treatment with *nickel* [673]. Such reaction applies not only to thiols but to sulfides and almost any sulfur-containing organic compound.

Divalent sulfur is a poison for most noble metal catalysts so that catalytic hydrogenation of sulfur-containing compounds poses serious problems (p. 10). However, allyl phenyl sulfide was *hydrogenated over tris(trisphenylphosphine)rhodium chloride* in benzene to give 93% yield of phenyl propyl sulfide [674].

Raney nickel and *nickel boride* somehow extract sulfur from sulfur-containing compounds forming nickel sulfide, and because they contain adsorbed hydrogen at their surfaces, even cause hydrogenation. The reaction is achieved by refluxing sulfur-containing compounds with a large excess of Raney nickel or nickel boride in ethanol and generally gives excellent yields [673].

Aliphatic and aromatic sulfides undergo desulfurization with *Raney nickel* [673], with *nickel boride* [673], with *lithium aluminum hydride* in the presence of cupric chloride [675], with *titanium dichloride* [676], and with *triethyl phosphite* [677]. In saccharides benzylthioethers were not desulfurized but reduced to toluene and mercaptodeoxysugars using *sodium* in liquid ammonia [678]. This reduction has general application and replaces catalytic hydrogenolysis, which cannot be used [637].

(76)

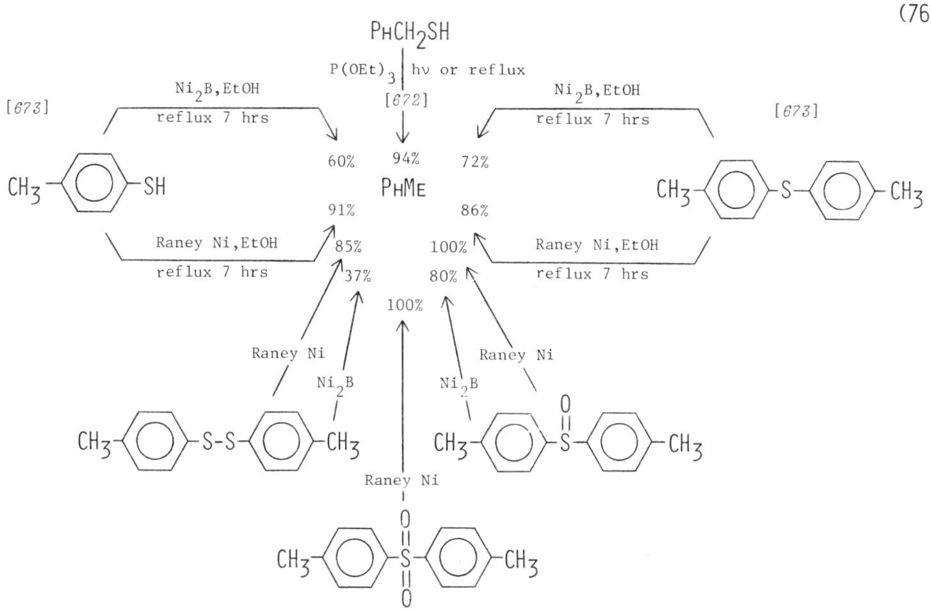

PHCH$_2$SH

[673] Ni$_2$B,EtOH reflux 7 hrs   P(OEt)$_3$ | hv or reflux [672]   Ni$_2$B,EtOH reflux 7 hrs [673]

60% 94% 72%

CH$_3$—⟨O⟩—SH   PHME   CH$_3$—⟨O⟩—S—⟨O⟩—CH$_3$

91% 86%

Raney Ni,EtOH reflux 7 hrs   85% 100% Raney Ni,EtOH reflux 7 hrs

37% 80%

100%

Raney Ni   Ni$_2$B   Ni$_2$B   Raney Ni

O

CH$_3$—⟨O⟩—S-S—⟨O⟩—CH$_3$ CH$_3$—⟨O⟩—S—⟨O⟩—CH$_3$

Raney Ni

O
‖
CH$_3$—⟨O⟩—S—⟨O⟩—CH$_3$
‖
O

Cyclic sulfides treated with *triethyl phosphite* eject sulfur and form a new ring less one member [677]. From compounds containing sulfur in three-membered rings (**thiiranes**) alkenes are formed in high yields [291, 294]. The same reaction can be achieved with *triphenyl phosphine* [291].

[291]

(77)

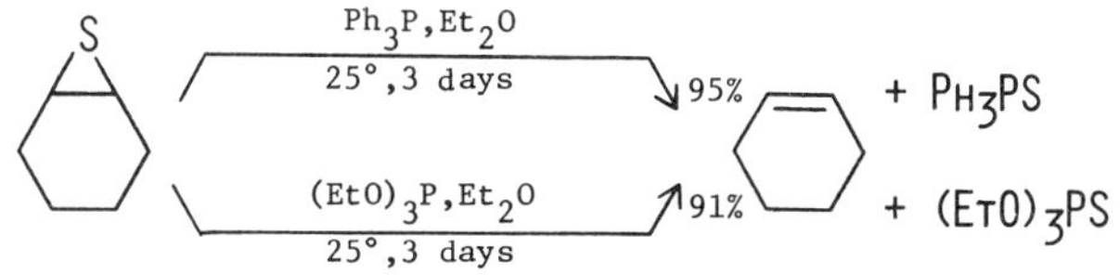

S△⬡   Ph$_3$P,Et$_2$O 25°,3 days   ↓95% ⬡ + PH$_3$PS

(EtO)$_3$P,Et$_2$O 25°,3 days   ↗91% ⬡ + (EtO)$_3$PS

Reductive cleavage of 1,3-dithiolanes is accomplished in low yields with sodium in liquid ammonia and gives either dithiols or mercapto sulfides depending on substituents in position two [679].

Alkyl thiocyanates were reduced with *lithium aluminum hydride* to mercaptans (yield 81%), and aryl thiocyanates to thiophenols (yield 94%) [680].

Disulfides can be either reduced to two thiols or desulfurized. The former reaction was achieved in high yields using *lithium aluminium hydride* [680, 681], *lithium triethylborohydride* [100] and *sodium borohydride* [682].

Partial desulfurization of disulfides to sulfides was accomplished by treatment with *tris(diethylamino)phosphine* in good yields [303]. 1,2-Dithiacyclohexane was thus quantitatively converted to thiophane (tetrahydrothiophene) at room temperature [303]. Complete desulfurization to hydrocarbons resulted when disulfides were refluxed in ethanol with *Raney nickel* or *nickel boride* (yields 86 and 72%, respectively) [673].

Reductive cleavage of disulfides is very easy and can be accomplished without affecting some other readily reducible groups such as nitro groups.

If, on the other hand, a nitro group must be reduced to an amino group without reducing the sulfidic bond, *hydrazine* is the reagent of choice [*683*].

Dialkyl and diaryl sulfoxides were reduced (deoxygenated) to sulfides by several methods. As a little surprise sulfides were obtained from sulfoxides by *catalytic hydrogenation* over 5% palladium on charcoal in ethanol at 80–90° and 65–90 atm in yields of 59–99% [*684*]. In the case of *p*-tolyl β-styryl sulfoxide deoxygenation was achieved even without reduction of the double bond giving 66% of *p*-tolyl β-styryl sulfide and only 16% of *p*-tolyl β-phenyl-ethyl sulfide [*684*].

Chemical deoxygenation of sulfoxides to sulfides was carried out by refluxing in aqueous-alcoholic solutions with *stannous chloride* (yields 62–93%) [*186*] (*Procedure 36*, p. 214), with *titanium trichloride* (yields 68–91%) [*203*], by treatment at room temperature with *molybdenum trichloride* (prepared by reduction of molybdenyl chloride MoOCl$_3$ with zinc dust in tetrahydrofuran) (yields 78–91%) [*216*], by heating with *vanadium dichloride* in aqueous tetra-hydrofuran at 100° (yields 74–88%) [*216*], and by refluxing in aqueous meth-anol with *chromium dichloride* (yield 24%) [*190*]. A very impressive method is the conversion of dialkyl and diaryl sulfoxides to sulfides by treatment in acetone solutions for a few minutes with 2.4 equivalents of *sodium iodide* and 1.2–2.6 equivalents of trifluoroacetic anhydride (isolated yields 90–98%) [*685*].

$$\underset{\text{PHCH}_2\overset{\displaystyle\text{O}}{\underset{\displaystyle\|}{\text{S}}}\text{CH}_2\text{PH}}{} \qquad\qquad\longrightarrow\qquad\qquad \text{PHCH}_2\text{SCH}_2\text{PH} \qquad (78)$$

[*684*] H$_2$/5% Pd(C),EtOH,86–90°,65–78 atm,4 days	90%
[*186*] SnCl$_2$·2 H$_2$O,HCl,MeOH,reflux 2 hrs	82%
[*203*] TiCl$_3$,H$_2$O,MeOH,CHCl$_3$,reflux	78%
[*216*] MoCl$_5$,THF,H$_2$O,Zn,25°,1 hr	89%
[*216*] VCl$_2$,H$_2$O,THF,100°,40 mm,8 hrs	82%
[*190*] CrCl$_2$,H$_2$O,MeOH,reflux 2 hrs	24%
[*685*] NaI,(CF$_3$CO)$_2$O,Me$_2$CO,25°,few minutes	93%

In addition to deoxygenation sulfoxides undergo reductive cleavage at the carbon–sulfur bond when heated in tetrahydrofuran with *aluminum amalgam*. Keto sulfoxides were thus converted to ketones in usually quantitative yields (the keto group remained intact) [*141*].

The sulfonyl group in **sulfones** resists catalytic hydrogenation. Double bonds in α,β-unsaturated sulfones are reduced by hydrogenation over palla-dium on charcoal (yield 94%) [*686, 687*] or over Raney nickel (yield 62%) without the sulfonyl group being affected [*686*]. In *p*-thiopyrone-1,1-dioxide both double bonds were reduced with *zinc* in acetic acid but the keto group and the sulfonyl group survived [*688*]. *Raney nickel* may desulfurize sulfones to hydrocarbons [*673*].

Reduction of sulfones to sulfides has been accomplished by *lithium aluminum hydride* [*687*], *diisobutyl aluminum hydride* [*689*] and sometimes by *zinc* dust in acetic acid [*687*]. Diisobutyl aluminum hydride (Dibal-H) was found to be superior to both lithium aluminum hydride and zinc. At least 2.2 mol of the hydride were necessary for the reduction carried out in mineral oil at room temperature or in refluxing toluene over 18–72 hours. Five-membered cyclic sulfones were reduced most readily, aliphatic sulfones least readily. Yields were 12–77% [689].

Reductions of five-membered cyclic sulfones with lithium aluminum hydride were run in refluxing ether and of other sulfones in refluxing ethyl butyl ether (92°) (yields 12–92%). Benzothiophene-1,1-dioxide was reduced at the double bond as well, giving 2,3-dihydrobenzothiophene in 79% yield after 18 hours of refluxing in ether [687].

An entirely different type of reduction occurs if sulfones are treated with *lithium aluminum hydride* in the presence of cupric chloride (1:2) [152], with *sodium* in liquid ammonia [690], with *aluminum amalgam* [141, 152] or, for aromatic ones, with *sodium amalgam* [691]. The carbon–sulfur bond is cleaved to give a sulfinic acid and a compound resulting from replacement of the sulfone group by hydrogen. Such hydrogenolysis can be achieved even without reducing a keto group in keto sulfones (2-methylsulfonylcyclohexanone on heating at 65° with aluminum amalgam in aqueous tetrahydrofuran afforded 89% yield of cyclohexanone [141]), and without reducing a double bond conjugated with aromatic rings in α,β-unsaturated sulfones [152]. Similar carbon–sulfur bond cleavage was observed with alkyl aryl sulfoximines [153].

[152]   (79)

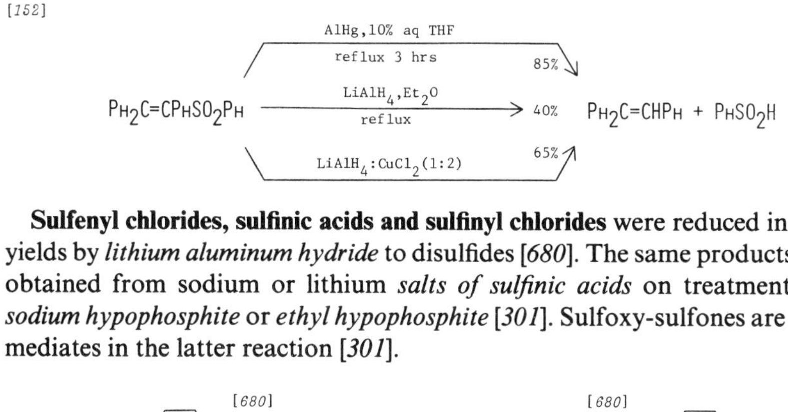

Sulfenyl chlorides, sulfinic acids and sulfinyl chlorides were reduced in good yields by *lithium aluminum hydride* to disulfides [680]. The same products were obtained from sodium or lithium *salts of sulfinic acids* on treatment with *sodium hypophosphite* or *ethyl hypophosphite* [301]. Sulfoxy-sulfones are intermediates in the latter reaction [301].

(80)

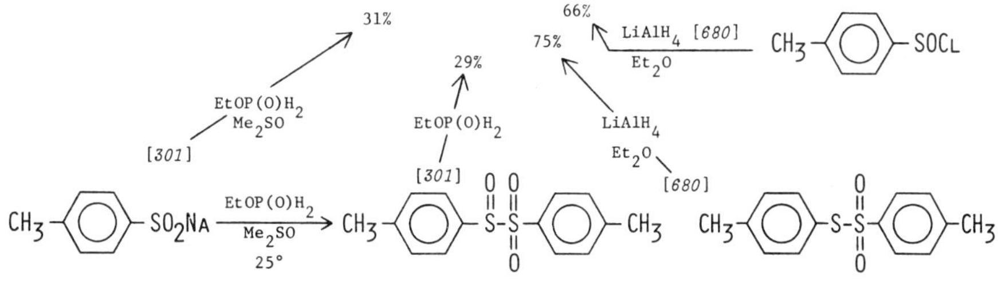

Sulfonic acids are completely resistant to any reductions, and other functions contained in their molecules can be easily reduced, e.g. nitro groups with *iron* [692].

Chlorides of sulfonic acids can be reduced either partially to sulfinic acids, or completely to thiols. Both reductions are accomplished in high yields with *lithium aluminum hydride*. An inverse addition technique at a temperature of −20° is used for the preparation of sulfinic acids, while the preparation of thiols is carried out at the boiling point of ether [693].

More reliable reagents for the **preparation of sulfinic acids** are *zinc* [694, 695], *sodium sulfide* [249] and *sodium sulfite* [252]. These reagents not only stop the reduction at the stage of the sulfinic acids (in the form of their salts) but do not reduce other functions present in the molecules. In the reduction of anthraquinone-1,5-disulfonyl chloride with sodium sulfide below 40° anthraquinone-1,5-disulfinic acid was obtained in 83.5% yield [249], and *p*-cyanobenzenesulfonyl chloride was reduced to *p*-cyanobenzenesulfinic acid in 87.4% yield [252].

Complete reduction of sulfonyl chlorides to thiols can be achieved by *lithium aluminum hydride* [680, 693], with *zinc* [696] and with *hydriodic acid* generated *in situ* from iodine and red phosphorus [230]. *m*-Nitrobenzenesulfonyl chloride, however, was reduced not to the thiol but to bis(*m*-nitrophenyl)disulfide by hydriodic acid in 86–91% yield [697].

(81)

$$CH_3-\langle O \rangle-SO_2CL \longrightarrow CH_3-\langle O \rangle-SO_2H \quad CH_3-\langle O \rangle-SH$$

[680] LiAlH$_4$,Et$_2$O		90%
[693] LiAlH$_4$,Et$_2$O,−20°	93%	
[693] LiAlH$_4$,Et$_2$O,reflux		89%
[695] Zn dust,70–90°;NaOH,Na$_2$CO$_3$;H$^\oplus$	64%	
[230] P,I$_2$,AcOH,reflux 2–3 hrs		90%

Esters of sulfonic acids give different products depending on the reducing agent, on the structure of the parent sulfonic acid, and especially on the structure of the hydroxy compound, alcohol or phenol.

Catalytic hydrogenation over Raney nickel converted benzenesulfonates of both alcohols and phenols to **parent hydroxy compounds,** benzene and nickel sulfide. *p*-Toluenesulfonates of alcohols are reduced similarly while *p*-toluenesulfonates of phenols gave nickel *p*-toluenesulfinates and **aromatic hydrocarbons.** The yields of the hydroxy compounds range from 25 to 96% [698].

Similarly, *lithium aluminum hydride* gives different products. β-Naphthyl *p*-toluenesulfonate affords *p*-thiocresol and β-naphthol, and phenyl methanesulfonate gives methyl mercaptan and phenol. On the other hand, propyl *p*-toluenesulfonate yields *p*-toluenesulfonic acid and propane, and cetyl methanesulfonate and cetyl *p*-toluenesulfonate give hexadecane in 92% and 96% yields, respectively [680].

While lithium aluminum hydride frequently leads to mixtures of hydrocarbons and alcohols from alkyl *p*-toluenesulfonates, *lithium triethylborohy-dride* (Super Hydride®) in tetrahydrofuran reduces *p*-toluenesulfonates of primary and secondary alcohols to the corresponding *hydrocarbons* at room temperature in high yields [*699*]. Also *sodium borohydride* in dimethyl sulfox-ide gives hydrocarbons: dodecane and cyclododecane from dodecyl and cy-clododecyl *p*-toluenesulfonates in 86.5% and 53.9% yields, respectively [*504*].

Replacement of sulfonyloxy groups by hydrogen is also achieved by refluxing the methanesulfonyl and *p*-toluenesulfonyl esters of alcohols with *sodium iodide and zinc dust* in wet dimethoxyethane. Evidently iodine displaces sul-fonyloxy group and is replaced by hydrogen by means of zinc. Yields range from 26% to 65% [*700*] (*Procedure 33*, p. 213).

Cleavage of the sulfonyl esters to the **parent alcohols** is accomplished in yields of 60–100% by treatment of the *p*-toluenesulfonates with 2–6 equiva-lents of *sodium naphthalene* in tetrahydrofuran at room temperature (yields 60–100%). Sodium naphthalene is prepared by stirring sodium with an equi-valent amount or a slight excess of naphthalene in tetrahydrofuran for 1 hour at room temperature under an inert gas [*701*]. Benzenesulfonates and bromo-benzenesulfonates are also cleaved to the parent alcohols while alkyl methanesulfonates are reduced also to hydrocarbons [*701*].

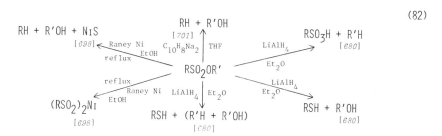

(82)

In saccharides *sodium amalgam* has been used for regeneration of sugars from their sulfonates but in low yields only [*702*].

Like sulfonyl esters, **sulfonamides** can be **reduced** either **to sulfinic** acids, or **to thiols.** Heating of *p*-toluenesulfonamide with fuming *hydriodic acid* and phosphonium iodide at 100° afforded 85% of *p*-thiocresol [*703*].

Substituted sulfonamides are reductively cleaved to amines and either sulfinic acids, or thiols, or even parent hydrocarbons of the sulfonic acid [*704*]. For the regeneration of amines *sodium naphthalene* proves to be very useful since it gives excellent yields (68–96%, isolated) under very gentle conditions (room temperature, 1 hour) [*705*] (*Procedure 28*, p. 212). Other reagents usually require refluxing. Thus *sodium* in refluxing amyl alcohol gave 78–93.5% yields, *zinc* dust in refluxing acetic and hydrochloric acid 2.7–87% yields, and *stannous chloride* in the same system 61.8–96% yields of the amines [*704*]. The other reduction products are toluene in the reductions with sodium, and *p*-thiocresol (74%) in the reductions with zinc [*704*].

Reduction of both sulfoesters and sulfonamides has as its main objective

(83)

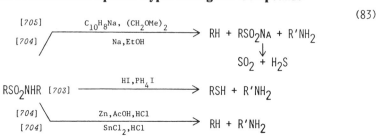

preparation of hydrocarbons, alcohols, phenols and amines. Reduction products of the sulfur-containing parts of the molecules are only corollaries.

REDUCTION OF AMINES AND THEIR DERIVATIVES

Like saturated alcohols, **saturated amines** are fairly resistant to reduction. The carbon–nitrogen bond has been hydrogenolyzed in the catalytic hydrogenation of 1,1-dimethylethylenimine which, on treatment with *hydrogen* over Raney nickel at 60° and 4.2 atm, afforded *tert*-butylamine in 75–82% yield [*706*]. Evidently the strain of the aziridine ring contributed to the relative ease of the hydrogenolysis since other cyclic amines, such as piperidine, required heating to 200–220° for their hydrogenolysis to alkanes [*707*].

Primary amines can be reduced to their parent compounds by treatment with *hydroxylamine-O-sulfonic acid* and aqueous sodium hydroxide at 0°. Substituted hydrazines and diimides are believed to be the reaction intermediates in the ultimate replacement of the primary amino group by hydrogen in yields 49–72% [708].

Unsaturated amines are hydrogenated at the multiple bonds by *catalytic hydrogenation* over any catalyst. The double bond in indole was saturated in catalytic hydrogenation over platinum dioxide in ethanol containing fluoroboric acid and indoline was obtained in greater than 85% yield [*456*]. **Allylic amines** such as allylpiperidine are also reduced by *sodium* in liquid ammonia in the presence of methanol (yield 75%) [*709*].

Vinylamines (enamines) are reduced by *alane, mono-* and *dichloroalane* to saturated amines, and hydrogenolyzed to amines and alkenes [*710*]. Reduction is favored by dichloroalane while hydrogenolysis is favored by alane. Alane, chloroalane and dichloroalane gave the following results with 1-*N*-pyrrolidinylcyclohexene: *N*-pyrrolidinylcyclohexane in 13, 15 and 22% yield, and pyrrolidine and cyclohexene in 80, 75 and 75% yields, respectively [*710*]. Saturated amines were also obtained by treatment of enamines with *sodium borohydride* [*711*], *with sodium cyanoborohydride* [*103, 712*] (*Procedure 22*, p. 210) and by heating for 1–2 hours at 50–70° with 87% or 98% *formic acid* (yields 37–89%) [*320*].

Aromatic amines are hydrogenated in the rings by *catalytic hydrogenation*. Aniline yields 17% of cyclohexylamine and 23% of dicyclohexylamine over platinum dioxide in acetic acid at 25° and 125 atm [*8*]. A better yield (90%) of cyclohexylamine was obtained by hydrogenation over nickel at 175° and 180 atm [*43*]. Reduction to saturated amines is also achieved by *lithium in ethyl-amine*: dimethylaniline gave dimethylcyclohexylamine in 44% yield [*713*]. Reduction with *sodium* in liquid ammonia in the presence of ethanol converted dimethylaniline and dimethyltoluidines to dihydro and even tetrahydro derivatives or products of their hydrolysis, cyclohexenones and cyclohexanones, in 58–72% yields [*714*].

In naphthylamines reduction with sodium in refluxing pentanol affected exclusively the substituted ring, giving 51–57% of 2-amino-1,2,3-4-tetrahydronaphthalene [*715*].

N-Alkyl or *N*-aryl derivatives of benzylamine are readily hydrogenolyzed to toluene and primary or secondary amines by *catalytic hydrogenation* [*637*], the best catalysts being palladium oxide [*716*] or palladium on charcoal [*717*, *718*]. The same type of cleavage takes place in 3-dimethylaminomethylindole, which gives 3-methylindole in 83.9% yield on treatment with hydrogen over palladium on charcoal in ethanol at room temperature and atmospheric pressure [*719*]. Hydrogenolysis over palladium is easier than over platinum and takes place preferentially to saturation of double bonds [*719*].

Quaternary ammonium salts, especially iodides, in the aromatic series were converted to tertiary amines. *Electrolysis* of trimethylanilinium iodide using a lead cathode gave 75% of benzene and 77% of trimethylamine, and electrolysis of methylethylpropylanilinium iodide gave 47.5% yield of methylethylpropylamine. The cleavage always affects the bond between nitrogen and the benzene ring [*720*]. Thus, 2,3-dimethylbenzyltrimethyl ammonium iodide was reduced by *sodium amalgam* to 1,2,3-trimethylbenzene in 85–90% yield [*721*].

If the quaternary nitrogen is a member of a ring, the ring is cleaved. 3-Benzyl-2-phenyl-*N*,*N*-dimethylpyrrolidinium chloride was cleaved by *hydrogenation* over Raney nickel at 20–25° almost quantitatively to 2-benzyl-4-dimethylamino-1-phenylbutane [*722*]. Reduction of methylpyridinium iodide (and its methyl homologs) with *sodium aluminum hydride* gave 24–89% yields of 5-methylamino-1,3-pentadiene (and its methyl homologs) in addition to *N*-methyl dihydro- and tetrahydropyridine [*448*].

Since amines are fairly resistant to reduction **halogens in halogenated amines** behave toward reducing agents as if the amino group were not present. Occasionally, however, the amino group may help hydrogenolysis of halogen. In *o*-aminobenzotrifluoride all three fluorine atoms were replaced by hydrogen in 80–100% yield on refluxing with *lithium aluminum hydride* [*630*]. This rather exceptional hydrogenolysis is due to the electron-releasing effect of the amino group which favors nucleophilic displacement by hydride ions. Similarly, **nitroso groups** and **nitro groups in amines** are reduced independently. *p*-Nitrosodimethylaniline gave 64% of *N*,*N*-dimethyl-*p*-phenylenediamine on reduction with *chromous chloride* [*190*], *o*-nitroaniline gave 74–85% of *o*-phenylenediamine on refluxing with *zinc* in aqueous ethanolic sodium hydroxide [*723*], and 5-amino-2,4-dinitroaniline yielded 49–52% of 1,2,5-triamino-4-nitrobenzene on reduction with *sodium sulfide* [*724*].

As in the hydrogenation of glycols, in vigorous hydrogenations of **hydroxy amines** a carbon–carbon bond cleavage may occur along with hydrogenolysis of carbon–oxygen and carbon–nitrogen bonds.

N-Ethyl-*N*-(2-hydroxyethyl)aniline afforded 52.5% of *N*-ethyl-*N*-[2-hydroxyethyl]cyclohexylamine and 14% of *N*-methyl-*N*-ethylcyclohexylamine upon *hydrogenation* over Raney nickel at 150° and 65 atm in ethanol [725].

Cleavage of a carbon–nitrogen bond in a position β to a keto group takes place during catalytic hydrogenation of Mannich bases and gives ketones in yields of 51–96% [726]. The carbon–nitrogen bond next to a carbonyl group in α-amino ketones is cleaved by *zinc* [727] (pp. 118, 119).

Cleavage of the carbon–nitrogen bond occurs in **benzyloxycarbonylamino compounds** as a result of decarboxylation of the corresponding free carbamic acids resulting from hydrogenolysis of benzyl residues [728, 729, 730] (p. 151).

In *N*-**nitrosamines** the more polar nitrogen–oxygen bond is usually reduced preferentially to the nitrogen–nitrogen bond with *lithium aluminum hydride* so that the products are disubstituted *asym*-hydrazines. *N*-Nitroso-*N*-methylaniline gives 77% of *N*-methyl-*N*-phenylhydrazine, *N*-nitrosodicyclohexylamine 48% of *N*,*N*-dicyclohexylhydrazine, and *N*-nitrosopiperidine 75% of *N*,*N*-pentamethylenehydrazine [731]. *N*-Nitrosodimethylamine is reduced to *N*,*N*-dimethylhydrazine with lithium aluminum hydride in 78% yield [732], and with *zinc* in acetic acid in 77–83% yield [733]. Reduction of *N*-nitrosodiphenylamine with an equimolar amount of *lithium aluminum hydride* or by using the 'inverse technique' yielded 90% of *N*,*N*-diphenylhydrazine. When 1 mol of *N*-nitrosodiphenylamine was added to 4 mol of lithium aluminum hydride, cleavage of the nitrogen–nitrogen bond took place giving diphenylamine [731].

(84)

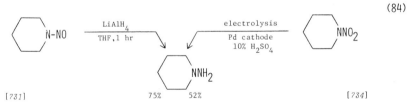

$$[731] \qquad\qquad 75\% \quad 52\% \qquad\qquad [734]$$

N-**nitramines** too were reduced to disubstituted hydrazines. *Electrolysis* in 10% sulfuric acid over copper or lead cathodes reduced *N*-nitrodimethylamine to *N*,*N*-dimethylhydrazine (yield 69%), *N*-nitro-*N*-methylaniline to *N*-methyl-*N*-phenylhydrazine (yield 54%), *N*-nitropiperidine to *N*,*N*-pentamethylenehydrazine (yield 52%) [734], and nitrourea to semicarbazide (yield 61–69%) [735].

N-**Amine oxides** can be reduced (deoxygenated) to tertiary amines. Such a reaction is very desirable, especially in aromatic nitrogen-containing heterocycles where conversion to amine oxides makes possible electrophilic substitution of the aromatic rings in different positions than it occurs in the parent heterocyclic compounds. The reduction is very easy and is accomplished by *catalytic hydrogenation* over palladium [736, 737], by *borane* [738], by *iron* in acetic acid (yields 80–100%) [*170*], with *titanium trichloride* (yield 71–97%) [*199*], or by *chromous chloride* (yields 39–96%) [*191*].

In nitroamine oxides the nitro group may be reduced preferentially but usually both functions are affected. 5-Ethyl-2-methyl-4-nitropyridine *N*-oxide is converted quantitatively to 4-amino-5-ethyl-2-methylpyridine by hydrogen over 30% palladium on charcoal in ethanolic solution [*737*]. The outcome of the hydrogenation of a nitroamine oxide may be influenced by reaction conditions [*736*].

[*736*] (85)

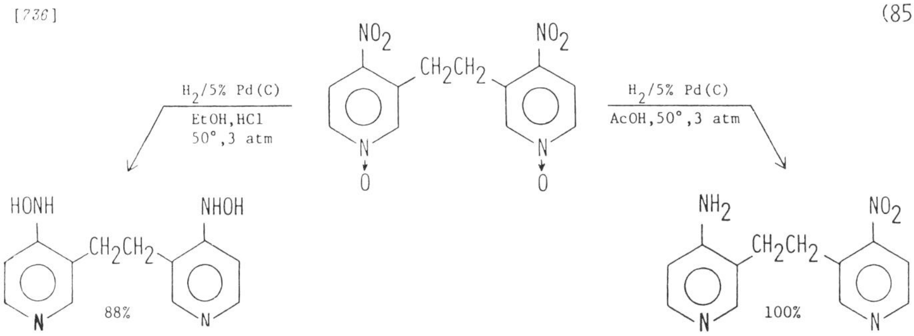

Iron in acetic acid at 100° reduced 4-nitropyridine oxide quantitatively to 4-aminopyridine [*170*], and 3-bromo-4-nitropyridine oxide to 3-bromo-4-aminopyridine in 80% yield [*170*]. *Chromous chloride*, on the other hand, failed in the deoxygenation of nitropyridine oxides and nitroquinoline oxides [*191*].

A bond between nitrogen and oxygen was hydrogenolyzed in a **cyclic derivative of hydroxylamine**, 2-methyl-6-[β-pyridyl]-1,2-oxazine, which on treatment with *zinc* dust in acetic acid gave, on ring cleavage, 84.5% of 1-hydroxy-4-methylamino-1-[β-pyridyl]butane [*739*].

Hydrazo compounds were hydrogenolyzed to primary amines by *catalytic hydrogenation* over palladium [*740*]. Since hydrazo compounds are intermediates in the reductive cleavage of azo compounds to amines it is very likely that all the reducing agents converting azo compounds to amines cleave also the hydrazo compounds (p. 96).

Hydrazines are hydrogenolyzed catalytically over Raney nickel [*741*] and other catalysts. Hydrogenolysis of hydrazones and azines is discussed elsewhere (pp. 34, 106, 133, 134).

Azo compounds can be reduced either to hydrazo compounds or else cleaved to two molecules of amines.

Conversion of azo compounds to hydrazo compounds is achieved by controlled *catalytic hydrogenation* [*572, 740, 742*], by treatment with *diimide* (generated *in situ* from di-potassium azodicarboxylate and acetic acid) [*264, 743*], by *sodium amalgam* (yield 90%) [*744*], by *aluminum amalgam* [*148*] and by *zinc* in alcoholic ammonia (yield 70–85%) [*745*]. *Hydrogen sulfide* has been used for selective reduction of a nitro group to an amino group with conservation of both azo groupings in mono-*p*-nitro-*m*-phenylenebisazobenzene (yields 79%) [241]. This is exceptional since hydrogen sulfide, under slightly different conditions, reduces the azo group to the hydrazo group [746] as well as azo compounds to primary amines [246].

$$P_HN=NP_H \xrightarrow{\hspace{3cm}} P_HNHNHP_H \qquad 2\ P_HNH_2 \qquad (86)$$

[742]	1 H_2/(py)$_3$RhCl$_3$-NaBH$_4$,25°,1 atm	· +	
[740]	1 H_2/Pd,25°,1 atm,5 min.	+	
[572]	1 H_2/2% Pd(C),83% EtOH,25°	~100%	
[264]	KO$_2$CN=NCO$_2$K,DMSO,25°,20 hrs	75%	
[148]	AlHg,EtOH,100°,few min.	~100%	
[740]	H_2/Pd,25°,1 atm,4.5 hrs		+
[742]	H_2/(py)$_3$RhCl$_3$-NaBH$_4$,25°,1 atm		+
[748]	(EtO)$_2$P(S)SH (10 equiv),70°,24 hrs		40%

The **reductive cleavage of azo compounds** is accomplished in good yields by *catalytic hydrogenation* [740, 747], with *sodium hydrosulfite* [258] and with *O,O-dialkyl dithiophosphoric acid* (yields 23–93%) [748].

Reduction of diazo and **azido compounds** is discussed on pp. 75, 76.

REDUCTION OF ALDEHYDES

Reduction of aldehydes to primary alcohols is very easy and can be accomplished by a legion of reagents. The real challenge is to reduce aldehydes containing other functional groups selectively or to reduce the other function in preference to the aldehyde group.

Reduction of Aliphatic Aldehydes

Saturated aliphatic aldehydes are readily reduced to alcohols by *catalytic hydrogenation* over almost any catalyst: e.g. platinum oxide, especially in the presence of iron ions which accelerate the reduction of the carbonyl group [749], and Raney nickel (after washing away strong alkalinity which could cause side reactions) [45]. High yields of primary alcohols are obtained by reductions with *lithium aluminum hydride* [83] (*Procedure 14*, p. 207), *lithium borohydride* [750] and with *sodium borohydride* [751], as well as with more sophisticated complex hydrides such as *lithium trialkoxyaluminum hydrides* [752], *sodium bis(2-methoxyethoxy)aluminum hydride* [544], *tetrabutylammonium cyanoborohydride* [757] and others. B-3-Pinanyl-9-borabicyclo[3,3,1]nonane (prepared from (±)-α-pinene and 9-BBN) is suitable for asymmetric reduction of 1-deuteroaldehydes to chiral 1-deuteroalcohols (enantiomeric excess of 64–83%) [109]. Aliphatic aldehydes were also reduced by treatment with a suspension of *sodium* or *lithium hydride* in a solution of ferrous or ferric chloride in tetrahydrofuran [753], by *iron* and acetic acid [754], by refluxing with *sodium dithionite* (hydrosulfite) in aqueous dioxane or dimethylformamide [262], and by *Meerwein–Ponndorf* reaction using isopropyl alcohol and aluminum isopropoxide [309] (*Procedure 48*, p. 217).

Several reagents reduce aldehydes preferentially to ketones in mixtures of both. Very high selectivity was found in reductions using dehydrated aluminum oxide soaked with *isopropyl alcohol* and especially diisopropylcarbinol [755], or silica gel and *tributylstannane* [756]. The best selectivity was achieved with *lithium trialkoxyaluminum hydrides* at $-78°$. In the system hexanal/cyclohexanone the ratio of primary to secondary alcohol was 87:13 at 0° and 91.5:8.5 at $-78°$ with lithium tris(*tert*-butoxy)aluminum hydride [752], and 93.6:6.4 at 0° and 99.6:0.4 at $-78°$ with lithium tris(3-ethyl-3-pentyloxy)aluminum hydride [752].

(87)

	R=	RCHO $\xrightarrow{\hspace{4cm}}$ RCH$_2$OH	RCH$_2$OH
[544]	C_3H_7	NaAlH$_2$(OCH$_2$CH$_2$OMe)$_2$,C$_6$H$_6$,30-80°	97%
[751]	C_3H_7	NaBH$_4$,H$_2$O	85%
[45]	C_5H_{11}	H$_2$/Raney Ni(W6),25°,2 hrs	~100%
[752]	C_5H_{11}	LiAlH(OCMe$_3$)$_3$,THF,-78° or 0°	≥90%
[262]	C_5H_{11}	Na$_2$S$_2$O$_4$,H$_2$O,reflux 4 hrs	63%
[49]	C_6H_{13}	H$_2$/Ni,EtOH,Et$_3$N,25°,1 atm,14 hrs	88%
[83]	C_6H_{13}	LiAlH$_4$,Et$_2$O,reflux	86%
[750]	C_6H_{13}	LiBH$_4$,Et$_2$O	83%
[754]	C_6H_{13}	Fe,AcOH,H$_2$O,100°,6-7 hrs	75-81%
[756]	C_7H_{15}	SiO$_2$,Bu$_3$SnH,hexane,25°,1 hr	90%
[753]	C_7H_{15}	FeCl$_3$,LiH,THF,25°,24-48 hrs	79%
[757]	C_8H_{17}	Bu$_4$NBH$_3$CN,HMPA,25°,1 hr	84%
[755]	C_9H_{19}	Al$_2$O$_3$,Me$_2$CHOH,CCl$_4$,25°,2 hrs	84%

Reduction of saturated aliphatic aldehydes to alkanes was carried out by refluxing with amalgamated *zinc* and hydrochloric acid (*Clemmensen reduction*) [160, 758] (p. 28) or by heating with *hydrazine* and potassium hydroxide (*Wolff–Kizhner reduction*) [280, 759] (p. 34). Heptaldehyde gave heptane in 72% yield by the first and in 54% yield by the second method.

Other possibilities of **converting the aldehyde group to a methyl group** are **desulfurization of the mercaptal** (p. 104) and reduction of azines, hydrazones and tosylhydrazones (p. 106).

An interesting reduction of aldehydes takes place on treatment with a reagent prepared from *titanium trichloride and potassium* [206] or magnesium [207] in tetrahydrofuran: propionaldehyde gave a 60% yield of a mixture of 30% *cis*- and 30% *trans*-3-hexene [207].

A method for the **conversion of unsaturated aliphatic aldehydes** to **saturated aldehydes** is a gentle *catalytic hydrogenation*. Palladium is more selective than nickel. Hydrogenation over sodium borohydride-reduced palladium in methanol at room temperature and 2 atm reduced crotonaldehyde to butyraldehyde but did not hydrogenate butyraldehyde [31]. Nickel prepared by reduction with sodium borohydride was less selective: it effected reduction of crotonaldehyde to butyraldehyde but also reduction of butyraldehyde to butyl alcohol, though at a slower rate [31]. Hydrogenation of 2,2,dimethyl-

4-pentenal over 5% palladium on alumina at 25° and 2 atm gave 88.5% of 2,2-dimethylpentanal while over Raney nickel at 125° and 100 atm 2,2-dimethylpentanol was obtained in 94% yield [760]. 3-Cyclohexenecarboxaldehyde afforded 81% yield of cyclohexanecarboxaldehyde on hydrogenation over 5% palladium on charcoal at 75–80° and 14 atm [761]. Reduction using cobalt hydrocarbonyl HCo(CO)₄ at 25° and 1 atm converted crotonaldehyde to butyraldehyde (yield 80%) with only a trace of butyl alcohol [762]. A very selective reduction of a double bond in an unsaturated aldehyde was achieved by homogeneous catalytic hydrogenation (*Procedure 8*, p. 206) and by *catalytic transfer of hydrogen* using 10% palladium on charcoal and triethylammonium formate. In citral, only the α,β-double bond was reduced, yielding 91% of citronellal (3,7-dimethyl-6-octenal) [317a].

Unsaturated aliphatic aldehydes were selectively **reduced to unsaturated alcohols** by specially controlled *catalytic hydrogenation*. Citral treated with hydrogen over platinum dioxide in the presence of ferrous chloride or sulfate and zinc acetate at room temperature and 3.5 atm was reduced only at the carbonyl group and gave geraniol (3,7-dimethyl-2,6-octadienol) [59], and crotonaldehyde on hydrogenation over 5% osmium on charcoal gave crotyl alcohol [763].

Complex hydrides can be used for the selective reduction of the carbonyl group although some of them, especially lithium aluminum hydride, may reduce the α,β-conjugated double bond as well. Crotonaldehyde was converted to crotyl alcohol by reduction with *lithium aluminum hydride* [83], *magnesium aluminum hydride* [577], *lithium borohydride* [750], *sodium borohydride* [751], *sodium trimethoxyborohydride* [99], *diphenylstannane* [114] and *9-borabicyclo[3,3,1]nonane* [764]. A dependable way to convert α,β-unsaturated aldehydes to unsaturated alcohols is the *Meerwein–Ponndorf reduction* [765].

Reduction of unsaturated aliphatic aldehydes to saturated alcohols was effected by catalytic hydrogenation over nickel catalysts [31, 760] and by *electroreduction* [766].

(88)

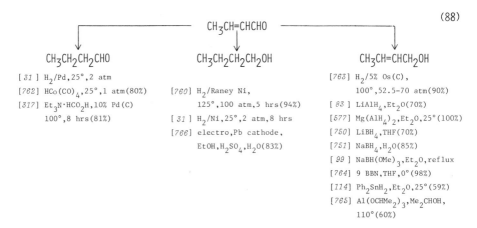

CH₃CH=CHCHO

CH₃CH₂CH₂CHO

[31] H₂/Pd,25°,2 atm
[762] HCo(CO)₄,25°,1 atm(80%)
[317] Et₃N·HCO₂H,10% Pd(C)
        100°,8 hrs(81%)

CH₃CH₂CH₂CH₂OH

[760] H₂/Raney Ni,
        125°,100 atm,5 hrs(94%)
[31] H₂/Ni,25°,2 atm,8 hrs
[766] electro,Pb cathode,
        EtOH,H₂SO₄,H₂O(83%)

CH₃CH=CHCH₂OH

[763] H₂/5% Os(C),
        100°,52.5–70 atm(90%)
[83] LiAlH₄,Et₂O(70%)
[577] Mg(AlH₄)₂,Et₂O,25°(100%)
[750] LiBH₄,THF(70%)
[751] NaBH₄,H₂O(85%)
[99] NaBH(OMe)₃,Et₂O,reflux
[764] 9 BBN,THF,0°(98%)
[114] Ph₂SnH₂,Et₂O,25°(59%)
[765] Al(OCHMe₂)₃,Me₂CHOH,
        110°(60%)

Reduction of Aromatic Aldehydes

Reduction of aromatic aldehydes can give three kinds of product: primary alcohols of benzylic type, methylated aromatics, or methylated alicyclic (or heterocyclic) compounds with fully hydrogenated rings. Since benzyl-type alcohols are easily hydrogenolyzed, reduction to the alcohol sometimes requires careful control or special choice of reducing agent to prevent reduction of the aldehyde group to a methyl group. Total reduction requires rather energetic hydrogenation and occurs only after the hydrogenolysis of the carbon–oxygen bond.

Conversion of aromatic aldehydes to alcohols is very easy and can be accomplished by a variety of ways. *Catalytic hydrogenation* over almost any catalyst gives good yields of alcohols provided it is monitored to prevent deeper hydrogenation. Platinum and palladium catalysts tend to hydrogenolyze the alcohols to methyls and even to hydrogenate the rings if used in acidic media. Some reviewers of hydrogenations consider platinum better than palladium while others have found the opposite. At any rate, good yields of alcohols were obtained with either catalyst, especially when the hydrogenation was stopped after absorption of the amount of hydrogen required for the hydrogenation to the alcohol [*30, 749, 767, 768*]. Addition of ferrous chloride in hydrogenations using platinum oxide proved advantageous [*749*].

Nickel, Raney nickel and copper chromite are other catalysts suitable for hydrogenation of aldehydes to alcohols with little if any further hydrogenolysis. Benzaldehyde was hydrogenated to benzyl alcohol over nickel [*43*], Raney nickel [*45*] and copper chromite [*50*] in excellent yields. In the last-

(89)

PhCHO	⟶	PhCH$_2$OH	PhCH$_3$
[*767*]	H$_2$/Pt,EtOH,20°,2 atm,5 hrs	(~100%)	
[*767*]	H$_2$/Pt,AcOH,2C°,2 atm,10 hrs		80%
[*49*]	H$_2$/Ni,EtOH,25°,1 atm,1.6 hrs	91%	
[*71*]	HCO$_2$H/Cu,200°	56%	18%
[*753*]	NaH/FeCl$_3$,THF,25°,24 hrs	85%	
[*83*]	LiAlH$_4$,Et$_2$O,exothermic	85%	
[*770*]	LiAlH$_4$/AlCl$_3$,Et$_2$O,reflux 30 min	60%	
[*750*]	LiBH$_4$,Et$_2$O,exothermic	91%	
[*99*]	NaBH(OMe)$_3$,Et$_2$O,reflux 4 hrs	78%	
[*771*]	Bu$_4$NBH$_4$,CH$_2$Cl$_2$,0°→25°,24 hrs	91%	
[*109*]	B-3-Pinanyl-9-BBN,enantiomeric excess	70%	
[*772*]	Et$_3$SiH/BF$_3$,CH$_2$Cl$_2$,0°,11 min		52%
[*756*]	Bu$_3$SnH,SiO$_2$,cyclohexane	81%	
[*114*]	Ph$_2$SnH$_2$,Et$_2$O,25°	62%	
[*773*]	Li/NH$_3$,THF,tert-BuOH or NH$_4$Cl		~90%
[*262*]	Na$_2$S$_2$O$_4$,H$_2$O,dioxane	84%	
[*774*]	N$_2$H$_4$,KOH,80-100°		79%
[*755*]	Al$_2$O$_3$,Me$_2$CHOH,CCl$_4$,25°,2.5 hrs	77%	
[*765*]	Al(OCHMe$_2$)$_3$,Me$_2$CHOH,reflux	55%	

mentioned case the benzyl alcohol was accompanied by 8% of toluene [50]. Reduction of benzaldehyde by *hydrogen transfer* using formic acid and copper gave 56% of benzyl alcohol and 18% of toluene [71]. Salicyl aldehyde was reduced by nascent hydrogen generated by dissolving *Raney nickel alloy* (50% nickel and 50% aluminum) in 10% aqueous sodium hydroxide. At a temperature of 10–20° the product was saligenin (salicyl alcohol) (yield 77%), whereas at 90° a 76% yield of *o*-cresol was obtained [769].

Chemical reduction of aromatic aldehydes to alcohols was accomplished with *lithium aluminum hydride* [83], *alane* [770], *lithium borohydride* [750], *sodium borohydride* [751], *sodium trimethoxyborohydride* [99], *tetrabutylammonium borohydride* [771], *tetrabutylammonium cyanoborohydride* [757], *B-3-pinanyl-9-borabicyclo[3.3.1]nonane* [109], *tributylstannane* [756], *diphenylstannane* [114], *sodium dithionite* [262], *isopropyl alcohol* [755], *formaldehyde* (crossed Cannizzaro reaction) [313] and others.

The chiral reagent prepared *in situ* from (+)-α-pinene and 9-borabicyclo-[3.3.1]nonane reduced benzaldehyde-*1-d* to benzyl-*1-d* alcohol in 81.6% yield and 90% enantiomeric excess [109].

(90)

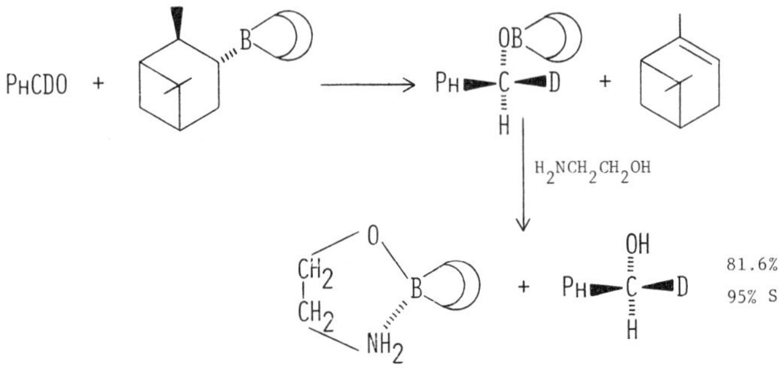

Most of the above reagents do not reduce aromatic aldehydes beyond the stage of alcohols. Reagents which reduce via carbonium ion intermediates – *alane* [770] and *triethylsilane* with boron trifluoride [772], for example – tend to reduce the carbonyl group to the methyl group, especially in aldehydes containing electron-releasing groups which stabilize the carbonium intermediates. On the contrary, electron-withdrawing substituents favor reduction to alcohols only.

(91)

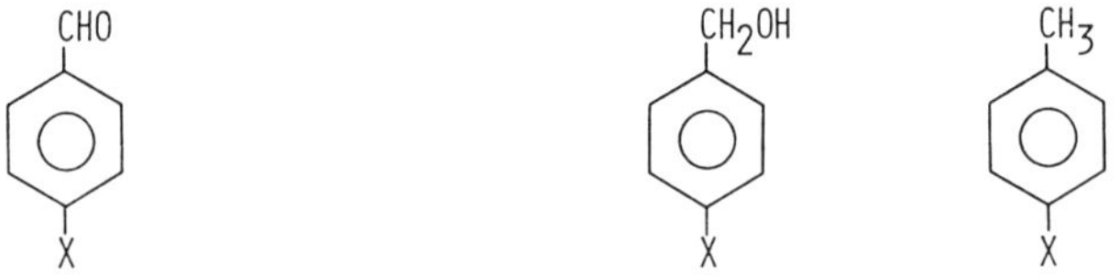

				CH_2OH	CH_3
[770]	X = H	$LiAlH_4/AlCl_3$, Et_2O, reflux	60%		
[772]		Et_3SiH/BF_3			52%
[770]	X = MeO	$LiAlH_4/AlCl_3$, reflux			79%
[772]		Et_3SiH/BF_3			100%
[772]	X = NO_2	Et_3SiH/BF_3	100%		

Reduction of the carbonyl group in aromatic aldehydes **to the methyl group** was achieved by *catalytic hydrogenation* over palladium in acetic acid (yield 86%) [*775*], or by energetic hydrogenation over nickel [*43*] or copper chromite [*776*]. Fluorene-1-carboxaldehyde gave 86% of 1-methylfluorene on hydrogenation over palladium on charcoal in acetic acid at room temperature and atmospheric pressure [*775*], and fural yielded 90–95% of α-methylfuran over copper chromite at 200–230° at atmospheric pressure [*776*].

Triethylsilane in the presence of boron trifluoride [*772*] or trifluoroacetic acid [*777*] also reduced the aldehyde group to a methyl group.

Electrolytic reduction using a lead cathode in 20% sulfuric acid converted pyridine α-carboxaldehyde to a mixture of 41% of α-picoline, 25% of α-pipecoline and 11% of 2-methyl-1,2,3,6-tetrahydropyridine [*443*].

Reduction of benzaldehyde and *p*-alkylbenzaldehydes to the corresponding hydrocarbons was carried out by *lithium* in liquid ammonia and tetrahydrofuran in the presence of *tert*-butyl alcohol or ammonium chloride (yields 90–94%) [*773*].

Like any aldehydes aromatic aldehydes undergo *Clemmensen reduction* [*758, 778*] and *Wolff-Kizhner* reduction [*759, 774*] and give the corresponding methyl compounds, generally in good yields. The same effect is accomplished by conversion of the aldehydes to *p*-toluenesulfonyl hydrazones followed by reduction with *lithium aluminum hydride* (p. 106).

Reduction of **aromatic aldehydes to pinacols** using sodium amalgam is quite rare. Equally rare is conversion of aromatic aldehydes **to alkenes** formed by deoxygenation and coupling and accomplished by treatment of the aldehyde with a reagent obtained by reduction of *titanium trichloride with lithium* in dimethoxyethane. Benzaldehyde thus afforded *trans*-stilbene in 97% yield [*206, 209*].

Hydrogenation of the nuclei in aromatic aldehydes is possible by *catalytic hydrogenation* over noble metals in acetic acid and is rather slow [*767*].

A typical example of an **unsaturated aromatic aldehyde** is cinnamaldehyde. Hardly any other aldehyde has become so extensively employed for testing new reagents for their selectivity. The α,β-double bond is conjugated not only with the aldehyde carbonyl group but also with the aromatic ring, which makes it readily reducible. Consequently the development of a selective method for reducing cinnamaldehyde to hydrocinnamaldehyde or to cinnamyl alcohol is a great challenge.

Reduction of the double bond only was achieved by *catalytic hydrogenation* over palladium prepared by reduction with sodium borohydride. This catalyst does not catalyze hydrogenation of the aldehyde group [*31*]. Also sodium borohydride-reduced nickel was used for conversion of cinnamaldehyde to hydrocinnamaldehyde [31]. Homogeneous hydrogenation over tris(triphenylphosphine)rhodium chloride gave 60% of hydrocinnamaldehyde and 40% of ethylbenzene [*56*]. Raney nickel, by contrast, catalyzes **total reduction** to hydrocinnamyl alcohol [*45*]. Total reduction of both the double bond and the carbonyl group was also accomplished by *electrolysis* [766] and by refluxing with *lithium aluminum hydride* in ether [576, 609].

(92)

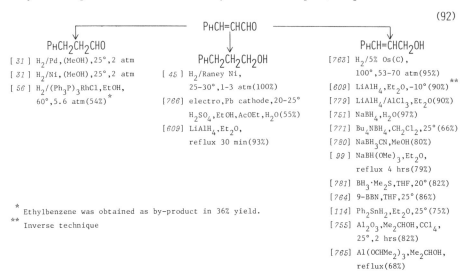

PHCH=CHCHO

PHCH₂CH₂CHO

[31] H₂/Pd,(MeOH),25°,2 atm
[31] H₂/Ni,(MeOH),25°,2 atm
[56] H₂/(Ph₃P)₃RhCl,EtOH,
       60°,5.6 atm(54%)*

PHCH₂CH₂CH₂OH

[45] H₂/Raney Ni,
       25-30°,1-3 atm(100%)
[766] electro,Pb cathode,20-25°
       H₂SO₄,EtOH,AcOEt,H₂O(55%)
[609] LiAlH₄,Et₂O,
       reflux 30 min(93%)

PHCH=CHCH₂OH

[763] H₂/5% Os(C),
      100°,53-70 atm(95%)
[609] LiAlH₄,Et₂O,-10°(90%)**
[779] LiAlH₄/AlCl₃,Et₂O(90%)
[751] NaBH₄,H₂O(97%)
[771] Bu₄NBH₄,CH₂Cl₂,25°(66%)
[780] NaBH₃CN,MeOH(80%)
[99] NaBH(OMe)₃,Et₂O,
       reflux 4 hrs(79%)
[781] BH₃·Me₂S,THF,20°(82%)
[764] 9-BBN,THF,25°(86%)
[114] Ph₂SnH₂,Et₂O,25°(75%)
[755] Al₂O₃,Me₂CHOH,CCl₄,
      25°,2 hrs(82%)
[765] Al(OCHMe₂)₃,Me₂CHOH,
      reflux(68%)

* Ethylbenzene was obtained as by-product in 36% yield.
** Inverse technique

Many more examples exist for **reduction of the carbonyl only.** Over an osmium catalyst [763] or platinum catalyst activated by zinc acetate and ferrous chloride [782] cinnamaldehyde was *hydrogenated* to cinnamyl alcohol. The same product was obtained by gentle reduction with *lithium aluminum hydride* at −10° using the inverse technique [609], by reduction with *alane* (prepared *in situ* from lithium aluminum hydride and aluminum chloride) [779], with *sodium borohydride* [751], with *tetrabutylammonium borohydride* [771], with *sodium trimethoxyborohydride* [99], with *sodium cyanoborohydride* [780], with *9-borabicyclo[3.3.1]nonane* [764], with *borane*-dimethyl sulfide complex [781], with *diphenylstannane* [114], and with dehydrated alumina soaked with *isopropyl alcohol* [755]. Reduction of furalacetaldehyde (β-(α-furyl)acrolein) with ethanol and aluminum ethoxide in xylene at 100° for 2 hours gave 60–70% β-(α-furyl)allyl alcohol [783].

On rare occasions the **aldehyde group in an α,β-unsaturated aromatic aldehyde** was *reduced to a methyl group.* 6-Benzyloxyindole-3-carboxaldehyde was transformed to 6-benzyloxy-3-methylindole by *sodium borohydride* in isopropyl alcohol in the presence of *10% palladium* on charcoal (yield 89%) [784].

Reduction of unsaturated aromatic aldehydes to unsaturated hydrocarbons poses a serious problem, especially if the double bond is conjugated with the benzene ring or the carbonyl or both. In *Clemmensen reduction* the α,β-unsaturated double bond is usually reduced [160], and in *Wolff–Kizhner reduction* a cyclopropane derivative may be formed as a result of decomposition of pyrazolines formed by intramolecular addition of the intermediate hydrazones across the double bonds [280]. The only way of converting unsaturated aromatic aldehydes to unsaturated hydrocarbons is the reaction of the aldehyde with *p*-toluenesulfonylhydrazide to form *p*-toluenesulfonylhydrazone and its subsequent reduction with *borane*. The reduction is accompanied by a double bond shift [*785, 786*].

Conversion of unsaturated aromatic aldehydes to saturated hydrocarbons can be realized by *Clemmensen reduction* [*160*].

REDUCTION OF DERIVATIVES OF ALDEHYDES

Halogenated aldehydes are usually reduced without the loss of the halogen [*774*]. In many reductions the rate of reduction of the carbonyl group is higher than that of the hydrogenolysis of most carbon–halogen bonds. Use of *alkali metals* and *zinc* for the reduction may involve some risk, especially if a reactive halogen is bonded in the position α to a carbonyl [*160*]. In such cases even *Meerwein-Ponndorf* reduction may lead to complications because of the reaction of the halogen with the alcoholic group in alkaline medium [*309*].

In **nitro aldehydes** both the nitro group and the aldehyde group are readily reduced by *catalytic hydrogenation*. It may be difficult, if not impossible to hydrogenate either function separately. More dependable methods are reduction by *alane* [*787*] or by *isopropyl alcohol* and aluminum isopropoxide (*Meerwein-Ponndorf*) [*788*] to *nitro alcohols*, and by *stannous chloride* [*789, 790*], *titanium trichloride* [*590*] or *ferrous sulfate* [*218*] to *amino aldehydes* (*Procedure 38*, p. 214).

$$(93)$$

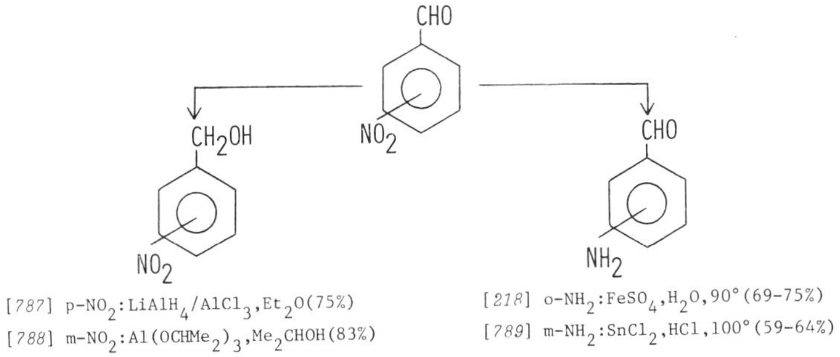

[*787*] p-NO$_2$:LiAlH$_4$/AlCl$_3$,Et$_2$O(75%)

[*788*] m-NO$_2$:Al(OCHMe$_2$)$_3$,Me$_2$CHOH(83%)

[*218*] o-NH$_2$:FeSO$_4$,H$_2$O,90° (69–75%)

[*789*] m-NH$_2$:SnCl$_2$,HCl,100° (59–64%)

Hydroxy aldehydes are reduced to hydroxy alcohols without complications. In benzyl ethers of phenol aldehydes the benzyl group was hydrogenolyzed in preference to the reduction of the carbonyl by *catalytic hydrogenation* over 10% palladium on charcoal at room temperature and 2 atm to give phenol aldehyde [*791*].

A peculiar reduction occurred on treatment of a **dialdehyde**, diphenyl-2,2'-dicarboxaldehyde, with tris(dimethylamino)phosphine at room temperature: phenanthrene-9,10-oxide was formed in 81–89% yield [*302*].

Acetals of aldehydes are usually stable to lithium aluminum hydride but are reduced to ethers with *alane* prepared *in situ* from lithium aluminum hydride and aluminum chloride in ether. Butyraldehyde diethyl acetal gave 47% yield of butyl ethyl ether, and benzaldehyde dimethyl acetal and diethyl acetal afforded benzyl methyl ether and benzyl ethyl ether in 88% and 73% yields, respectively [*792*].

2-Tetrahydrofuranyl and 2-tetrahydropyranyl alkyl ethers, being acetals, resist lithium aluminum hydride but are reduced by *alane* prepared from lithium aluminum hydride and aluminum chloride or boron trifluoride etherate. Exocyclic reductive cleavage gives tetrahydrofuran or tetrahydropyran and alcohols (yields 5–90%) whereas endocyclic reductive cleavage affords 4-hydroxybutyl or 5-hydroxypentyl alkyl ethers, respectively (yields 10–79%). Which of the cleavages takes place depends on the structure of the alkyl groups and is affected both by electronic and steric effects. Use of aluminum chloride gives better yields with tertiary alkyl groups while boron trifluoride etherate favors better yields with secondary and especially primary alkyl groups [793].

$$\underset{\underset{\displaystyle O \quad OR}{\diagdown \quad \diagup}}{\overset{\displaystyle (CH_2)_N}{CH_2 \quad CH}} \xrightarrow[\text{Reflux 2 hrs}]{\text{LiAlH}_4,\text{Et}_2O} \quad \underset{\underset{\displaystyle O}{\diagdown \diagup}}{\overset{\displaystyle (CH_2)_N}{CH_2 \quad CH_2}} \quad ROH \quad \underset{\underset{\displaystyle OH \quad OR}{}}{\overset{\displaystyle (CH_2)_N}{CH_2 \quad CH_2}} \quad (94)$$

[793]			ROH	
n=2, R=Me₃C;	AlCl₃			58%
	BF₃·Et₂O			16%
R=C₆H₁₃;	AlCl₃		40%	27%
	BF₃·Et₂O			66%
n=3, R=Me₃C;	AlCl₃			72%
	BF₃·Et₂O			46%
R=C₆H₁₃;	AlCl₃		~90%	11%
	BF₃·Et₂O			41%

Sulfur analogs, 2-tetrahydrofuranyl and 2-tetrahydropyranyl **thioethers,** were reduced by *alane* to alkyl 4- or alkyl 5-hydroxyalkyl thioethers resulting from the preferential reductive cleavage of the carbon–oxygen (rather than the carbon–sulfur) bond. Thus refluxing for 2 hours with alane in ether converted 2-alkylthiotetrahydrofurans to alkyl 4-hydroxybutyl thioethers in 63–72% yields, and 2-alkylthiotetrahydropyrans to alkyl 5-hydroxypentyl thioethers in 58–82% yields [794].

Cyclic five-membered **mono- and dithioacetals** were reduced with *calcium* in liquid ammonia to give β-alkoxy- and β-thioalkoxymercaptans, respectively: $RCH_2OCH_2CH_2SH$ and $RCH_2SCH_2CH_2SH$. Benzaldehyde ethylenedithioacetal, on the other hand, gave toluene (75%), and phenylacetaldehyde ethylenedithioacetal gave ethylbenzene (71% yield) [795].

Mercaptals (dithioacetals) and cyclic mercaptals (prepared from aldehydes and ethanedithiol or 1,2-propanedithiol) are easily **desulfurized** by refluxing with an excess of *Raney nickel* in ethanol, and give *hydrocarbons* in good yields [796, 797]. If the molecule of the mercaptal contains a function reducible by hydrogen, Raney nickel must be refluxed for several hours in acetone prior to the desulfurization to remove the hydrogen adsorbed on its surface [46]. Desulfurization of mercaptals can also be achieved by heating of the mercaptal with 3–4 equivalents of *tributylstannane* and a catalytic amount of azobis(isobutyronitrile) for 1.5 hours at 80° and distilling the mixture *in vacuo* [798] or on treatment of the mercaptal with *sodium* in liquid ammonia [799].

These reactions represent a gentle method for conversion of aldehydes to hydrocarbons in yields of 40-95% and are frequently used, especially in saccharide chemistry [796, 797, 798].

[794]   (95)

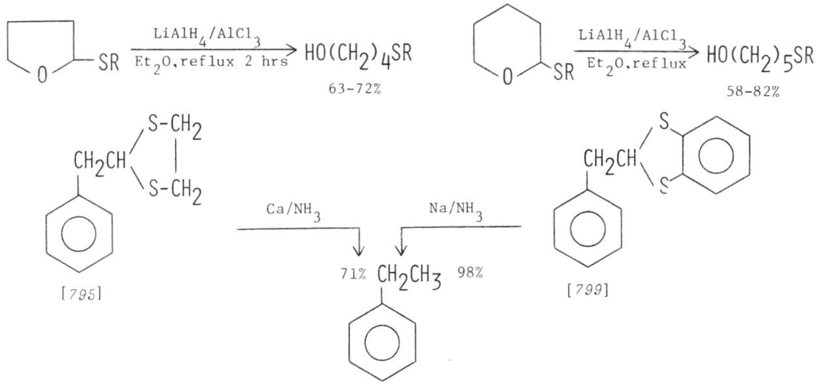

Cinnamaldehyde diacetate was reduced by heating at 100° with *iron* in 50% acetic acid to cinnamyl acetate in 60.7% yield [800].

2-Alkylaminotetrahydropyrans may be considered derivatives of aldehydes. They are reductively cleaved by *lithium aluminum hydride* to *amino alcohols*. Thus 2-(N-piperidyl)tetrahydropyran afforded, after refluxing for 2 hours with 2 mol of lithium aluminum hydride in ether, 5-piperidino-1-pentanol in 82% yield [801].

Alkyl or aryl aldimines (Schiff's bases) gave, on reduction, secondary amines. Reduction was carried out by *catalytic hydrogenation* over platinum [802] and over nickel in good yields [803, 804], by *lithium aluminum hydride* in refluxing ether or tetrahydrofuran, by *sodium aluminum hydride, lithium borohydride* and *sodium borohydride* [805], by *sodium amalgam* [806] and by *potassium-graphite* [807], generally in fair to excellent yields. In the reduction of N-cyclopropylbenzaldimine the energetic action of lithium aluminum hy-

(96)

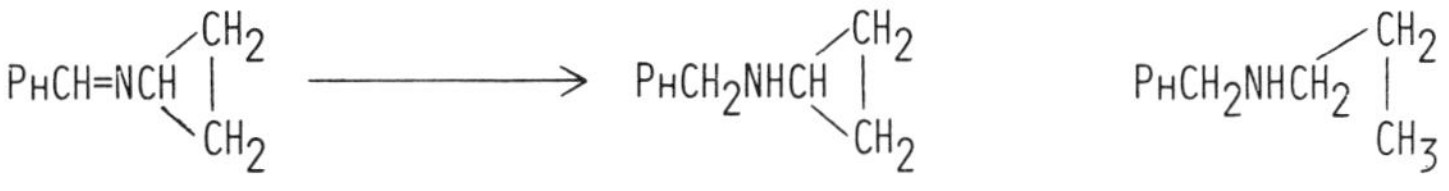

[805]		
H$_2$/Pt,EtOH,2 atm,12 hrs	72%	
LiAlH$_4$,Et$_2$O,reflux 12 hrs		80%
LiAlH$_4$,Et$_2$O,reflux 4 hrs	59%	
NaAlH$_4$,THF,reflux 24 hrs	23.5%	23.5%
LiBH$_4$,THF,reflux 24 hrs	52%	
NaBH$_4$,THF,reflux 24 hrs	67%	

dride not only reduced the carbon–nitrogen double bond but also cleaved the three-membered ring and yielded benzylpropylamine [805].

Aldimines derived from aromatic aldehydes suffered hydrogenolysis in *hydrogenation* over palladium at 117–120° at 20 atm and gave products in which the original aldehyde group was transformed to methyl group (yields 62–72%) [*808*].

Aldoximes yielded *primary amines* by *catalytic hydrogenation*: benzaldehyde gave benzylamine in 77% yield over nickel at 100° and 100 atm [*803*], with *lithium aluminum hydride* (yields 47–79%) [*809*], with *sodium* in refluxing ethanol (yields 60–73%) [*810*] and with other reagents. **Hydrazones** of aldehydes are intermediates in the Wolff–Kizhner reduction of the aldehyde group to a methyl group (p. 97) but are hardly ever reduced to amines.

p-**Toluenesulfonylhydrazones of aldehydes** prepared from aldehydes and *p*-toluenesulfonhydrazide are reduced by *lithium aluminum hydride* [*811, 812*], *sodium borohydride* [*785, 811*], *sodium cyanoborohydride* [*813*], or with *bis(benzyloxy) borane* prepared *in situ* from benzoic acid and borane–tetrahydrofuran complex [*786*]. The aldehyde group is converted to the methyl group. If sodium cyanoborohydride is used as the reducing agent the tosylhydrazones may be prepared *in situ* since the cyanoborohydride does not reduce the aldehydes prior to their conversion to the tosylhydrazones. Thus the above sequence represents a method for the reduction of aldehydes to hydrocarbons under very gentle reaction conditions. It is therefore suitable even for reduction of α,β-unsaturated aldehydes which cannot be reduced by Clemmensen or Wolff–Kizhner reductions. Rearrangement of the double bond accompanies the reduction of tosylhydrazones of α,β-unsaturated aldehydes [*785, 786*].

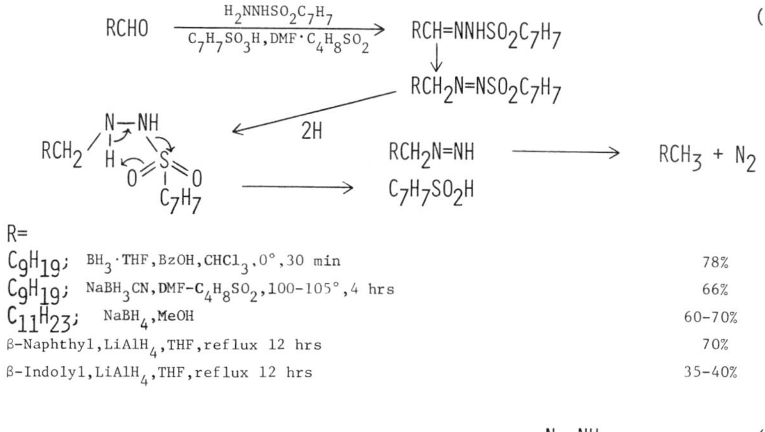

$$RCHO \xrightarrow[C_7H_7SO_3H, DMF \cdot C_4H_8SO_2]{H_2NNHSO_2C_7H_7} RCH=NNHSO_2C_7H_7 \qquad (97)$$

$$\downarrow$$

$$RCH_2N=NSO_2C_7H_7$$

$$RCH_2\overset{N-NH}{\underset{O=S=O}{H}} \xleftarrow{\quad 2H \quad}$$

$$RCH_2N=NH \longrightarrow RCH_3 + N_2$$

$$\longrightarrow C_7H_7SO_2H$$

R=

[*786*]	C_9H_{19};	$BH_3 \cdot THF, BzOH, CHCl_3, 0°, 30$ min	78%
[*813*]	C_9H_{19};	$NaBH_3CN, DMF-C_4H_8SO_2, 100-105°, 4$ hrs	66%
[*811*]	$C_{11}H_{23}$;	$NaBH_4, MeOH$	60–70%
[*812*]	β-Naphthyl, $LiAlH_4$, THF, reflux 12 hrs		70%
[*812*]	β-Indolyl, $LiAlH_4$, THF, reflux 12 hrs		35–40%

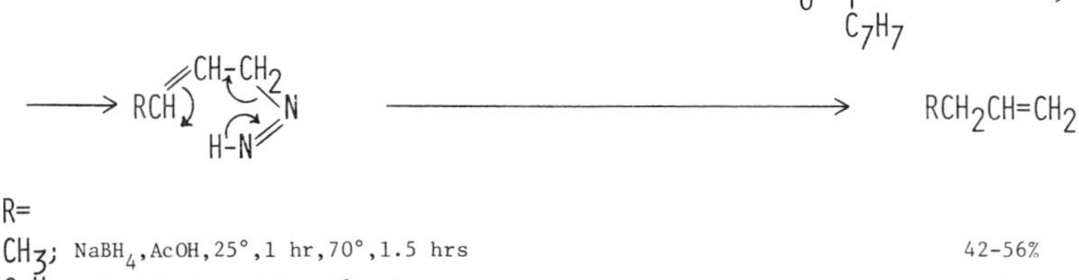

$$RCH=CHCH=NNHSO_2C_7H_7 \xrightarrow{\quad 2H \quad} RCH=CHCH_2\overset{N-NH}{\underset{O=S=O}{H}} \longrightarrow \qquad (98)$$

$$C_7H_7$$

$$\longrightarrow RCH\overset{CH=CH_2}{\underset{H-N}{N}} \longrightarrow RCH_2CH=CH_2$$

R=

[*785*]	CH_3;	$NaBH_4, AcOH, 25°, 1$ hr, $70°, 1.5$ hrs	42–56%
[*786*]	C_6H_5;	$BH_3 \cdot THF, BzOH, CHCl_3, 0°, 1$ hr	85%

REDUCTION OF KETONES

The immense number of reductions performed on ketones dictates subdivision of this topic into reductions of aliphatic, alicyclic, aromatic and unsaturated ketones. However, since differences in behavior toward reduction are very small between aliphatic and alicyclic ketones the section on reductions of alicyclic ketones includes only specific examples due to stereochemistry of the ring systems.

Reduction of Aliphatic Ketones

Reduction of saturated ketones to alcohols is very easy by many means but is distinctly slower than that of comparable aldehydes. It is frequently possible to carry out selective reductions of the two types (p. 97).

Alcohols were obtained by *catalytic hydrogenation*. While palladous oxide proved ineffective [*38, 814*], hydrogenation over platinum oxide and especially rhodium–platinum oxide [*38*], and over platinum, rhodium and ruthenium, was fast and quantitative at room temperature and atmospheric pressure [*38, 814*]. Under comparable conditions Raney nickel and nickel prepared by reduction of nickel acetate with sodium hydride in the presence of sodium alkoxides [*49*] have also been used and gave excellent yields. Normal nickel [*43*] as well as copper chromite [*50*] catalyst required high temperatures (100–150°) and high pressures (100–165 atm) and also gave almost quantitative yields. Moderate to good yields were obtained by reduction of ketones with 50% nickel–aluminum alloy in 10% aqueous sodium hydroxide [*769*]. This is somewhat surprising since the reduction was carried out under conditions favoring aldol condensation.

Transformation of ketones to alcohols has been accomplished by many hydrides and complex hydrides: by *lithium aluminum hydride* [*83*], by *magnesium aluminum hydride* [*89*], by *lithium tris(tert-butoxy)aluminum hydride* [*815*], by *dichloroalane* prepared from lithium aluminum hydride and aluminum chloride [*816*], by *lithium borohydride* [*750*], by *lithium triethylborohydride* [*100*], *by sodium borohydride* [*751, 817*], by *sodium trimethoxyborohydride* [*99*], by *tetrabutylammonium borohydride* [*771*] and *cyanoborohydride* [*757*], by chiral *diisopinocampheylborane* (yields 72–78%, optical purity 13–37%) [*818*], by *dibutyl-* and *diphenylstannane* [*114*], *tributylstannane* [*756*] and others (*Procedure 21*, p. 209).

As a consequence of the wide choice of hydride reagents the classical methods such as reduction with *sodium* in ethanol almost fell into oblivion [*819, 820*]. Nevertheless some old reductions were resuscitated. *Sodium dithionite* was found to be an effective reducing agent [*262*], and the reduction by *alcohols* [*309*] was modified to cut down on the temperature [*755*] or the time required [*821*], or to furnish chiral alcohols ('in good yields and excellent optical purity') by using optically active pentyl alcohol and its aluminum salt [*822*]. Formation of chiral alcohols by reduction of pro-chiral ketones is discussed at length in the sections on cyclic and aromatic ketones (p. 111). Another, better way for the preparation of optically active hydroxy compounds is biochemical reduction using *microorganisms*. Such reductions frequently apply to polyfunctional derivatives such as unsaturated ketones, hydroxyketones, keto acids, etc., and are dealt with in the appropriate sections (pp. 117, 118, 125 and 143).

(99)

$$RCOCH_3 \longrightarrow RCH(OH)CH_3$$

R=

[45]	CH_3;	H_2/Raney Ni,25-30°,1-3 atm,38 min	100%
[43]	CH_3;	H_2/Ni,125°,165 atm,1.5 hrs	88-100%
[50]	CH_3;	H_2/CuCr$_2$O$_4$,150°,100-150 atm,42 min	100%
[83]	C_2H_5;	LiAlH$_4$,Et$_2$O,reflux	80%
[750]	C_2H_5;	LiBH$_4$,Et$_2$O,reflux	77%
[751]	C_2H_5;	NaBH$_4$,MeOH-H$_2$O	87%
[818]	C_2H_5;	Diisopinocampheylborane,THF,-30°	73%*
[822]	C_2H_5;	EtMeCHCH$_2$OH,Al(OCH$_2$CHMeEt)$_3$,reflux 3 hrs	94.6%**
[769]	C_3H_7;	Ni-Al,NaOH,80-90°,1 hr	54%
[771]	$(CH_3)_3C$;	Bu$_4$NBH$_4$,CHCl$_3$,reflux 5.5 hrs	91%
[821]	$(CH_3)_3C$;	iso-PrOH,Al(O-iso-Pr)$_3$,reflux 1 hr	36%
[824]	$(CH_3)_2CHCH_2CH_2$;	electro,Cd cathode,dild. H$_2$SO$_4$,EtOH	83.5%
[441]	C_6H_{13};	LiAlH$_4$·4 C$_5$H$_5$N,25°,12 hrs	69%
[817]	C_6H_{13};	NaBH$_4$,H$_2$O,Al$_2$O$_3$	90%
[262]	C_6H_{13};	Na$_2$S$_2$O$_4$,DMF-H$_2$O,reflux 4 hrs	75%
[757]	C_9H_{19};	Bu$_4$NBH$_3$CN,HMPA,25°	84%
[819]	C_9H_{19};	Na,EtOH,reflux	75%

* Optical purity 16.5%

** Optically active

For the **reduction of aliphatic ketones to hydrocarbons** several methods are available: reduction with *triethylsilane and boron trifluoride* [772], *Clemmensen reduction* [160, 758] (p. 28), *Wolff-Kizhner reduction* [280, 281, 759] (p. 34), reduction of *p*-toluenesulfonylhydrazones with *sodium borohydride* [785], *sodium cyanoborohydride* [813] or *borane* [786] (p. 134), *desulfurization of dithioketals* (mercaptoles) [799, 823] (pp. 130, 131) and electroreduction [824].

(100)

$$RCOCH_3 \qquad RCH_2CH_3$$

R=

[759]	C_6H_{13};	N$_2$H$_4$·H$_2$O,HO(CH$_2$CH$_2$O)$_3$H,AcOH,130-180°, MeONa,HO(CH$_2$CH$_2$O)$_3$H,190-200°,0.5-1.5 hrs	66%
[786]	C_6H_{13};	TosNHNH$_2$,EtOH; BH$_3$·THF,BzOH,CHCl$_3$,1 hr	78%
[799]	C_8H_{17};	o-C$_6$H$_4$(SH)$_2$,TosOH; Na/NH$_3$	98%
[772]	C_9H_{19};	Et$_3$SiH/BF$_3$,CH$_2$Cl$_2$,1 hr	80%
[758]	C_9H_{19};	ZnHg,HCl,reflux 24 hrs	87%
[813]	C_9H_{19};	TosNHNH$_2$,DMF-C$_4$H$_8$SO$_2$,NaBH$_3$CN,100-105°,4 hrs	86%

Somewhat less frequent than the reductions of aliphatic ketones to secondary alcohols and to hydrocarbons are one-electron **reductions to pinacols.** These are accomplished by metals such as *sodium*, but better still by *magnesium* or *aluminum*. Acetone gave 43–50% yield of pinacol on refluxing with magnesium amalgam in benzene [*140*], and 45% and 51% yields on refluxing in methylene chloride or tetrahydrofuran, respectively [*825*].

An interesting **deoxygenation of ketones** takes place on treatment with *low valence state titanium*. Reagents prepared by treatment of titanium trichloride in tetrahydrofuran with lithium aluminum hydride [*205*], with potassium [*206*], with magnesium [*207*], or in dimethoxyethane with lithium [*206*] or zinc–copper couple [*206, 209*] convert **ketones to alkenes** formed by coupling of the ketone carbon skeleton at the carbonyl carbon. Diisopropyl ketone thus gave tetraisopropylethylene (yield 37%) [*206*], and cyclic and aromatic ketones afforded much better yields of symmetrical or mixed coupled products [*206, 207, 209*]. The formation of the alkene may be preceded by pinacol coupling. In some cases a pinacol was actually isolated and reduced by low valence state titanium to the alkene [*206*] (p. 118).

Reduction of Aromatic Ketones

Like aliphatic ketones, **aromatic and aromatic aliphatic ketones** are **reduced to alcohols** very easily. Since the secondary alcoholic group adjacent to the benzene ring is easily hydrogenolyzed, special precautions must be taken to prevent the reduction of the keto group to the methylene group, especially in catalytic hydrogenations.

An interesting example of hydrogenation with hydrogen in the absence of transition metal catalyst is reduction of benzophenone to benzhydrol with hydrogen in *tert*-butyl alcohol containing potassium *tert*-butoxide at 150–210° and 96–135 atm. Although the yields range from 47 to 98% the method is not practical because of its drastic conditions, and because of a cornucopia of more suitable reductions.

Catalytic hydrogenation of aromatic ketones was successful over a wide range of catalysts under mild conditions (25°, 1 atm). Of the noble metal family some metals more than others tend to hydrogenolyze the alcohol formed, and/or saturate the ring. Evaluation of some metal catalyst is based on hydrogenation of acetophenone as a representative [*38, 814*].

While palladium proved ineffective in reducing aliphatic and alicyclic ketones it is the catalyst of choice for reduction of acetophenone: it does not hydrogenate the ring, and in proper solvents reduction may stop at the stage of the alcohol without hydrogenolysis. Over platinum oxide in ethanol reduction to 1-phenylethanol is fast and proceeds further only slowly while in acetic acid up to 84% of completely hydrogenolyzed and saturated product, ethylcyclohexane, was obtained. Rhodium–platinum oxide in ethanol or acetic acid facilitates quantitative reduction to 1-cyclohexylethanol with a slight hydrogenolysis (6%) [*38*]. Use of palladium and platinum oxides in acetic acid with added hydrochloric acid enhances hydrogenation to ethylbenzene [*38*]. Compared with palladium, platinum favors more extensive hydrogenolysis, and rhodium favors both hydrogenolysis and hydrogenation of the ring [*814*].

In contrast to hydrogenation over noble metals hydrogenation of acetophenone over different nickel catalysts and over copper chromite results in the formation of 1-phenylethanol without hydrogenolysis [*43, 45, 49, 50*].

Hydrogenation of the acetophenone analogs, isomeric acetylpyridines, is more complicated. 2-Acetylpyridine hydrogenated over 5% palladium on charcoal in 95% ethanol gave, at room temperature and 3 atm, 79% yield of 2-(1-hydroxyethyl)pyridine, also obtained over Raney nickel. Hydrogenations over platinum oxide or rhodium afforded, in addition to 60–70% of the above alcohol, 23–25% of the starting material and varying amounts of 2-(1-hydroxy)piperidine resulting from the ring hydrogenation. 3-Acetylpyridine was hydrogenated over 5% palladium on charcoal in ethanol to a mixture of 70% of 3-acetyl-1,4,5,6-tetrahydropyridine and 7.2% of 3-acetylpiperidine. Treatment of 4-acetylpyridine with hydrogen over platinum oxide in 95% ethanol yielded 62% of 4-(1-hydroxyethyl)pyridine and 15% of a pinacol, 2,3-bis(4-pyridyl)-2,3-butanediol. The pinacol was also obtained in 60% yield in hydrogenation over palladium on charcoal or over rhodium on alumina, in 75–80% yield over platinum oxide, and in 90% yield over Raney nickel in alcohol [*826*]. This formation of a pinacol in catalytic hydrogenation is unique.

Surprisingly good results in the reduction of aromatic ketones were obtained in treating the ketones with a 50% nickel–aluminum alloy in 10–16% aqueous sodium hydroxide at temperatures of 20–90° (yields 65–90%) [*769, 827*].

Very dependable reduction of aromatic ketones to secondary alcohols is accomplished by hydrides and complex hydrides: *lithium aluminum hydride* [*83, 828*], *lithium tetrakis(N-dihydropyridyl)aluminate* (diaryl ketones only) [*441*], *lithium borohydride* [*750*], *lithium triethylborohydride* [*100*], *sodium bis(2-methoxyethoxy)aluminum hydride* [*544*], *sodium borohydride* [*751, 817*], *sodium trimethoxyborohydride* [*99*], *tetrabutylammonium borohydride* [*771*], *tetrabutylammonium cyanoborohydride* [*757*], *diphenylstannane* [*114*] and others, generally in good to excellent yields with practically no hydrogenolysis of the alcohol. As a consequence, reduction with metals almost ceased to be used, and other reducing reagents are used only sporadically. Some of these are *sodium dithionite* [*262*], *alcohols* (especially isopropyl alcohol) in Meerwein–Ponndorf reduction [*309*] or its modifications such as shorter periods of heating without the time-consuming removal of acetone [*821, 829*], and *aldehydes* (formaldehyde or salicyl aldehyde, suitable only for purely aromatic ketones) [*314*]. Reduction of aromatic ketones by *Grignard reagents* [*323*] lacks practical importance (Scheme 101, p. 111).

As in catalytic hydrogenation, in reduction with metals **alkyl pyridyl ketones** make a complex picture.

(101)

PhCOMe		PhCH(OH)Me
[49]	H_2/Ni-NaH,EtOH,25°,1 atm,2.5 hrs	92%
[45]	H_2/Raney Ni,25-30°,1-3 atm,22 min	100%
[43]	H_2/Ni,EtOH,175°,160 atm,8 hrs	60%
[50]	H_2/CuCr$_2$O$_4$,150°,100-150 atm,30 min	100%
[770]	LiAlH$_4$/AlCl$_3$,Et$_2$O,reflux 30 min	94%
[544]	NaAlH$_2$(OCH$_2$CH$_2$OMe)$_2$,C$_6$H$_6$,20-40°,6 min	95%
[99]	NaBH(OMe)$_3$,Et$_2$O,reflux 4 hrs	82%
[771]	Bu$_4$NBH$_4$,CHCl$_3$,reflux 2 hrs	91%
[769]	Ni-Al alloy,10% aq. NaOH,10-20°	75%
[822]	MeEtCHCH$_2$OH,Al(OCH$_2$CHMeEt)$_3$,reflux 10 hrs	80%
[262]	Na$_2$S$_2$O$_4$,DMF-H$_2$O,reflux 4 hrs	94%

2-Acetyl- and 4-acetylpyridines reduced by refluxing for 8 hours with amalgamated zinc in hydrochloric acid (*Clemmensen reduction*) gave 83% and 86% yield of pure 1-(2-pyridyl)ethanol and 1-(4-pyridyl)ethanol, respectively. On the other hand, reduction by refluxing for 8 hours with zinc dust in 80% formic acid yielded, respectively, 86% and 64% of pure 2-ethyl- and 4-ethyl-pyridine. Refluxing with zinc dust and acetic acid for 9 hours gave mixtures of the alcohols and alkylpyridines. 3-Acetylpyridine afforded 14% of 1-(3-pyridyl)ethanol by the first method, 65% of 3-ethylpyridine by the second method, and by refluxing with zinc dust in acetic acid mainly the alcohol and also a pinacol, 2,3-bis(3-pyridyl)-2,3-butanediol [830].

Special attention should be directed to reductions of pro-chiral ketones by *chiral reducing agents* that lead to chiral secondary alcohols. Such reagents were prepared for example by treatment of *(2S,3S)- and (2R,3R)-1,4-bis(diethylamino)-2,3-butanediol* with *lithium aluminum hydride* [831] or by adding *borane* [108] or *9-bora[3.3.1]bicyclononane* [109] to optically active α-pinene. Reductions with such *boranes* gave alcohols of varying optical purities. Treatment of acetophenone with the first reagent afforded 1-phenyl-ethanol in 65% yield and 9% optical purity, and treatment with the second reagent gave the alcohol in 78% of enantiomeric excess [818, 832].

Another way of preparing optically active alcohols 'in good yields and excellent optical purity' is based on reduction of pro-chiral ketones with the aluminum salt of optically active amyl alcohol (pentyl alcohol) [822].

Alcohols of highest enantiomeric excess were obtained by *biochemical reductions* using microorganisms. Acetophenone gave 90% yield of (−)-S-1-phenylethanol after incubation with *Cryptococcus macerans*. Other ketones were reduced in much lower conversions (2–41%) by *Sporobolomycetes pararoseus* [833]. α-, β- and γ-acetylpyridines were similarly reduced by *Cryptococcus macerans* to the corresponding (−)-S-1-pyridylethanols of 79–85% enantiomeric excess [834]. The S configuration predicted by Prelog's rule (if the ketone is placed with the larger group on the observer's left, the hydroxyl group formed is closer to the observer [835]) was confirmed by chemical means [834]. Treatment of 1,2-di(β-pyridyl)-2-methyl-1-propanone with *Bo-*

tryodiplodia theobromae Pat. afforded $(-)$-1,2-di(β-pyridyl)-2-methyl-1-pro-panol of 99% optical purity in 90% yield [*836*].

In ketones having a chiral cluster next to the carbonyl carbon reduction with *lithium aluminum hydride* gave one of the two possible diastereomers, *erythro* or *threo*, in larger proportions. The outcome of the reduction is determined by the approach of the reducing agent from the least hindered side (steric control of asymmetric induction) [*828*]. With lithium aluminum hydride as much as 80% of one diasteromer was obtained. This ratio is higher with more bulky reducing hydrides [*837*].

One-electron reduction of **aromatic ketones gives pinacols.** Exceptionally such a reduction was achieved by catalytic hydrogenation of 3-acetylpyridine [*826*] (p. 110). Common reducing agents for the synthesis of pinacols are metals: *aluminum amalgam* [*144*] (*Procedure 30*, p. 212), *zinc* in acetic acid [*838*] and magnesium with magnesium iodide [*839*]. An excellent yield of tetraphenylpinacol was obtained by *irradiation* of benzophenone dissolved in *isopropyl alcohol* [*308*].

Titanium in a low valence state, as prepared by treatment of solutions of titanium trichloride with potassium [*206*] or magnesium [*207*] in tetrahydro-furan or with lithium in dimethoxyethane [*206*], **deoxygenates ketones and effects coupling** of two molecules at the carbonyl carbon to form alkenes, usually a mixture of both stereoisomers. If a mixture of acetone with other ketones is treated with titanium trichloride and lithium, the alkene formed by combination of acetone with the other ketone predominates over the sym-metrical alkene produced from the other ketone [*206*] (*Procedure 39*, p. 215).

PhCOPh	$Ph_2C=CPh_2$	$\underset{HO\ OH}{Ph_2C-CPh_2}$	$PhCH_2Ph$ (102)
[*206*] $TiCl_3$,Li,$(CH_2OMe)_2$,reflux 16 hrs	96%		
[*838*] Zn,AcOH,heat		92%	
[*839*] Mg,MgI_2,Et_2O,C_6H_6		99.6%	
[*308*] Me_2CHOH,hν,3-5 hrs		93-94%	
[*281*] $N_2H_4 \cdot H_2O$,NaOH,$HO(CH_2CH_2O)_3H$			83.3%

Reduction of aromatic ketones to hydrocarbons occurs very easily as the carbonyl group is directly attached to an aromatic ring. In these cases reduc-tion produces benzylic-type alcohols which are readily hydrogenolyzed to hydrocarbons. This happens during catalytic hydrogenations as well as in chemical reductions.

Carbonyl groups adjoining an aromatic ring have been converted to methylene in good to excellent yields by *hydrogenation* over palladium and platinum catalysts in acetic and hydrochloric acid at room temperature and atmospheric pressure [*38, 840*] (p. 100), over Raney nickel [*428*] or over copper chromite [*428*] at 130-175° and 70-350 atm, and over molybdenum sulfide at 270° and 100 atm [*55*]. Aromatic ketones containing pyrrole [*428*] and pyridine rings [*826*] must not be hydrogenated under too vigorous conditions if the aromatic rings are to be preserved. In the case of 2,4-diacetyl-3,5-dimethylpyrrole temperature during the hydrogenation over Raney nickel or copper chromite should not exceed 180° [428].

Reduction of carbonyl to methylene in aromatic ketones was also achieved by *alane* prepared from lithium aluminum hydride and aluminum chloride [770], by *sodium borohydride* in trifluoroacetic acid [841], with *triethylsilane* in trifluoroacetic acid [335, 777], with *sodium* in refluxing ethanol [842], with *zinc* in hydrochloric acid [843] and with *hydrogen iodide* and *phosphorus* [227], generally in good to high yields.

However, most frequently used methods for reduction of aromatic ketones to hydrocarbons are, as in the case of other ketones, *Clemmensen reduction* [160, 161, 758, 843, 844] (*Procedure 31*, p. 213), *Wolff-Kizhner* reduction [280, 281, 282, 759, 774, 845] (*Procedure 45*, p. 216), or reduction of *p*-toluene-sulfonylhydrazones of the ketones with *lithium aluminum hydride* [811, 812] or with *borane and benzoic acid* [786].

Reduction of carbocyclic rings in aromatic ketones can be accomplished by *catalytic hydrogenation* over platinum oxide or rhodium–platinum oxide and takes place only after the reduction of the carbonyl group, either to the alcoholic group, or to a methylene group [38].

In **heterocyclic aromatic ketones** containing pyrrole and pyridine rings partial or total reduction of the ring may precede hydrogenation of the keto group. Hygrine was hydrogenated over platinum oxide in acetic acid at 20° and 1 atm to (*N*-methyl-2-pyrrolidinyl)acetone in 60% yield [432]. On the other hand, in 2,4-diacetyl-3,5-dimethylpyrrole hydrogenation of the pyrrole ring required temperatures of 180° for Raney nickel and 250° for copper chromite to give 2,4-diethyl-3,5-dimethylpyrrolidine in 70% and 50% yields, respectively [428]. In the hydrogenation of 3-acetylpyridine over 5% palladium on charcoal in ethanol the keto group survived the reduction to 3-acetyl-1,4,5,6-tetrahydropyridine (70%) and 3-acetylpiperidine (7.2%) [826] (p. 110). In 4-acetylquinoline the pyridine ring was reduced to the 1,2-dihydro- and 1,2,3,4-tetrahydro derivatives, but only after the conversion of the acetyl group to an ethyl group [319].

Reduction of Cyclic Ketones

All reducing agents used for reductions of aliphatic and aromatic ketones can be used for **reduction of cyclic ketones to secondary alcohols** (pp. 107 and 109). In fact, reduction of cyclic ketones is sometimes easier than that of both the above mentioned categories [262]. What is of additional importance in the reductions of cyclic ketones is **stereoselectivity** of the reduction and stereochemistry of the products.

According to the older literature *catalytic hydrogenation* tends to favor *cis* isomers (where applicable) while reduction with *metals* gives *trans* isomers [846]. But even catalytic hydrogenation may give either *cis* isomers, when carried out over platinum oxide [847], or *trans* isomers, if Raney nickel is used

[847]. An acidic medium favors alcohols with an axial hydroxyl; an alkaline medium favors alcohols with an equatorial hydroxyl [848].

Similarly reductions with metal hydrides, metals and other compounds may give predominantly one isomer. The stereochemical outcome depends strongly on the structure of the ketone and on the reagent, and may be affected by the solvents.

For example, reduction of 2-alkylcycloalkanones with *lithium aluminum hydride* in tetrahydrofuran gave the following percentage proportions of the less stable *cis*-2-alkylalkanol (with axial hydroxyl): 2-methylcyclobutanol 25%, 2-methylcyclopentanol 21%, 2-methylcyclohexanol 25%, 2-methylcycloheptanol 73%, and 2-methylcyclooctanol 73% (the balance to 100% being the other, *trans*, isomer) [837].

Reduction of 2-methylcyclohexanone gave the following percentage proportions of the less stable *cis*-2-methylcyclohexanol (with axial hydroxyl): with *lithium aluminum hydride* 25%, with *diborane* in tetrahydrofuran 26%, with *di-sec-amylborane* in tetrahydrofuran 79%, with *dicyclohexylborane* in diglyme 94%, and with *diisopinocampheylborane* in diglyme 94% (the balance to 100% being the *trans* isomer) [837].

Different stereoselectivities caused by solvent effects are demonstrated in the reduction of dihydroisophorone (3,3,5-trimethylcyclohexanone) with *sodium borohydride* which gave less stable *trans*-3,3,5-trimethylcyclohexanol (with axial hydroxyl) by reduction in anhydrous isopropyl alcohol (55–56%), in anhydrous *tert*-butyl alcohol (55%), in 65% aqeuous isopropyl alcohol (59.5%), in anhydrous ethanol (67%), and in 71% aqueous methanol (73%) (the balance to 100% being the more stable *cis* isomer with equatorial hydroxyl) [849].

(103)

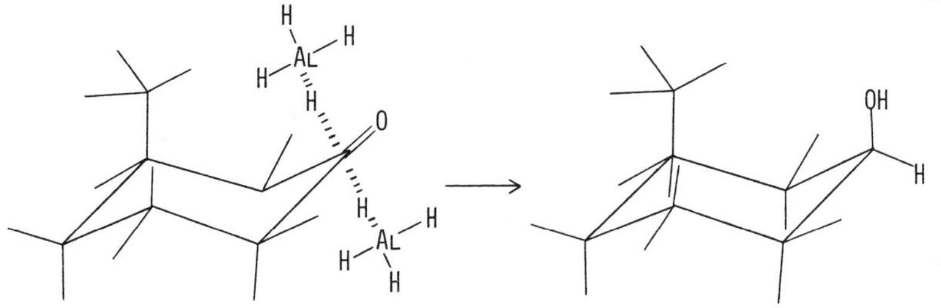

The structure of the cyclic ketone is of utmost importance. Reduction of cyclic ketone by complex hydrides is started by a nucleophilic attack at the carbonyl function by a complex hydride anion. The approach of the nucleophile takes place from the less crowded side of the molecule (**steric approach or steric strain control**) leading usually to the less stable alcohol. In ketones with no steric hindrance (no substituents flanking the carbonyl group or bound in position 3 of the ring) usually the more stable (equatorial) hydroxyl is generated (**product development or product stability control**) [850, 851, 852, 853]. The contribution of the latter effect to the stereochemical outcome of the reduction has been disputed [850]. Since the reductions with hydrides and complex hydrides are very fast it is assumed that the transition states develop early in the reaction, and consequently the steric approach is more important than the stability of the product. What is sometimes difficult to estimate is what causes stronger steric hindrance: interactions between the incoming nucleophile and axial substituents in position 3 of the ring or torsional strain between the nucleophile and a hydrogen or a substituent flanking the carbonyl carbon. In other words, the determination of which side of the molecule is more accessible is at times difficult.

(104)

Percentage of equatorial (trans) alcohol

Reference	[816]	[850]	[852]	[854]	[855]
LiAlH$_4$,Et$_2$O	89	90-91			
LiAlH$_4$,THF			42	92	
LiAlH(iso-Bu)$_2$Bu,Et$_2$O,C$_6$H$_{14}$					61
LiAlH(iso-Bu)$_2$sec-Bu,Et$_2$O,C$_6$H$_{14}$					58
LiAlH(iso-Bu)$_2$tert-Bu,Et$_2$O,C$_6$H$_{14}$					51
LiAlH(OMe)$_3$,THF		59	36		
LiAlH(OCMe$_3$)$_3$,THF		90	46	89.7	
AlHCl$_2$(LiAlH$_4$/AlCl$_3$),Et$_2$O	80-99.5				
NaBH$_4$,MeOH		85-86			
NaBH(OMe)$_3$,MeOH		75-76			
NaBH(OCHMe$_2$)$_3$,iso-PrOH		75-80			
iso-PrOH-Al(O-iso-Pr)$_3$	79				

The effect of steric hindrance can be nicely demonstrated in the reduction of two bicyclic ketones, norcamphor and camphor. The relatively accessible norcamphor yielded on reduction with complex hydrides predominantly (the less stable) *endo* norborneol while sterically crowded camphor was reduced by the same reagents predominantly to the less stable *exo* compound, isobor-neol [837]. From the numerous examples shown it can be deduced that the stereoselectivity increases with increasing bulkiness (with some exceptions), and that it is affected by the nucleophilicity of the reagent and by the solvent.

Similar stereoselectivity was noticed in reductions of ketones by *alkali metals* in liquid ammonia and alcohols. 4-*tert*-Butylcyclohexanone gave almost exclusively the more stable equatorial alcohol, norcamphor 68-91% of the less stable *endo*-norborneol, and camphor a mixture of borneol and isoborneol [860].

Stereoselective reductions of cyclic ketones have immense importance in the chemistry of steroids where either α or β epimers can be obtained. A few

(105)

Percentage of axial (<u>trans</u>) alcohol

Reference	[816]	[856]	[850]	[854]	[849]
LiAlH$_4$,Et$_2$O	55	52–55	58–63		
LiAlH$_4$,THF		72–74		87.8	
LiAlH(OMe)$_3$,Et$_2$O		75			
LiAlH(OMe)$_3$,THF		92	98		
LiAlH(OEt)$_3$,Et$_2$O		83			
LiAlH(OCHMe$_2$)$_3$,Et$_2$O		54			
LiAlH(OCHMe$_2$)$_3$,THF		69			
LiAlH(OCMe$_3$)$_3$,Et$_2$O		63–73			
LiAlH(OCMe$_3$)$_3$,THF		88	96		
AlHCl$_2$(LiAlH$_4$/AlCl$_3$),Et$_2$O	85*,0**	85–86			
NaBH$_4$,MeOH			94		
NaBH$_4$,71% aq. MeOH					73
NaBH$_4$,EtOH					67
NaBH$_4$,iso-PrOH					55–56
NaBH$_4$,65% aq. iso-PrOH					59.5
NaBH$_4$,tert-BuOH					55
NaBH(OMe)$_3$,MeOH			81		
NaBH(OMe)$_3$,iso-PrOH					65
NaBH(OCHMe$_2$)$_3$,iso-PrOH			80–83		
iso-PrOH,Al(O-iso-PrOH)$_3$	6				

* Kinetic product. ** Thermodynamic product.

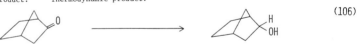

(106)

Percentage of less stable <u>endo</u> alcohol

Reference	[816]	[837]	[852]	[857]	[858]
LiAlH$_4$,Et$_2$O	81				
LiAlH$_4$,THF		90	89		
LiAlH(OMe)$_3$,THF			98		
LiAlH(OEt)$_3$,THF			85		
LiAlH(OCMe$_3$)$_3$,THF			93		
AlHCl$_2$(LiAlH$_4$/AlCl$_3$),Et$_2$O	96*,11**				
LiBH(sec-Bu)$_3$,THF				99.6	
LiBH(CHMeCHMe$_2$)$_3$,THF				>99.5	
B$_2$H$_6$,THF		98			
BH(CHMeCHMe$_2$)$_2$,THF		92			
BH(C$_6$H$_{11}$)$_2$,THF		94			
Li/NH$_3$,EtOH					47
Na/NH$_3$,EtOH					40
K/NH$_3$,EtOH					43
Ca/NH$_3$,EtOH					90
iso-PrOH,Al(O-iso-Pr)$_3$	20				

* Kinetic product. ** Thermodynamic product.

(107)

Percentage of less stable <u>exo</u> alcohol

Reference	[816]	[837]	[852]	[857]	[858]	[859]
LiAlH$_4$,Et$_2$O	90					
LiAlH$_4$,THF		91	92			
LiAlH(OMe)$_3$,THF			99			
LiAlH(OCMe$_3$)$_3$,THF			93			
AlHCl$_2$(LiAlH$_4$/AlCl$_3$),Et$_2$O	73					
LiBH(sec-Bu)$_3$,THF				99.6		
LiBH(C$_6$H$_{11}$)$_3$,THF				99.3		
B$_2$H$_6$,THF		52				
BH(CHMeCHMe$_2$)$_2$,THF		65				
BH(C$_6$H$_{11}$)$_2$,THF		93				
Li/NH$_3$,EtOH					21-23	
Na/NH$_3$,EtOH					42	
K/NH$_3$,EtOH					60-70	
Ca/NH$_3$,EtOH					28	
iso-PrOH,Al(O-iso-Pr)$_3$	29					70

examples demonstrate that while *catalytic hydrogenation* [49] or *catalytic hydrogen transfer (Henbest reduction)* [861] favor formation of α epimers reduction with *hydrides* and *complex hydrides* is much more stereoselective and gives predominantly β epimers [757, 816].

(108)

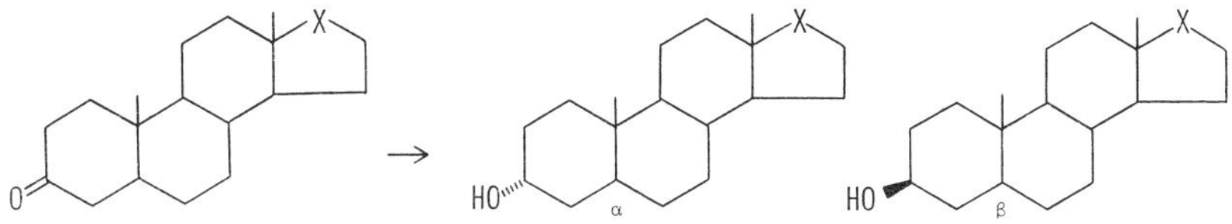

X = CHC$_8$H$_{17}$ (3-Cholestanone)

[49] H$_2$/Ni(NaH),25°,1 atm,3.8 hrs(96%)	67%	33%
[816] LiAlH$_4$,Et$_2$O	9%	91%
[816] AlHCl$_2$(LiAlH$_4$/AlCl$_3$),Et$_2$O	0-17%	83-100%
[757] Bu$_4$NBH$_3$CN,HMPA,25°(94.5%)	17%	83%
[816] iso-PrOH,Al(O-iso-Pr)$_3$,reflux	16%	84%

X = CO (Androstane-3,17-dione)

[49] H$_2$/Ni(NaH),25°,1 atm,70 min(97%)	47%	53%
[861] H$_2$IrCl$_6$,(MeO)$_3$P,90% iso-PrOH, reflux 94 hrs (Henbest reduction)	94%	2%

Cyclic ketones were reduced to **optically active cyclic alcohols** by biochemical methods. Incubation of 2,3-benzosuberone with *Cryptococcus macerans* gave 27% conversion to the optically active (−)-S-2,3-benzo-1-cycloheptanol. Other hydroaromatic ketones were reduced, although in low conversions (2–

41%), to S-alcohols (according to Prelog's rule, p. 111) [833]. Biochemical reductions are especially important in steroids.

(109)

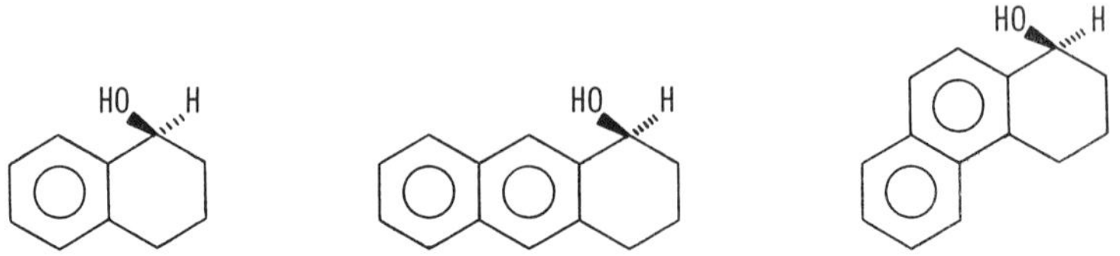

[833]

| (+)-S | (+)-S | (−)-S |
| 15% (Recovered 80%) | 13% (Recovered 36%) | 6% (Recovered 77%) |

Reductions of cyclic ketones to cycloalkanes, to pinacols and to alkenes are accomplished by the same reagents which are used for similar reductions of aliphatic and aromatic ketones (pp. 109 and 112).

(110)

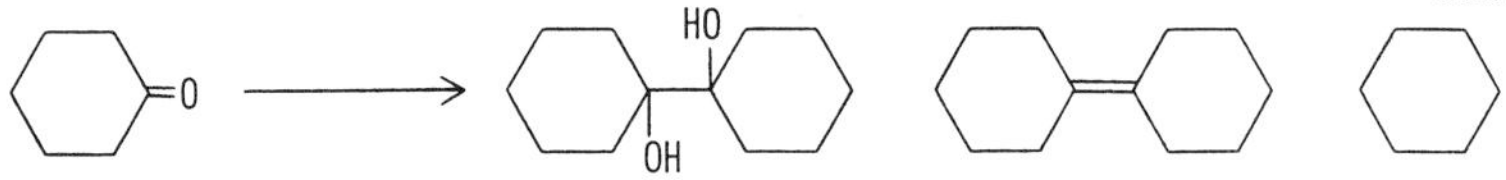

[825]	AlHg,CH$_2$Cl$_2$,reflux 1-4 hrs	55%		
[207]	TiCl$_3$,Mg,THF	45%		
[206]	TiCl$_3$,Li,(CH$_2$OMe)$_2$,reflux 16 hrs		79%	
[206]	TiCl$_3$,K,THF,reflux 16 hrs		85%	
[68]	ZnHg,HCl			50%
[281]	N$_2$H$_4$·H$_2$O,KOH,HO(CH$_2$CH$_2$O)$_3$H, reflux 1 hr			80.4%
[811]	H$_2$NNHC$_7$H$_7$;NaBH$_4$,dioxane			70-80%

Reduction of the carbonyl group to methylene is carried out by *Clemmensen reduction* [160, 758], by *Wolff-Kizhner reduction* [280, 282], or by its modifications: decomposition of hydrazones with potassium *tert*-butoxide in dimethyl sulfoxide at room temperature in yields of 60-90% [845], or by reduction of *p*-toluenesulfonylhydrazones with *sodium borohydride* (yields 65-80%) [811] (p. 134).

In keto steroids the reductions were also achieved by *electrolysis* in 10% sulfuric acid and dioxane using a divided cell with lead electrodes (yields 85-97%) [862], by specially activated *zinc* dust in anhydrous solvent (ether or acetic anhydride saturated with hydrogen chloride) (yields 50-87%) [155, 863], and by the above mentioned reduction of tosylhydrazones with *sodium borohydride* (yields 60-75%) [811].

Heterocyclic ketones having sulfur or nitrogen atoms in the ring in a position β to the carbonyl group are cleaved during Clemmensen reduction between the α-carbon and the hetero atom. Recyclization during the reaction leads to ring contractions [159, 864]. Reduction of *N*-alkyl-α-pyridones, which are amides in disguise, is discussed elsewhere (p. 170).

For the reduction to **pinacols** *aluminum amalgam* [825] or *low valence state titanium chloride* [207] were used. Under different conditions the titanium

(111)

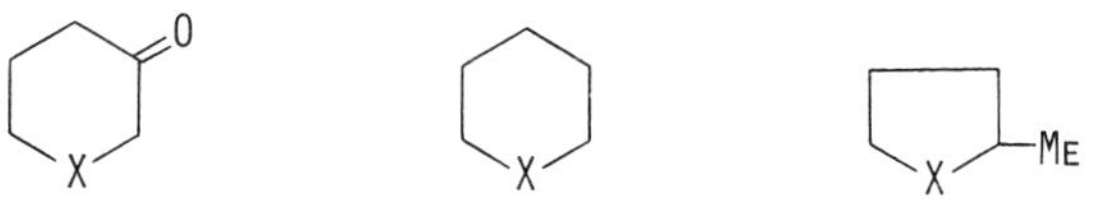

X = S   [864]   ZnHg,HCl,reflux 1 hr   21%
X = NME   [159]   ZnHg,HCl,reflux 12 hrs   60%
X = NME   [159]   N$_2$H$_4$·H$_2$O,KOH
                  HO(CH$_2$CH$_2$O)$_3$H,reflux 1 hr   45%

reagent deoxygenates the ketones and couples them at the carbonyl carbon to alkenes [206].

Reduction of Unsaturated Ketones

Unsaturated ketones of all kinds can be converted to saturated ketones, to unsaturated alcohols, to saturated alcohols, to alkenes and to alkanes. If the double bond is conjugated with the carbonyl group reduction usually takes place more readily and may give, in addition to the above products, ε-diketones formed by coupling at β-carbons.

Unsaturated ketones having double bonds not conjugated with the carbonyl group can be **reduced to saturated ketones** by controlled *catalytic hydrogenation*. However, if another conjugated double bond is present in the molecule of the ketone, it may be reduced preferentially. In 3-ethylenedioxy-5,16-pregnadiene-20-one only the conjugated double bond in position 16 was reduced by hydrogen over 10% palladium on calcium carbonate in ethyl acetate (yield 91%) [865], and in 4,6,22-ergostatriene-3-one only the double bond in the position γ,δ to the carbonyl was reduced over 5% palladium on charcoal in methanol containing potassium hydroxide (yield 70%) [866]. The isolated double bonds in the above compounds were not hydrogenated provided the reduction was stopped after absorption of 1 mol of hydrogen. The same results were obtained in heterogeneous hydrogenation of eremophilone (5,10-dimethyl-3-isopropenyl-1-oxo-1,2,3,4,5,6,7,10-octahydronaphthalene) over palladium catalysts. However, homogeneous hydrogenation using tris(triphenylphosphine)rhodium chloride in benzene gave 94% yield of 13,14-dihydroeremophilone in which the conjugated double bond survived [651].

α,β-Unsaturated ketones (enones) were **reduced to saturated ketones** by catalytic hydrogenation provided it was stopped after absorption of 1 mol of hydrogen [814]. Platinum [767], platinum oxide [34, 867], platinum on carbon [814], palladium on carbon [814, 868], rhodium on carbon [814], and tris(triphenylphosphine)rhodium chloride [56, 869] were used as the catalysts. *Nickel–aluminum alloy* in 10% aqueous sodium hydroxide [769] and zinc-reduced nickel in aqueous medium [870] also reduced only the conjugated double bonds.

In cyclic α,β-unsaturated ketones such as 2-oxo-2,3,4,5,6,7,8,10-octahydronaphthalene, where the reduction may lead to two stereoisomers, hydro-

genation over 10% palladium on charcoal in aqueous ethanol gave 93% of
cis- and 7% of *trans*-2-decalone in the presence of hydrochloric acid, and 62%
of *cis*- and 38% of *trans*-2-decalone in the presence of sodium hydroxide [868].
A whole spectrum of ratios of the two stereoisomers was obtained over
different catalysts in solvents of different polarities and different pHs [21,
868].

Lithium aluminum hydride reduces preferentially the carbonyl function (p.
98) but *alanes* prepared by reactions of aluminum hydride with two equivalents of isopropyl or *tert*-butyl alcohol or of diisopropylamine reduce the
conjugated double bonds with high regioselectivity in quantitative yields
[871] (p. 121).

Triphenylstannane reduced the α,β double bond in dehydro-β-ionone in
84% yield [872]. Complex *copper hydrides* prepared *in situ* from lithium
aluminum hydride and cuprous iodide in tetrahydrofuran at 0° [873], or
from lithium trimethoxyaluminum hydride or sodium bis(methoxy-
ethoxy)aluminum hydride and cuprous bromide [874] in tetrahydrofuran at 0° reduced the α,β double bonds selectively in yields from 40 to
100%. Similar selectivity was found with a complex *sodium bis(iron
tetracarbonyl)hydride* $NaHFe_2(CO)_8$ [875].

Reduction of α,β-unsaturated to saturated ketones was further achieved by
electrolysis in a neutral medium using copper or lead cathodes (yields 55–
75%) [766], with *lithium* in propylamine (yields 40–65%) [876], with
potassium-graphite clathrate C_8K (yields 57–85%) [807], and with *zinc* in
acetic acid (yield 87%) [688]. Reduction with amalgamated zinc in hydrochloric acid (*Clemmensen reduction*) usually reduces both functions [877].

Biochemical reduction of α,β-unsaturated ketones using microorganisms
(best *Beauveria sulfurescens*) takes place only if there is at least one hydrogen
in the β-position and the substituents on α-carbons are not too bulky. The
main product is the saturated ketone, while only a small amount of the
saturated alcohol is formed, especially in slightly acidic medium (pH 5–5.5).
The carbonyl is attacked from the equatorial side. Results of biochemical
reduction of 5-methylcyclohex-2-en-1-one are illustrative of the biochemical
reduction by incubation with *Beauveria sulfurescens*: after 24 hours 74% of
the enone was reduced to 3-methylcyclohexanone and 26% to 3-methylcy-
clohexanol containing 55% of *cis* and 45% of *trans* isomer. After 48 hours the
respective numbers were 70% and 30%, and 78% and 22%, respectively [878].

Reduction of unsaturated ketones to unsaturated alcohols is best carried out
with *complex hydrides*. α,β-Unsaturated ketones may suffer reduction even at
the conjugated double bond [764, 879]. Usually only the carbonyl group is
reduced, especially if the inverse technique is applied. Such reductions are
accomplished in high yields with *lithium aluminum hydride* [879, 880, 881,
882], with *lithium trimethoxyaluminum hydride* [764], with *alane* [879], with
diisobutylalane [883], with *lithium butylborohydride* [884], with *sodium borohydride* [751], with *sodium cyanoborohydride* [780, 885] with *9-borabicyclo
[3.3.1]nonane* (9-BBN) [764] and with *isopropyl alcohol* and aluminum isopro-

poxide (*Meerwein–Ponndorf reduction*) [*309, 886*]. Many α,β-unsaturated cyclopentenones and cyclohexenones and their homologs are reduced by the above reagents [*764, 780, 781, 879, 884*].

$$\text{P}_H\text{CH=CHCOR} \quad (112)$$

$$\text{P}_H\text{CH}_2\text{CH}_2\text{COR} \qquad\qquad \text{P}_H\text{CH=CHCHR} \atop \text{OH}$$

R = M$_E$

[*56*] H$_2$/(Ph$_3$P)$_3$RhCl,C$_6$H$_6$-EtOH,	[*881*] LiAlH$_4$,Et$_2$O (~100%)
60°,4.5-5.5 atm,8-12 hrs (80%)	[*884*] LiAlH$_3$Bu,C$_6$H$_{14}$-PhMe,-78° (98%)
[*769*] Ni-Al,10% NaOH,10-20° (78%)	[*780*] NaBH$_3$CN,MeOH,HCl,25°,1.5 hr (77%)
[*871*] AlH[N(CHMe$_2$)$_2$]$_2$,THF,0° (96%)	[*780*] NaBH$_3$CN,HMPA,H$_2$SO$_4$,25°,1 hr (58%)
[*766*] Pb cathode,EtOH,AcOEt,1 amp (75%)	[*886*] Me$_2$CHOH,Al(OCHMe$_2$)$_3$,reflux (96%)
[*807*] C$_8$K,THF,25°,10 min (85%)	

R = P$_H$

[*867*] H$_2$/PtO$_2$,AcOEt,25°,1 atm (81-95%)	[*882*] LiAlH$_4$,Et$_2$O,35° (65%)
[*871*] AlH(OCMe$_3$)$_2$,THF,0° (98%)	[*884*] LiBH$_3$Bu,C$_6$H$_{14}$-PhMe,-78°,2 hrs (99%)
[*873*] LiAlH$_4$/CuI,THF,0° (101%)	[*114*] Ph$_2$SnH$_2$,Et$_2$O,25° (75%)
[*874*] NaAlH$_2$(OCH$_2$CH$_2$OMe)$_2$,THF,BuOH,	
-20°,1 hr (65%)	

Reduction of unsaturated ketones to saturated alcohols is achieved by *catalytic hydrogenation* using a nickel catalyst [*49*], a copper chromite catalyst [*50, 887*] or by treatment with a *nickel–aluminum alloy* in sodium hydroxide [*888*]. If the double bond is conjugated, complete reduction can also be obtained with some hydrides. 2-Cyclopentenone was reduced to cyclopentanol in 83.5% yield with *lithium aluminum hydride* in tetrahydrofuran [*764*], with *lithium tris(tert-butoxy)aluminium hydride* (88.8% yield) [*764*], and with *sodium borohydride* in ethanol at 78° (yield 100%) [*764*]. Most frequently, however, only the carbonyl is reduced, especially with application of the inverse technique (p. 21).

$$(113)$$

$$\xrightarrow[\text{175°, 100-150 atm}]{[\text{\textit{50}}]\ \text{H}_2/\text{CuCr}_2\text{O}_4} \text{P}_H\text{CH=CHCOM}_E \xrightarrow[\substack{\text{NaBH}_4,\text{AcOH} \\ 25°,1\ \text{hr};70°,3\ \text{hrs}}]{[\textit{785}]\ \text{H}_2\text{N}\cdot\text{NHC}_7\text{H}_7}$$

P$_H$CH$_2$CH$_2$CHM$_E$ 100%
OH

54% P$_H$CH$_2$CH=CHM$_E$

Reduction of α,β-unsaturated ketones to unsaturated hydrocarbon is rather rare, and is almost always accompanied by a shift of the double bond. Such reductions are accomplished in good to high yields by treatment of the *p*-toluenesulfonylhydrazones of the unsaturated ketones with *sodium borohydride* [*785*], *borane* [*786*] or *catecholborane* [*889*], or by *Wolff–Kizhner reduction* or its modifications [*890*]. However, **complete reduction to saturated hydrocarbons** may also occur during Wolff–Kizhner reduction [*891*] as well as during Clemmensen reduction [*160*].

One-electron reduction of α,β-unsaturated ketones yields, instead of pinacols, ε-**diketones** formed by coupling of semireduced species at the β-carbons

[*892*]. The practical usefulness of this reaction is minimized by low yields (16% in case of benzalacetone).

Ketones containing acetylenic bonds were reduced selectively at the triple bond by *catalytic hydrogenation*. Over 5% palladium on calcium carbonate in pyridine (which decreases the activity of the catalyst) 17-ethynyltestosterone was reduced in 95% yield to 17-vinyltestosterone, while over palladium on charcoal in dioxane 80% of 17-ethyltestosterone was obtained; the carbonyl in position 3 and the conjugated 4,5-double bond remained intact [*386*].

Ketones containing triple bonds in the α,β-positions are **reduced to the corresponding unsaturated alcohols** with *sodium cyanoborohydride* or *tetrabutylammonium cyanoborohydride* in 64–89% yields [*780*]. Thus 4-phenyl-3-butyn-2-one gave 4-phenyl-3-butyn-2-ol [*780*]. If the same ketone was converted to its *p*-toluenesulfonylhydrazone and this was reduced with *bis(benzyloxy)borane*, 1-phenyl-1,2-butadiene was obtained in 21% yield [*786*].

Reduction of α,β-acetylenic ketones with *chiral borane NB-Enanthrane*® prepared by addition of 9-borabicyclo[3.3.1]nonane to the benzyl ether of nopol yielded optically active acetylenic alcohols in 74–84% yields and 91–96% enantiomeric excess [*110*]. Another way to optically active acetylenic alcohols is reduction with a reagent prepared from lithium aluminum hydride and (2S,3R)-(+)-4-dimethylamino-3-methyl-1,2-diphenyl-2-butanol. At −78° mainly R alcohols were obtained in 62–99% yield and 34–90% enantiomeric excesses [*893*].

[*110*]   (114)

$$C_5H_{11}COC\!\equiv\!CH \xrightarrow[\text{THF},25°,48\text{ hrs}]{\text{NB-Enanthrane}} C_5H_{11}\!-\!\overset{\displaystyle OH}{\underset{\displaystyle H}{C}}\!-\!C\!\equiv\!CH$$

REDUCTION OF DERIVATIVES OF KETONES

The following section deals with ketones containing substituents in the rest of their molecules, and with functional derivatives of ketones.

Reduction of Substitution Derivatives of Ketones

Halogenated ketones can be **reduced to halogenated alcohols** with complex hydrides. Although these reagents are capable of hydrogenolyzing the carbon–halogen bond, reduction of the carbonyl group takes preference. 1,3-Dichloroacetone was reduced with *lithium aluminum hydride* to 1,3-dichloro-2-propanol in ether at −2° in 77% yield, even though an excess of the hydride was used [*894*], and 4-bromo-2-butanone yielded 4-bromo-2-butanol on reduction with *sodium borohydride* [*895*]. In the case of reduction of halogenated ketones containing reactive halogens, the inverse technique and mild conditions are advisable.

Catalytic hydrogenation in basic medium may hydrogenolyze halogen even without reducing the keto group: 8-chloro-5-methoxy-1-tetralone gave 53–56% yield of 5-methoxy-1-tetralone on treatment with hydrogen over 10% palladium on charcoal in ethanol containing triethylamine at room temperature and atmospheric pressure [*896*]. A particularly reactive halogen like the bromine in 1-bromo-1,1-dibenzoylethane was replaced quantitatively by hydrogen on refluxing for 5 minutes with a solution of *potassium iodide* and dilute *hydrochloric acid* in acetone [*229*]. Steroidal α-iodo ketones are reduced to ketones with chromous chloride [*188*] (*Procedure 37*, p. 214).

Meerwein–Ponndorf reduction of 2-bromocyclohexanol gave 2-bromo-cyclohexanol (30%) and cyclohexanol (33%) [*311*]. Treatment of α-**halo cyclohexanones** with *hydrazine* in alkaline medium affords **unsaturated hydrocarbons.** The reaction is carried out by refluxing the α-halo ketone with hydrazine and potassium acetate in dimethoxyethane or cyclohexene. Cyclohexene is used as an acceptor of hydrogen produced by the decomposition of hydrazine to diimide and further to nitrogen and hydrogen in order to prevent the formation of saturated hydrocarbons. This reaction gives good yields only with ketones containing the carbonyl group in a six-membered ring and has found use in transformations of steroids [*897*].

[*897*]   (115)

$$X= \quad F \quad CL \quad BR \quad I$$
Yield   71%   68%   62%   54%

Vinylic bromine in 2-bromocoprosten-1-one-3 was hydrogenolyzed in high yield by refluxing with *zinc* in ethanol without the double bond or carbonyl group being affected [*898*]. A trichloromethyl ketone, treated with zinc in acetic acid, gave a methyl ketone. With amalgamated zinc in hydrochloric acid even the carbonyl was reduced [*520*]. Bromine in α-bromo ketones is selectively replaced by hydrogen by means of *chromous chloride* [*899, 900*] or *vanadous chloride* (yields 80–98%) [*214*].

The **reduction of ketones containing nitro groups to nitro alcohols** is best carried out by borohydrides. 5-Nitro-2-pentanone was converted to 5-nitro-2-pentanol in 86.6% yield by reduction with *sodium borohydride* at 20–25°. Other nitro ketones gave 48.5–98.7% yields, usually higher than were obtained by *Meerwein–Ponndorf reduction* [*901*]. 2-Acetamido-3-(*p*-nitrophenyl)-1-hydroxypropan-3-one was reduced with *calcium borohydride* at −30° to 70% of *threo*- and 10% of *erythro*-2-acetamido-3-(*p*-nitrophenyl)propane-1,3-diol while sodium borohydride afforded a mixture of the above isomers in 25% and 47% yields, respectively [*902*].

The **reduction of a dinitro ketone to an azo ketone** is best achieved with glucose. 2,2'-Dinitrobenzophenone treated with *glucose* in methanolic sodium hydroxide at 60° afforded 82% of dibenzo[c,f][1,2]diazepin-11-one whereas *lithium aluminum hydride* yielded 24% of bis(o-nitrophenyl)methanol [315].

Conversion of aromatic nitro ketones with a nitro group in the ring into amino ketones has been achieved by means of *stannous chloride*, which reduced 4-chloro-3-nitroacetophenone to 3-amino-4-chloroacetophenone in 91% yield [178]. A more dependable reagent for this purpose proved to be *iron* which, in acidic medium, reduced *m*-nitroacetophenone to *m*-aminoacetophenone in 80% yield and *o*-nitrobenzophenone to *o*-aminobenzophenone in 89% yield (stannous chloride was unsuccessful in the latter case) [903]. Iron has also been used for the reduction of *o*-nitrochalcone, 3-(o-nitrophenyl)-1-phenyl-2-propen-1-one, to 3-(o-aminophenyl)-1-phenyl-2-propen-1-one in 80% yield [583].

Catalytic hydrogenation over palladium in acetic acid and sulfuric acid at room temperature and 2.5 atm reduced nitroacetophenones and their homologs and derivatives all the way through the alkyl anilines in yields of 78.5–95% [904].

By reduction combined with hydrolysis, 5-nitro-2-heptanone was converted to 2,5-heptanedione on treatment with *titanium trichloride* in aqueous glycol monomethyl ether in 85% yield [905].

α-Diazoketones yield different products depending on their structures and on the reducing agents. Alicyclic α-diazo ketones such as camphor, on *hydrogenation* over palladium oxide after absorption of 1 mol of hydrogen, gave the corresponding hydrazone (camphorquinone hydrazone, 83.5%) [906]. Aromatic and aliphatic-aromatic α-diazo ketones, over the same catalyst after absorption of 2 mol of hydrogen, yielded α-amino ketones (which condensed to 2,5-disubstituted pyrazines) [906, 907]. In the presence of copper oxide or in hydrochloric acid nitrogen was replaced by hydrogen in hydrogenation with palladium oxide or palladium on charcoal, and ketones were formed [906, 907]. The same result was obtained (without the presence of copper oxide) by reduction with *hydrogen iodide* (yields 68–100%) [231, 908]. Also *ethanol* on irradiation caused replacement of nitrogen by hydrogen giving up

(116)

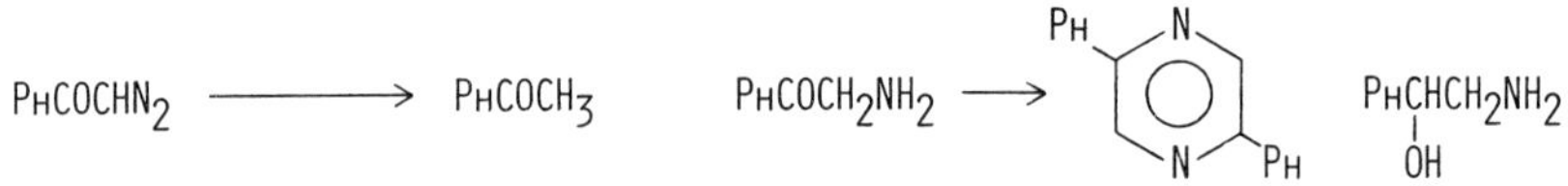

[906]	H$_2$/PdO,AcOEt		70%
	25°,1 atm		
[906]	H$_2$/PdO,AcOEt,AcOH	55%	
[906]	H$_2$/PdO,AcOH,HCl	30%	
[907]	H$_2$/Pd(C),EtOH	50%	
[907]	H$_2$/Pt	15%	45%
[907]	AlHg,HCl,or Zn,HCl	21%	
[907]	LiAlH$_4$,Et$_2$O		93%

to 85% yield of ω-(p-toluenesulfonyl)acetophenone from ω-(p-toluenesulfonyl)diazoacetophenone [*307*]. *Lithium aluminum hydride* in ether reduced ω-diazoacetophenone to 2-amino-1-phenylethanol in 93% yield [*906*].

In **aromatic ketones containing diazonium groups,** *hypophosphorous acid* replaces nitrogen by hydrogen. Diazotized 4-amino-3,5-dichloroacetophenone thus afforded 3,5-dichloroacetophenone in 80% yield [*288*].

In **azido ketones** both functions were reduced with *lithium aluminium hydride* in refluxing ether: ω-azidoacetophenone gave 49.5% yield of 2-amino-1-phenylethanol [*600*].

Reduction of Hydroxy and Amino Ketones

Hydroxy ketones and hydroxy-α,β-unsaturated ketones in the steroids such as estrone and testosterone, respectively, can be **reduced to diols** *biochemically*. Estrone acetate gave 68% yield of α-estradiol on incubation with baker's yeast at room temperature after five days [*909*]. Testosterone was reduced by bacteria to the saturated hydroxy ketones, etiocholan-17-ol-3-one and androstan-17-ol-3-one, and further to the diols, *epi*-etiocholane-3,17-diol and the epimeric isoandrostane-3,17-diol, both in low yields [*329*].

Bacillus polymyxa in a hydrogen atmosphere reduced (R,S)-acetoin to *erythro-* and *threo*-butane-1,3-diol in 100% yield [*910*], and *Saccharomyces cerevisiae* converted 3,3-dimethyl-1-hydroxybutan-2-one to ($-$)-R-3,3-dimethylbutane-1,2-diol in 66% yield [*911*].

Acyloins were converted to mixtures of stereoisomeric **vicinal diols** by catalytic hydrogenation over copper chromite [*912*]. More frequently they were reduced to **ketones**: by *zinc* (yield 77%) [*913, 914*], by zinc amalgam (yields 50-60%) [*915*], by *tin* (yields 86-92%) [*173*], or by *hydriodic acid*: by refluxing with 47% hydriodic acid in glacial acetic acid (yields 70-90%) [*916*], or by treatment with red phosphorus and iodine in carbon disulfide at room temperature (yields 80-90%) [*917*] (*Procedure 41*, p. 215). Since acyloins are readily accessible by reductive condensation of esters (p. 152) the above reductions provide a very good route to ketones and the best route to macrocyclic ketones [*913*].

(117)

$(CH_2)_8 \overset{CO}{\underset{CHOH}{\big|}}$	$(CH_2)_8 \overset{CHOH}{\underset{CHOH}{\big|}}$	$(CH_2)_8 \overset{CO}{\underset{CH_2}{\big|}}$	$(CH_2)_8 \overset{CH_2}{\underset{CH_2}{\big|}}$
[*912*] $H_2/CuCr_2O_4$,150°,135 atm	48-52% <u>cis</u> 27-32% <u>trans</u>		
[*913*] Zn wool,HCl		77%	
[*913*] Zn wool,HgCl$_2$,HCl,AcOH,reflux 88 hrs			79%
[*916*] HI,AcOH,reflux 2 hrs		90%	

In **steroidal α-hydroxy ketones** with the keto group in position 20 and the hydroxylic group in position 17, the hydroxy, and better still, acetoxy group is replaced by hydrogen on refluxing for 24 hours with *zinc* dust and acetic acid. 3β,17α-Diacetoxyallopregnan-20-one gave 89% yield of 3β-acetoxyal-lopregnane while its 17β-epimer's yield was only 46% [918].

Complete reduction of acyloins to hydrocarbons was accomplished by the *Clemmensen reduction*. Sebacoin afforded cyclodecane in 79% yield [913], and benzoin gave 1,2-diphenylethane in 84% yield [758].

In **ketones containing oxirane rings** in positions α,β to their carbonyls only the oxiranes were deoxygenated to α,β-unsaturated ketones by refluxing with *zinc* dust in acetic acid for 3 hours [646] or on treatment with *chromous chloride* [192, 650]. The keto groups remain intact, even if there are other keto groups with conjugated double bonds in the same molecule. Accordingly 16,17-epoxycorticosterone-21-acetate was converted to 16,17-dehydro-corticosterone-21-acetate in 94% yield [192]. α,β-Epoxy ketones may also be converted to allylic alcohols. On heating with *hydrazine* hydrate to 90° and refluxing for 15 minutes, 4β,5-epoxy-3-coprostanone gave 3-coprostene-5-ol in 68% yield. This reaction can be carried out even at room temperature using alcoholic solutions of α,β-epoxy ketones, 2-3 equivalents of hydrazine hydrate and 0.2 equivalent of acetic acid, and gives good yields of the rearranged allylic alcohols [919].

[919]   (118)

Ketones containing sulfur or nitrogen atoms bound to α-carbons suffer carbon–sulfur or carbon–nitrogen bond cleavage under the conditions of the *Clemmensen reduction* [159, 864] (p. 118). A ketosulfone was reduced to a sulfone-alcohol with *zinc* in refluxing 80% acetic acid in 70% yield [920].

Reduction of Diketones and Quinones

The reduction of diketones is very complex. They can be partially reduced to ketols (hydroxy ketones) or ketones, or completely reduced to hydrocarbons. Depending on the mutual distance of the two carbonyl groups and reagents used, carbon–carbon bond cleavage may occur and may be followed by recyclizations or rearrangements. Some reactions may result in the formation of alkenes. Quinones react in their own specific way.

α-Diketones (1,2-diketones) **are reduced to α-hydroxy ketones or ketones.** Diacetyl (butane-2,3-dione) was quantitatively converted to acetoin (3-hydroxy-2-butanone) by heating at 100° with granulated *zinc* in dilute sulfuric acid [921]; benzils were quantitatively converted to benzoins by refluxing with zinc dust in aqueous dimethyl formamide [164], by treatment with *titanium trichloride* or *vanadous chloride* [215], by heating with *benzpinacol* [922], and by treatment of 0° with *hydrogen sulfide* in piperidine–dimethylformamide [237] (*Procedure 42*, p. 216). Cyclotetradecane-1,2-dione was reduced to 2-

hydroxycyclotetradecanone by refluxing for 1 hour with triethyl phosphite in benzene (yield 78%) [296].

(119)

PhCOCOPh   PhCHCOPh   PhCH$_2$COPh   PhCH$_2$CH$_2$Ph
                                   OH

Ref	Conditions	PhCHCOPh/OH	PhCH$_2$COPh	PhCH$_2$CH$_2$Ph
[164]	Zn dust, 80% aq DMF, reflux 5-6 hrs	93%		
[215]	TiCl$_3$,THF,25°	88%		
[215]	VCl$_2$,THF,25°	100%		
[237]	H$_2$S,C$_5$H$_{11}$N-DMF,0°,1 hr	100%		
[922]	Ph$_2$C(OH)CPh$_2$(OH),160-70°	85%		
[237]	H$_2$S,C$_5$H$_5$N-MeOH,0°,4 hrs		100%	
[923]	1. N$_2$H$_4$·H$_2$O; 2. dil NaOH,reflux 7.5 hrs		73%	
[923]	1. 2N$_2$H$_4$; 2. 4 N KOH in MeOH, reflux 40 hrs			~100%

Reduction of diketones to either **hydroxy ketones** or **diols** can be accomplished by different biochemical reductions [327, 834] (Procedure 50, p. 218).

Ketones are obtained from α-diketones by reduction with *hydrogen sulfide* in a pyridine–methanol solution [237], by refluxing with 47% *hydriodic acid* in acetic acid (yield 80%) [916], and by *decomposition of monohydrazones* with alkali [923]. Reduction of α-diketones to *hydrocarbons* is achieved by *decomposition of bis-hydrazones* by alkali [923].

The Clemmensen **reduction of β-diketones** (1,3-diketones) is rather complicated. The first step in the reaction of 2,4-pentanedione with *zinc amalgam* is an intramolecular pinacol reduction leading to a cyclopropanediol. Next the cyclopropane ring is opened in the acidic medium, and a rearrangement followed by a reduction gives the final product, a ketone, with a changed carbon skeleton [924, 925]. The ketone is usually accompanied by small amounts of the corresponding hydrocarbon [924] or an α-hydroxy ketone [925].

(120)

RCOCH$_2$COMe ⟶ ... R = Me: 35-50% R = Ph: 65%

[924] ZnHg,HCl,reflux

γ-Diketones (1,4-diketones) are reduced to hydroxy ketones by *catalytic hydrogenation* using ruthenium on silica gel at 20° and 6.2 atm. 1,4-Cyclohexanedione thus gave 70% yield of 4-hydroxycyclohexanone [926]. *Clemmensen reduction* is very complex. In addition to products of partial reduction of hexane-2,5-dione, 2-hexanone (11%), 2-hexanol (32%) and a mixture of cis- and trans-2-hexen-4-ol (11% and 24%, respectively), a small amount (3%) of 3-methyl-2-pentanone was isolated [927]. This compound resulted from a rearrangement of the initially formed 1,2-dimethyl-1,2-cyclobutanediol to 1-acetyl-1-methylcyclopropane that ultimately rearranged to 3-methyl-2-pen-

tanone [*927*]. 1-Phenyl-1,4-pentanedione was reduced to 5-phenyl-2-pentan-
one (yield 76%) and 1-phenylpentane (12%).

In the Clemmensen reduction of 1,4-cyclohexanedione, all the products
isolated from the reduction of 2,5-hexanedione were found in addition to
2,5-hexanedione (20%) and 2-methylcyclopentanone (6%). The presence of
the two latter compounds reveals the mechanism of the reduction. In the first
stage the carbon–carbon bond between carbons 2 and 3 ruptured, and the
product of the cleavage, 2,5-hexanedione, partly underwent aldol condensation, partly its own further reduction [*927*]. The cleavage of the carbon–
carbon bond in 1,4-diketones was noticed during the treatment of 1,2-diben-
zoylcyclobutane which afforded, on short refluxing with zinc dust and zinc
chloride in ethanol, an 80% yield of 1,6-diphenyl-1,6-hexanedione [*156*].

An interesting reaction takes place when diketones with the keto groups in
positions 1,4 or more remote are refluxed in dimethoxyethane with *titanium
dichloride* prepared by reduction of titanium trichloride with a zinc–copper
couple. By deoxygenation and intramolecular coupling, cycloalkenes with up
to 22 members in the ring are obtained in yields of 50–95%. For example, 1-
methyl-2-phenylcyclopentene was prepared in 70% yield from 1-phenyl-1,5-
hexanedione, and 1,2-dimethylcyclohexadecene in 90% yield from 2,17-octa-
decanedione [*206, 210*].

(121)

$$PhCO(CH_2)_2COPh \qquad [206] \qquad MeCO(CH_2)_{20}COMe$$

$$TiCl_3/Zn-Cu$$
$$(CH_2OMe)_2$$
$$reflux\ 21\ hrs$$

(structures: cyclobutene with two Ph, 87%; macrocycle $(CH_2)_{20}$ with C–Me, C–Me, 83%)

In **enediones** in which two carbonyl groups of a diketone are linked by an
ethylenic bond *tin* [*174*] and *chromous chloride* [*196*] reduce only the double
bond, and none of the conjugated carbonyl groups. A double bond conjugated with one carbonyl group only is not reduced. Refluxing cholest-4-ene-
3,6-dione with chromous chloride in tetrahydrofuran yielded 49% of 5β-
cholestane-3,6-dione, and a similar reduction of cholesta-1,4-diene-3,6-one
gave 5β-cholest-1-ene-3,6-dione [*196*].

Quinones constitute one of the **most easily reducible systems.** They can be
reduced to hydroquinones, to ketones, to diols, or to hydrocarbons.

p-Benzoquinone and its derivatives are *catalytically hydrogenated* to **hydroquinones** under very mild conditions. At room temperature and atmospheric
pressure hydrogenation of *p*-benzoquinone stopped after absorption of just
1 mol of hydrogen when platinum in acetic acid or platinum or palladium in
ethanol were used as catalysts. The reduction over palladium was faster than
over platinum. However, platinum proved more efficient since, in the presence
of a mineral acid, the hydrogenation proceeded further to the stage of cyclo-

hexanol [*928*]. Palladium proved to be the best catalyst for the hydrogenation of *p*-benzoquinone to hydroquinone [*928*].

Lithium aluminum hydride reduced *p*-benzoquinone to hydroquinone (yield 70%) [*576*] and anthraquinone to anthrahydroquinone in 95% yield [*576*]. *Tin* reduced *p*-benzoquinone to hydroquinone in 88% yield [*174*] (*Procedure 35*, p. 214). *Stannous chloride* converted tetrahydroxy-*p*-benzoquinone to hexahydroxybenzene in 70–77% yield [*929*], and 1,4-naphthoquinone to 1,4-dihydroxynaphthalene in 96% yield [*180*]. Other reagents suitable for reduction of quinones are *titanium trichloride* [*930*], *chromous chloride* [*187*], *hydrogen sulfide* [*248*], *sulfur dioxide* [*250*] and others. Yields are usually good to excellent. Some of the reagents reduce the quinones selectively in the presence of other reducible functions. Thus hydrogen sulfide converted 2,7-dinitrophenanthrene quinone to 9,10-dihydroxy-2,7-dinitrophenanthrene in 90% yield [*248*].

(122)

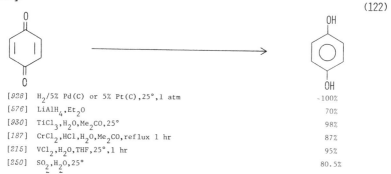

[*928*]	H$_2$/5% Pd(C) or 5% Pt(C),25°,1 atm	~100%
[*576*]	LiAlH$_4$,Et$_2$O	70%
[*930*]	TiCl$_3$,H$_2$O,Me$_2$CO,25°	98%
[*187*]	CrCl$_2$,HCl,H$_2$O,Me$_2$CO,reflux 1 hr	87%
[*215*]	VCl$_2$,H$_2$O,THF,25°,1 hr	95%
[*250*]	SO$_2$,H$_2$O,25°	80.5%

Anthraquinone is usually reduced to anthrahydroquinone but refluxing with *tin*, hydrochloric acid and acetic acid transformed anthraquinone to anthrone in 62% yield [*175*].

Further reduction of quinones – acquisition of four or more hydrogens per molecule – was achieved with *lithium aluminum hydride* which reduced, in yields lower than 10%, 2-methyl-1,4-naphthoquinone to 1,2,3,4-tetrahydro-1,4-dihydroxy-2-methylnaphthalene and to 1,2,3,4-tetrahydro-4-hydroxy-1-keto-2-methylnaphthalene [*931*]. *Lithium aluminum hydride* [*931*], *sodium borohydride, lithium triethylborohydride* and *9-borabicyclo[3.3.1]nonane* [*100*] converted anthraquinone to 9,10-dihydro-9,10-dihydroxyanthracene in respective yields of 67, 65, 77 and 79%.

Complete deoxygenation of quinones to hydrocarbons is accomplished in yields of 80–85% by heating with a mixture of *zinc*, zinc chloride and sodium chloride at 210–280° [*932*]. Refluxing with *stannous chloride* in acetic and hydrochloric acid followed by refluxing with *zinc* dust and 2 N sodium hydroxide reduced 4′-bromobenzo[5′.6′:1.2]anthraquinone to 4′-bromobenzo[5′.6′:1.2]anthracene in 95% yield [*181*], and heating with iodine, phosphorus and 47% *hydriodic acid* at 140° converted 2-chloroanthraquinone to 2-chloroanthracene in 75% yield [*222*]. Also *aluminum* in dilute sulfuric acid can be used for reductions of the same kind [*151*].

Reduction of Ketals and Thioketals

Ketals of acetone and cyclohexanone with methyl, butyl, isopropyl and cyclohexyl alcohols are **hydrogenolyzed to ethers and alcohols** by *catalytic hydrogenation*. While platinum and ruthenium are inactive and palladium only partly active, 5% rhodium on alumina proves to be the best catalyst which, in the presence of a mineral acid, converts the ketals to ethers and alcohols in yields of 70–100% [*933*].

The reduction is believed to be preceded by an acid-catalyzed reversible cleavage of the ketals to alcohols and unsaturated ethers which are subsequently hydrogenated. Mineral acid is essential. Best yields and fastest reductions are found with ketals of secondary alcohols. The hydrogenation proceeds at 2.5–4 atm at room temperature with ketals of secondary alcohols, and at 50–80° with ketals of primary alcohols. Acetone and cyclohexanone diisopropyl ketals gave 75% and 90% yields of diisopropyl and cyclohexyl isopropyl ether at room temperature after 1 and 2.5 hours, respectively [*933*].

The reagent of choice for the reduction of ketals to ethers is *alane* prepared *in situ* from lithium aluminum hydride and aluminum chloride in ether. At room temperature ethers are obtained in 61–92% yields [*792, 934*]. Cyclic ketals prepared from ketones and 1,2- or 1,3-diols afford on hydrogenolysis by alanes alkyl β- or γ-hydroxyalkyl ethers in 83–92% yields [*792*].

Cyclic five-membered **monothioketals** are **desulfurized by Raney nickel** mainly **to their parent ketones** (53–55% yields) and several by-products [*935*]. After stirring for 2 hours at 25° in benzene with W-2 Raney nickel, 4-*tert*-butylcyclohexanone ethylene monothioketal afforded 53–55% of 4-*tert*-butylcyclohexanone, 12–14% of *cis*-4-*tert*-butylcyclohexyl ethyl ether, 6–11% of 4-*tert*-butylcyclohexanone diethyl ketal, 16% of 4-*tert*-butylcyclohex-1-enyl ethyl ether, and 7–12% of 4-*tert*-butylcyclohexene [*935*].

Alane formed by the reaction of lithium aluminum hydride and aluminum chloride in ether cleaves exclusively the carbon-oxygen bond in cyclic monothioketals derived from ketones and mercaptoethanol, and on refluxing in 100% excess for 2 hours produces β-hydroxyethyl sulfides (yields 66–91%); on prolonged heating with the reagent these β-hydroxyethyl sulfides are further reduced to the corresponding ethyl sulfides (thioethers) (yields 28–81%) [*936*].

Reduction of cyclic five-membered ethylene monothioketals with *calcium* in liquid ammonia cleaves the bond between carbon and sulfur and yields **alkyl β-mercaptoethyl ethers** (7–88%) [*795*]. Cyclic five-membered **ethylene dithioketals** (ethylenemercaptoles) afford, analogously, **alkyl β-mercaptoethyl thioethers** (yields 85%) [*795*].

Most frequently **mercaptoles (dithioketals) are desulfurized to hydrocarbons** by *Raney nickel* [*796, 797*]. As in the case of aldehydes, conversion of ketones to mercaptoles followed by desulfurization to hydrocarbons represents the most gentle reduction of the carbonyl group to methylene. The desulfurization is accomplished by refluxing of the mercaptole with a large excess of

(123)

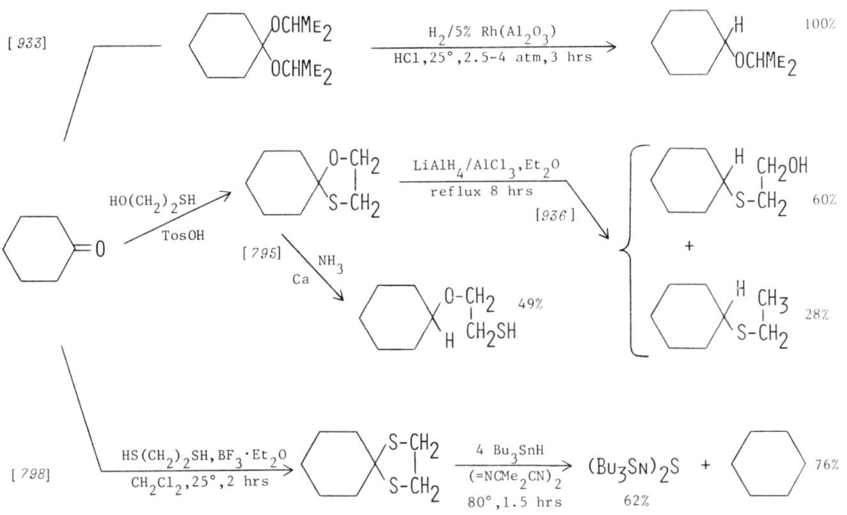

Raney nickel in ethanol or other solvents. If reducible functions are present in the molecule of the mercaptole, Raney nickel must be stripped of hydrogen by refluxing with acetone for a few hours before use [46].

Nickel prepared by reduction of nickel chloride with sodium borohydride was used for desulfurization of diethyl mercaptole of benzil. Partial desulfurization using 2 mol of nickel per mol of the mercaptole gave 71% yield of ethylthiodesoxybenzoin while treatment with a 10-fold molar excess of nickel over the mercaptole gave 61% yield of desoxybenzoin (benzyl phenyl ketone) [937].

(124)

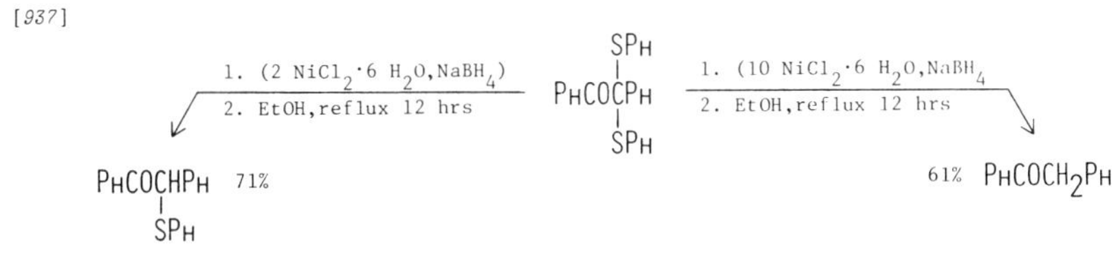

Mercaptoles of ketones are best prepared by treatment of ketones with ethanedithiol or 1,3-propanedithiol in the presence of anhydrous zinc chloride or boron trifluoride etherate. Many desulfurizations have been carried out with these cyclic mercaptoles, especially in steroids. Yields of the hydrocarbons range from 50 to 95% [797].

Alternative desulfurizations can be achieved using *tributylstannane* or hydrazine. Reaction with the former reagent is carried out by heating the mercaptole for 1.5 hours at 80° with 3–4 equivalents of tributylstannane and azobis(isobutyronitrile) as a catalyst and distilling the product and the by-product, bis(tributyltin) sulfide, *in vacuo*. Yields are 74–95% [798].

The reaction with *hydrazine* consists of heating the mercaptole with 3–5 parts by volume of hydrazine hydrate, 1.5–2.5 parts by weight of potassium hydroxide and 8–20 parts by volume of diethylene or triethylene glycol to 90–190° until the evolution of nitrogen ceases. Times required are 0.5–3 hours and yields range from 60% to 95% [*938*].

Reduction of Ketimines, Ketoximes and Hydrazones

Ketimines are reduced to amines very easily by *catalytic hydrogenation*, by complex hydrides and by formic acid. They are intermediates in reductive amination of ketones (p. 134). An example of the reduction of a ketimine is conversion of 3–aminocarbonyl-2,3-diphenylazirine to the corresponding aziridine by *sodium borohydride* (yield 73%), by *potassium borohydride* (yield 71%) and by *sodium bis(2-methoxyethoxy)aluminum hydride* (yield 71%) [*939*].

[*939*]   (125)

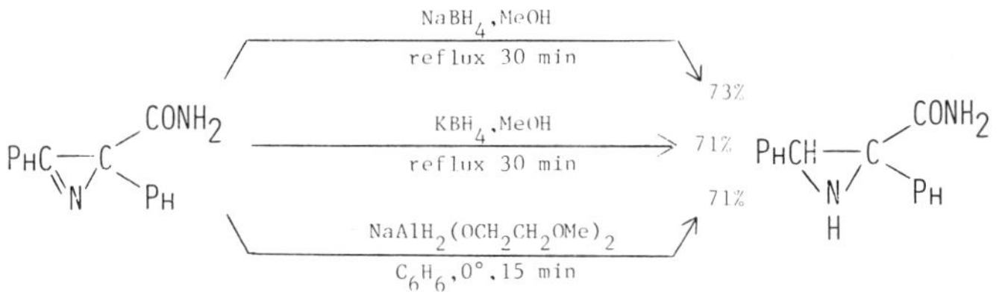

Reduction of ketoximes is one of the most useful ways **to primary amines.** Many methods are available: *Catalytic hydrogenation* over 5% rhodium on alumina converted cycloheptanone oxime to cycloheptylamine in 80% yield at 20–60° and 0.75–1 atm [*940*], and treatment with hydrogen and Raney nickel at 25–30° and 1–3 atm gave cyclohexylamine from cyclohexanone oxime in 90% yield [*45*]. 2-Alkylcyclohexanone oximes yielded *cis-* or *trans-*2-alkylcyclohexylamines depending on the method and reaction conditions used. Catalytic hydrogenation over platinum in acetic acid or over palladium in ethanol and hydrochloric acid afforded *cis*-2-alkylcyclohexylamine while hydrogenation over Raney nickel in ammonia or reduction with *sodium* in ethanol yielded *trans*-2-alkylcyclohexylamine, and reduction with sodium amalgam a mixture of both isomers [*464, 941*]. Similarly, hydrogenation of 11-oximino-5β-pregnane-3α,17α,20β-triol over platinum oxide in acetic acid at 60° and 56 atm gave 86% of β-, and reduction with sodium in refluxing propanol 75% of α-11-amino-5β-pregnane-3α,17α,20β-triol [*942, 943*]. Hydrogenation of methyl ethyl ketoxime to *sec*-butylamine in 76% yield was achieved by treatment with *hydrazine* hydrate and *Raney nickel* W4 at room temperature for 4 hours [*944*].

Reduction of cyclohexanone oxime with *lithium aluminum hydride* in tetrahydrofuran gave cyclohexylamine in 71% yield [*809*], and reduction of ketoximes with *sodium* in methanol and liquid ammonia [*945*] or in boiling ethanol [*946*] afforded alkyl amines, usually in good to high yields. *Stannous* chloride in hydrochloric acid at 60° reduced the dioxime of 9,10-phenanthra-

(126)

Ref	Conditions	Yield
[45]	Raney Ni,EtOH,25-30°,1-3 atm,45 min	90%
[944]	$N_2H_4 \cdot H_2O$,Raney Ni,25°,4 hrs	65%
[809]	$LiAlH_4$,Et_2O,reflux 30 min	71%
[945]	Na,NH_3,MeOH	91%

(127)

	Ref	Conditions	Yield (2-NH2)	Yield (other)
R=Me;	[941]	H_2/Pt,AcOH	70-75%	Yield not indicated
	[941]	Na,ROH		
R=Et;	[464]	H_2/Pd(C),EtOH,25°,1 atm	13%	
	[464]	H_2/Raney Ni,130°,83 atm		79%
	[464]	Na,EtOH,reflux		74.5%

quinone to 9,10-diaminophenanthrene in 90% yield [185]. *Titanium trichloride* behaves differently: in methanol or dioxane solutions in the presence of sodium acetate it regenerates the parent ketones in 90% yields [200].

Oximes of α,β-unsaturated ketones, on reduction with *lithium aluminum hydride*, depending on the structure of the ketoxime and on the reaction conditions, yield unsaturated amines, saturated amines, and sometimes aziridines in fair yields [947].

Whereas *diborane* in tetrahydrofuran reduces oximes only at 105-110°, oxime ethers and oxime esters are reduced to amines and alcohols at room temperature in good yields. For example the *p*-nitrobenzoyl ester of cyclohexanone oxime gave a 67% yield of cyclohexylamine and 81% yield of *p*-nitrobenzyl alcohol [948].

Nitro group in an oxime is reduced preferentially to the oximino group with *ammonium sulfide* [240]. In a monoxime of a diketone, the oximino group is reduced to amino group to the exclusion of the carbonyl group by *catalytic hydrogenation* over platinum oxide in methanolic hydrochloric acid: 9-keto-10-oximino-1,2,3,4-tetrahydrophenanthraquinone afforded 10-amino-9-keto-1,2,3,4-tetrahydrophenanthrene in 78% yield [949].

Hydrazones of ketones may be **reduced to hydrazines, to amines,** and **to hydrocarbons,** or reconverted **to parent ketones.**

The phenylhydrazone of acetone gave on *hydrogenation* over colloidal platinum a 90% yield of *N*-isopropyl-*N*-phenylhydrazine [950], and the semicarbazone of benzil on *electroreduction* a 70% yield of *N*-aminocarbonyl-*N'*-(1,2-diphenyl-2-ketoethyl)hydrazine [951].

The phenylhydrazone of *N*-acetylisopelletierine (*N*-acetyl-2-piperidylacetone) was hydrogenolyzed over platinum oxide in acetic acid at 25° and 3 atm to 1-(*N*-acetyl-2-piperidyl)-2-aminopropane in 92% yield [952], and the phenylhydrazone of levulinic acid was reduced with *aluminum amalgam* to 4-aminovaleric acid in 60% yield [*154*].

Hydrazones treated with alkalis decompose to nitrogen and hydrocarbons [*845, 923*] (*Wolff-Kizhner reduction*) (p. 34), and *p*-toluenesulfonylhydrazones are reduced to hydrocarbons by *lithium aluminum hydride* [*812*], *sodium borohydride* [*785*] or *sodium cyanoborohydride* [*813*]. *Titanium trichloride* hydrogenolyzes the nitrogen–nitrogen bond in phenylhydrazones and forms amines and ketimines which are hydrolyzed to the parent ketones. Thus 2,4-dinitrophenylhydrazone of cycloheptanone afforded cycloheptanone in 90% yield [*202*].

REDUCTIVE ALKYLATION (REDUCTIVE AMINATION)

Treatment of aldehydes or ketones with ammonia, primary or secondary amines in reducing media is called reductive alkylation (of ammonia or amines) or reductive amination (of aldehydes or ketones). Reducing agents are most frequently hydrogen in the presence of catalysts such as platinum, nickel or Raney nickel [*953*], complex borohydrides [*103, 954, 955*], formaldehyde or formic acid [*322*].

Reductive alkylation of ammonia should give primary amines, reductive alkylation of primary amines secondary amines, and reductive alkylation of secondary amines tertiary amines. In reality, secondary and even tertiary amines are almost always present to varying extents since the primary amines formed in the reaction of the carbonyl compounds with ammonia react with the carbonyl compounds to give secondary amines, and the secondary amines similarly afford tertiary amines according to Scheme 128. In addition, secondary amines may be formed, especially at higher temperatures, by additional reactions shown in Scheme 129. Depending on the ratios of the carbonyl compounds to ammonia or amines, different classes of amines predominate.

(128)

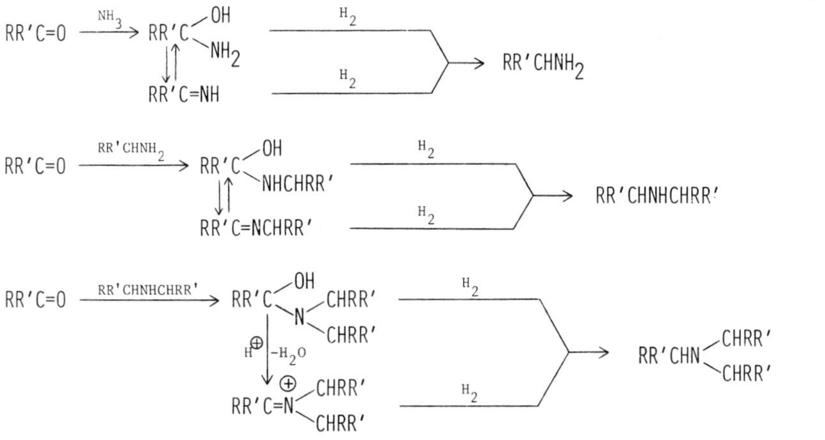

R,R' =   H, Alkyl, Aryl

For example, hydrogenation of benzaldehyde with 1 mol of ammonia gave 89.4% of primary and 7.1% of secondary amine while the reaction with 0.5 mol of ammonia afforded 11.8% of the primary and 80.8% of the secondary amine [*956*].

(129)

$$RCH_2NH_2 + RCH_2NH_2 \xrightarrow[\text{heat}]{\text{Pd or Ni}} RCH_2NHCH_2R + NH_3$$

$$[\textit{803}] \begin{cases} \xrightarrow{RCH_2NH_2} NH_3 + RCH=NCH_2R \xrightarrow[67\%]{H_2} RCH_2NHCH_2R \\ \\ \xrightarrow{2\ RCH=NH} NH_3 + RCH(N=CHR)_2 \xrightarrow{3\ H_2} \begin{matrix} \nearrow 96\% \\ \searrow 94\% \end{matrix} \begin{matrix} \\ RCH_2NH_2 \end{matrix} \end{cases}$$

$$RCH=NH$$

Consequently, by choosing proper conditions, especially the ratios of the carbonyl compound to the amino compound, very good yields of the desired amines can be obtained [*322, 953*]. In *catalytic hydrogenations* alkylation of amines was also achieved by alcohols under the conditions when they may be dehydrogenated to the carbonyl compounds [*803*]. The reaction of aldehydes and ketones with ammonia and amines in the presence of hydrogen is carried out on catalysts: platinum oxide [*957*], nickel [*803, 958*] or Raney nickel [*956, 959, 960*]. Yields range from low (23–35%) to very high (93%). An alternative route is the use of complex borohydrides: *sodium borohydride* [*954*], *lithium cyanoborohydride* [*955*] and *sodium cyanoborohydride* [*103*] in aqueous-alcoholic solutions of pH 5–8.

(130)

$$PHCHO \begin{cases} \xrightarrow[\begin{subarray}{c}[\textit{956}]\ H_2/\text{Raney Ni},40-70°,90\ \text{atm},30\ \text{min}\\ 0.5\ NH_3,\text{EtOH};\end{subarray}]{1\ NH_3,\text{EtOH};} 89.4\%\ PHCH_2NH_2 \\ \\ \xrightarrow{} 80.8\%\ (PHCH_2)_2NH \\ \\ \xrightarrow{[\textit{955}]\ \ EtNH_2;\ LiBH_3CN,MeOH,25°,72\ hrs} 80\%\ PHCH_2NHEt \\ \\ \xrightarrow{[\textit{954}]\ PHNH_2;\ NaBH_4,\text{aq. EtOH,AcOH,AcONa},0°} 83\%\ PHCH_2NHPH \end{cases}$$

(131)

$$ME_2CO \begin{cases} \xrightarrow{[\textit{958}]\ 1\ NH_3,\text{EtOH};\ Ni,15-20°,1\ \text{atm}} 62\%\ ME_2CHNH_2 \\ \\ \xrightarrow[25°,1-2\ \text{atm},6-10\ hrs]{[\textit{957}]\ 0.77\ H_2N(CH_2)_2OH;\ EtOH;\ H_2/PtO_2} 94-95\%\ ME_2CHNH(CH_2)_2OH \\ \\ \xrightarrow{[\textit{954}]\ PHNH_2;\ NaBH_4,\text{aq. EtOH,AcOH,AcONa},0°} 91\%\ ME_2CHNHPH \end{cases}$$

Reductive alkylation can also be accomplished by heating carbonyl compounds at 150–250° with 4–5 mol of *ammonium formate, formamide*, or *formates or formamides* prepared by heating primary on secondary amines with formic acid at 180–190° (*Leuckart reaction*) [*322*]. An excess of 85–90% formic acid is frequently used. Formyl derivatives of primary or secondary amines are sometimes obtained as products and have to be hydrolyzed to the corresponding amines by refluxing with 30% sodium hydroxide or with 10%, 20% or concentrated hydrochloric acid. The reaction requires 4–30 hours and the yields range from some 25% to almost 100% [321, 322, 961] (*Procedure 49*, p. 218).

(132)

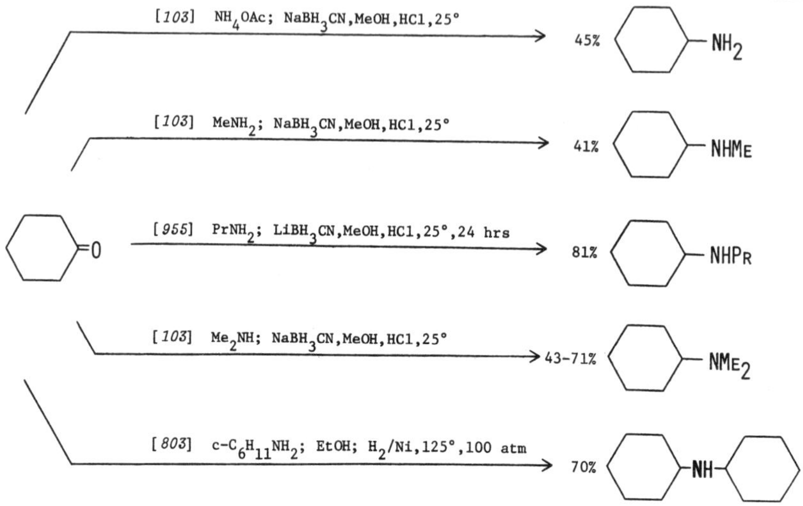

(133)

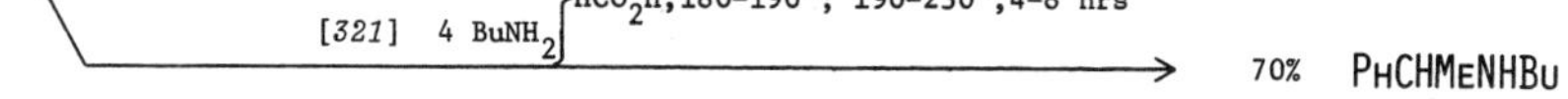

Reductive methylation is achieved by the reaction of *formaldehyde* with ammonium chloride. It is carried out by heating the components at 100–120° and gives mono-, di- and trimethylamine in high yields (*Eschweiler reaction*) [312, 962]. No catalyst is needed; part of the formaldehyde provides the necessary hydrogen while the other part is oxidized to formic acid. The same reaction can be applied to methylation of primary and secondary amines [962]. Reductive alkylation can also be accomplished by reducing mixtures of amines with acids which are first reduced to aldehydes (p. 171).

REDUCTION OF CARBOXYLIC ACIDS

The carboxylic group in carboxylic acids can be reduced to an aldehyde group, to an alcoholic group and even to a methyl group. Unsaturated acids and aromatic acids can be reduced at the multiple bonds or aromatic rings, respectively, without or with concomitant reduction of the carboxylic groups. Other functions in the molecules of carboxylic acids may or may not be affected by different reducing agents. Most frequently, products of the reduction of carboxylic acids are alcohols.

Reduction of Aliphatic Carboxylic Acids

Saturated aliphatic acids are **converted to aldehydes** by *aminoalanes* prepared *in situ* from alane and two molecules of a secondary amine. The best reagent is obtained by adding 2 mol of *N*-methylpiperazine to 1 mol of alane in tetrahydrofuran at 0–25°. Refluxing hexanoic acid, octanoic acid or palmitic acid for 6 hours with the solution of the aminoalane afforded the corresponding aldehydes in respective yields of 63%, 69% and 77% [*963*]. Aliphatic acids containing 5–14 carbon atoms were reduced to the corresponding aldehydes in 61–84% yields by treatment with *lithium* in methylamine followed by hydrolysis of the intermediates *N*-methylaldimines [*964*]. The carboxylic group is also reduced to the aldehyde group by *electrolysis* using lead electrodes [*965*] and on reduction with sodium amalgam (p. 139).

More frequent than the reduction to aldehydes is **reduction of carboxylic acids to alcohols.** *Catalytic hydrogenation* requires special catalysts, high temperatures (140–420°) and high pressures (150–990 atm). Under such vigorous conditions the resulting alcohols are accompanied by esters resulting from esterification of the alcohols with the parent acids. The catalysts used for such hydrogenations are copper and barium chromate [*966*], ruthenium dioxide or ruthenium on carbon [*41*], and, especially, rhenium heptoxide [*42*] and rhenium heptasulfide [*54, 967*]. Although yields of alcohols are in some cases very high, this method is of limited use.

The reduction of free acids to alcohols became practical only after the advent of complex hydrides. *Lithium aluminum hydride* reduces carboxylic acids to alcohols in ether solution very rapidly in an exothermic reaction. Because of the presence of acidic hydrogen in the carboxylic acid an additional equivalent of lithium aluminum hydride is needed beyond the amount required for the reduction. The stoichiometric ratio is 4 mol of the acid to 3 mol of lithium aluminum hydride (Equation 12, p. 18). Trimethylacetic acid was reduced to neopentyl alcohol in 92% yield, and stearic acid to 1-octadecanol in 91% yield. Dicarboxylic sebacic acid was reduced to 1,10-decanediol even if less than the needed amount of lithium aluminum hydride was used [*968*].

Another reagent, *sodium bis(2-methoxyethoxy)aluminum hydride* (Vitride), was used to reduce nonanoic acid to 1-nonanol in refluxing benzene in 92% yield [*969*]. The same reagent converts sodium or bromomagnesium salts of acids to alcohols: sodium stearate to 1-octadecanol at 80° in 96% yield, and bromomagnesium octanoate to 1-octanol at 80° in 85% yield [*970*].

Sodium borohydride does not reduce the free carboxylic group, but *borane* prepared from sodium borohydride and boron trifluoride etherate in tetrahydrofuran converts aliphatic acids to alcohols at 0–25° in 89–100% yields

[*971*]. The reagent is suitable for selective reduction of a free carboxyl group in the presence of an ester group (p. 163).

Carboxylic acids containing double bonds are easily **converted to saturated acids** by *catalytic hydrogenation* over common catalysts. If a new chiral center is generated in the reduction process, homogeneous hydrogenation over a chiral catalyst gave 40–45% enantiomeric excess of one enantiomer [*19*].

Unsaturated carboxylic acids with double bonds conjugated with the carboxyl are reduced at the double bond by *sodium amalgam* [*972*]. Sorbic acid (2,4-hexadienoic acid) containing a conjugated system of two double bonds was reduced by *catalytic hydrogenation* over palladium or Raney nickel preferentially at the more distant double bond, giving 83–90% of 2-hexenoic acid [*973*]. Sodium amalgam adds hydrogen to the 1,4- as well as the 1,6-position and affords a mixture of 2- and 3-hexenoic acid [*972*]. Maleic and fumaric acids were converted to succinic acid with *chromous sulfate* in aqueous solution at room temperature in 30 and 60 minutes in 86% and 91% yields, respectively [*974*].

(134)

[972] 3% NaHg, AcOH 15-20°

CH$_3$CH=CHCH$_2$CH$_2$CO$_2$H 25.6%

CH$_3$CH$_2$CH=CHCH$_2$CO$_2$H 31.4%

CH$_3$CH=CHCH=CHCO$_2$H [968]

LiAlH$_4$, Et$_2$O

CH$_3$CH=CHCH=CHCH$_2$OH 92%

Lithium aluminum hydride **reduces exclusively the carboxyl** group, even in an unsaturated acid with α,β-conjugated double bonds. Sorbic acid afforded 92% yield of sorbic alcohol [*968*], and fumaric acid gave 78% yield of *trans*-2-butene-1,4-diol [*975*]. If, however, the α,β-conjugated double bond of an acid is at the same time conjugated with an aromatic ring it is reduced (p. 141).

Carboxylic acids containing triple bonds are **converted to *cis* olefinic acids by** *catalytic hydrogenation* over Raney nickel [*358*], and **to trans acids** by reduction with *sodium* in liquid ammonia [*358*]. Acetylenedicarboxylic acid afforded fumaric acid in 94% yield on treatment with *chromous sulfate* at room temperature [*195*].

(135)

H$_2$/Raney Ni, EtOH [358] 25°, 1 atm

C$_2$H$_5$C≡C(CH$_2$)$_3$CO$_2$H

1. NaOH, MeOH 2. Na/NH$_3$ [358]

C$_2$H$_5$\C=C/(CH$_2$)$_3$CO$_2$H H/ \H 100%

85% C$_2$H$_5$\C=C/H H/ \(CH$_2$)$_3$CO$_2$H

Lithium aluminum hydride reduces acetylenic acids containing conjugated triple bonds to olefinic alcohols. Acetylenedicarboxylic acid gave, at room temperature after 16 hours, 84% yield of *trans*-2-butene-1,4-diol [*975*].

Reduction of Aromatic Carboxylic Acids

Aromatic carboxylic acids can be reduced to aldehydes or alcohols. In addition, a carboxylic group linked to an aromatic ring can be converted to methyl, and the aromatic ring can be partially or totally hydrogenated.

Aldehydes are obtained in 86% and 75% yields, respectively, from benzoic acid on refluxing for 6 hours and from nicotinic acid on standing at room temperature for 24 hours with *bis(N-methylpiperazino)alane* in tetrahydrofuran [963]. Reduction of 3-fluorosalicylic acid with 2% *sodium amalgam* in aqueous solution containing sodium chloride, boric acid and *p*-toluidine gave, at 13-15°, a Schiff base which on hydrolysis with hydrochloric acid and steam distillation afforded 3-fluorosalicylaldehyde in 57% yield [136]. The purpose of *p*-toluidine is to react with the aldehyde as it is formed and protect it from further reduction.

$$PhCO_2H \longrightarrow PhCHO \qquad\qquad PhCH_2OH \qquad (136)$$

		PhCHO	PhCH$_2$OH
[968]	LiAlH$_4$,Et$_2$O		81%
[969]	NaAlH$_2$(OCH$_2$CH$_2$OMe)$_2$,C$_6$H$_6$,80°		97%
[963]	AlH(-NNMe)$_2$,THF,reflux 3 hrs	86%	4%
[971]	NaBH$_4$,BF$_3$·Et$_2$O,THF,0°→25°,1 hr		89%

Reduction of aromatic carboxylic acids to alcohols can be achieved by hydrides and complex hydrides, e.g. *lithium aluminum hydride* [968], *sodium aluminum hydride* [88] and *sodium bis(2-methoxyethoxy)aluminum hydride* [544, 969, 970], and with *borane* (diborane) [976] prepared from sodium borohydride and boron trifluoride etherate [971, 977] or aluminum chloride [738, 978] in diglyme. Sodium borohydride alone does not reduce free carboxylic acids. Anthranilic acid was reduced to the corresponding alcohol by *electroreduction* in sulfuric acid at 20-30° in 69-78% yield [979].

Reduction of aromatic carboxylic acids having hydroxy or amino groups in *ortho* or *para* positions is accompanied by *hydrogenolysis* when carried out with *sodium bis(2-methoxyethoxy)aluminum hydride* in xylene at 141-142°: *o*- and *p*-cresols [980] and *o*- and *p*-toluidines [981] respectively were obtained in yields ranging from 71% to 92%. *meta*-Substituted benzoic acids under the same conditions gave *m*-hydroxybenzyl alcohol [980] and *m*-aminobenzyl alcohol [981] in 72% yields. Conversion of *aromatic acids to hydrocarbons* was accomplished with *trichlorosilane*. A mixture of 0.1 mol of an acid and 0.6 mol of trichlorosilane in 80 ml acetonitrile was refluxed for 1 hour, then treated under cooling with 0.264 mol of tripropylamine and refluxed for 16 hours. After dilution with 850 ml of ether, removal of tripropylamine hydrochloride and evaporation of the ether, the residue was refluxed for 20 hours with 1 mol of potassium hydroxide in 120 ml of 80% aqueous methanol to give 78-94% yield of the corresponding hydrocarbon [982]. Hydrogenolysis to hydrocarbons may also occur in catalytic hydrogenations over nickel or copper chromite under very drastic conditions.

Saturation of the aromatic rings in aromatic carboxylic acids takes place during energetic *hydrogenation*. Gallic acid (3,4,5-trihydroxybenzoic acid) hydrogenated over 5% rhodium on alumina in 95% ethanol at 90–100° and 150 atm gave, after 8–12 hours, 45–51% yield of *all-cis*-3,4,5-trihydroxy-cyclohexanecarboxylic acid. Palladium, platinum, rhodium on other supports, and Raney nickel were less satisfactory as catalysts [983].

Sodium in liquid ammonia and ethanol reduced benzoic acid to 1,4-dihydrobenzoic acid. Reduction of *p-toluic* acid was more complicated and afforded a mixture of *cis*- and *trans*-1,2,3,4-tetrahydro-*p*-toluic acid and *cis*- and *trans*-1,4-dihydrotoluic acid. *m*-Methoxybenzoic acid yielded 1,2,3,4-tetrahydro-5-methoxybenzoic acid, and 3,4,5-trimethoxybenzoic acid gave 1,4-dihydro-3,5-dimethoxybenzoic acid in 87% yield (after hydrogenolysis of the methoxy group *para* to carboxyl) [984]. In the case of 4'-methoxy-biphenyl-4-carboxylic acid, sodium in isoamyl alcohol at 130° reduced completely only the ring with the carboxylic group, thus giving 92% yield of 4-(*p*-methoxyphenyl)cyclohexanecarboxylic acid [985].

1-Naphthoic acid was reduced with sodium in liquid ammonia to 1,4-dihydro-1-naphthoic acid which, after heating on a steam bath with 20% sodium hydroxide for 30 minutes, isomerized to 3,4-dihydro-1-naphthoic acid (yield 63%) [399]. 2-Naphthoic acid treated with 4 equivalents of *lithium* in liquid ammonia and ethanol gave 69% yield of 1,2,3,4-tetrahydro-2-naphthoic acid. With 7 equivalents of lithium, 1,2,3,4,5,8-hexahydro-2-naphthoic acid was obtained in 82% yield [986].

Partial reduction of the aromatic ring is especially easy in anthracene-9-carboxylic acid which was reduced to 9,10-dihydroanthracene-9-carboxylic acid with 2.5% *sodium amalgam* in aqueous sodium carbonate at 10° in 80% yield [987]. Aromatic carboxylic acids with hydroxyl groups in the *ortho* positions suffer ring cleavage during reductions with sodium in alcohols and are converted to dicarboxylic acids after fission of the intermediate β-keto acids.

Unsaturated aromatic carboxylic acids with double bonds conjugated with both the carboxyl and the aromatic ring undergo easy **saturation of the double bond** by *catalytic hydrogenation* over colloidal palladium [1067], Raney nickel [45], copper chromite [50] or tris(triphenylphosphine)rhodium chloride [56], generally in high to quantitative yields. Homogeneous catalytic hydrogenation using a chiral diphosphine–rhodium catalyst converted atropic acid (2-phenylacrylic acid) to S-hydratropic acid (2-phenylpropionic acid) in quan-

(137)

$PhCH=CHCO_2H$	$\xrightarrow{\hspace{2cm}}$	$PhCH_2CH_2CO_2H$	$PhCH_2CH_2CH_2OH$
[45]	H_2/Raney Ni,25°–30°,1–3 atm,10 min	100%	
[50]	H_2/CuCr$_2$O$_4$,175°,100–150 atm,18 min	100%	
[56]	H_2/(Ph$_3$P)$_3$RhCl,EtOH,60°,4–6 atm,8–12 hrs	85%	
[991]	Electroredn.,NaOH,Hg cathode,30v,5–10 amp	80–90%	
[68]	1. NaOH; 2. NaHg	80%	
[974]	CrSO$_4$,DMF–H$_2$O,25°,40 hrs	89%	
[968]	LiAlH$_4$,Et$_2$O		85%

titative yield and 63% optical purity [*988*]. Double bonds were also saturated by treatment with *nickel–aluminum alloy* in 10% sodium hydroxide at 90–100° (yields 80–95%) [*989, 990*]. Cinnamic acid was reduced to hydrocinnamic acid not only by catalytic hydrogenation but also by *electroreduction* [*991*], by reduction with *sodium amalgam* [*68*] (*Procedure 29*, p. 212) and with *chromous sulfate* [*974*]. *Lithium aluminum hydride* reduces the above acids completely, at the carboxyl as well as at the double bond [*968*].

Acetylenic aromatic acids having the triple bond flanked by carboxyl and an aromatic ring were partially reduced to olefinic aromatic acids by *chromous sulfate* in aqueous dimethylformamide at room temperature in high yields. Phenylpropiolic acid afforded *trans*-cinnamic acid in 91% yield [*195*]. Its sodium salt in aqueous solution gave on *catalytic hydrogenation* over colloidal platinum at room temperature and atmospheric pressure 80% yield of *cis*-cinnamic acid if the reaction was stopped after absorption of 1 mol of hydrogen. Otherwise phenylpropanoic acid was obtained in 75–80% yield [*992*].

REDUCTION OF CARBOXYLIC ACIDS CONTAINING SUBSTITUENTS OR OTHER FUNCTIONAL GROUPS

Carboxylic acids containing halogens are easily **reduced to halogenated alcohols** by alanes or boranes. Lithium aluminum hydride could endanger more reactive halogens (p. 63). *Borane* in tetrahydrofuran converted chloroacetic acid to chloroethanol at 0–25° in 30 minutes in quantitative yield, and 2-bromododecanoic acid to 2-bromododecanol in 1 hour in 92% yield without hydrogenolyzing the reactive halogens in positons α to carboxyls [*971*]. 3-Bromopropanoic acid was reduced to 3-bromopropanol with an ethereal solution of *alane* (prepared from lithium aluminum hydride and aluminum chloride) at 35° in 50% yield [*993*], and with *lithium aluminum hydride* in ether by inverse technique at −15° in 26% yield [*993*]. Alane in tetrahydrofuran at 10° converted 3-chloropropanoic acid to 3-chloropropanol in 89% yield (61% isolated), and 3-bromobutanoic acid to 3-bromobutanol in 87% yield, both in 15 minutes [*994*]. *p*-Chlorobenzoic acid afforded *p*-chlorobenzyl alcohol on reduction with borane in diglyme in 64–88% yields [*738, 977*]. Partial replacement of one atom of iodine took place during reduction of 3,4,5-triiodobenzoic acid which afforded 3,5-diiodobenzyl alcohol in 60% yield on treatment with lithium aluminum hydride. Similar reduction occurred with other polyiodinated benzene derivatives [*995*].

o-Chlorobenzoic acid and *p*-bromobenzoic acid were transformed into *o*-chlorotoluene and *p*-bromotoluene in 94% yields on treatment with *trichlorosilane* and tripropylamine [*982*] (p. 139). On the other hand, *o*-, *m*- and *p*-chlorobenzoic acids were converted to benzoic acid in respective yields of 91%, 64% and 82% by *catalytic hydrogenation* over Raney nickel in methanolic potassium hydroxide at room temperature [*536*], and *o*-bromobenzoic acid was reduced quantitatively to benzoic acid over colloidal palladium in aqueous sodium hydroxide [*996*]. In *N*-acetyl-β-(2-bromo-3-benzofuryl)alanine the bromine on the furan ring was replaced by hydrogen over Raney nickel at room temperature in 91% yield [*548*]. In **halogenated un-saturated acids** having chlorine or bromine linked to sp^2 carbons, catalytic hydrogenation may cause replacement of the halogen without saturation of the double bond. β-Chlorocrotonic acid was converted to crotonic acid by hydrogenation over 10% palladium on barium sulfate at room temperature and atmospheric pressure [*996*]. On the other hand, on hydrogenation over 10% palladium on charcoal at room temperature and atmospheric pressure dichloromaleic acid afforded predominantly succinic acid [*68*]. Vinylic fluorines in fluorobutenedioic acids are hydrogenolyzed with surprising ease. Products are always saturated acids, with or without fluorine [*66*].

(138)

		meso		
[*66*]	10% Pd(C),Et$_2$O,−70°	37.5%	41.2%	2.5%
	10% Pd(C),Et$_2$O,25°	19.0%	70.0%	11.0%
	10% Pd(C),H$_2$O,25°	0	67.0%	33.0%
	5% Rh(C),H$_2$O,25°	0	11.0%	89%

Refluxing with *zinc* in ethanol reduced α-bromocinnamic acid to cinnamic acid in 80% yield [*997*]. Allylic chlorines in γ,γ,γ-trichlorocrotonic acid were partly or completely hydrogenolyzed by *zinc* and *sodium amalgam* [*519*]. Hydrogenolysis of allylic bromine in α,β-unsaturated esters with zinc in acetic acid gave predominantly β,γ-unsaturated esters in 65–97% yields [*998*].

The biochemical reduction of α,β-unsaturated β-haloaliphatic acids by means of *Clostridium kluyveri* yielded halogen-free saturated acids. The same products were obtained from saturated β-halo acids. However, the same microorganism converted α,β-unsaturated α-halo acids to saturated α-halo acids with R configuration. Yields of reduction of α-fluoro-, α-chloro- and α-bromocrotonic acid ranged from 30% to 100% [*330*].

In **carboxylic acids containing nitro groups** *borane* reduced only the carboxyl group [*738, 977*] whereas *ammonium sulfide* reduced only the nitro group [*239*].

(139)

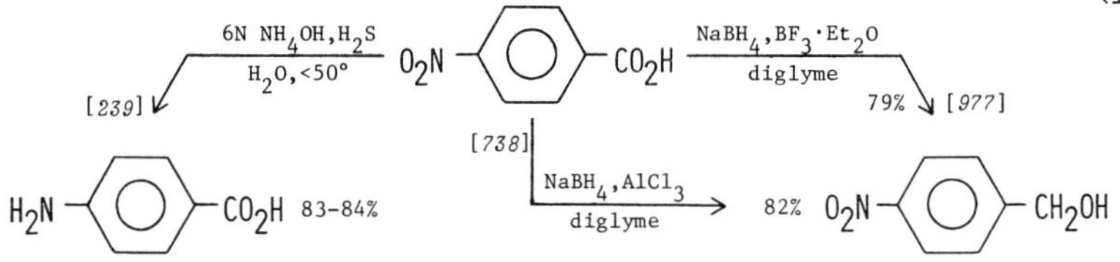

The **diazonium group** in *o*-carboxybenzenediazonium salts was *replaced by hydrogen* using *sodium borohydride* in methanol [*593*], or *ethanol* and ultraviolet irradiation [*306*]. Yields of benzoic acid were 77% and 92.5%, respectively.

The **sulfidic bond** in o,o'-dicarboxydiphenyl disulfide was cleaved by *zinc* in refluxing acetic acid giving o-carboxythiophenol in 71–84% yield [999].

In **2,3-epoxybutyric acid** *sodium borohydride* opened the epoxide ring without affecting the carboxyl. Varying ratios of 2- and 3-hydroxybutyric acid were obtained depending on the reaction conditions. Sodium borohydride in alkaline solution gave 18% of α- and 82% of β-hydroxybutyric acid while in the presence of lithium bromide the two isomers were obtained in 60:40 percentage ratio [1000].

In **keto acids,** carboxyl was reduced preferentially to the carbonyl with *borane* in tetrahydrofuran. 4-Phenyl-4-ketobutanoic acid afforded 4-phenyl-4-ketobutanol in 60% yield [971].

Keto acids were **reduced** more frequently **to hydroxy acids or to acids.** On *catalytic hydrogenation* over Raney nickel cyclohexanone-4-carboxylic acid gave 86% yield of *cis*-4-hydroxycyclohexanecarboxylic acid and on reduction with 4% *sodium amalgam* 65–70% yield of the *trans* product [846]. 4-Keto-4-phenylbutanoic acid was hydrogenated over palladium on barium sulfate to 4-hydroxy-4-phenylbutanoic acid which lactonized to a γ-lactone (yield 73%) [1001]. The same compound was obtained in 73% yield on treatment with *triethylsilane* and trifluoroacetic acid [777]. In 4-ketodecanoic acid the keto group was reduced *biochemically* using baker's yeast to a hydroxy group which lactonized to give 85% yield of the coresponding γ-lactone [1002]. Phenylglyoxylic acid was transformed by heating at 100° with amalgamated *zinc* in hydrochloric acid to mandelic acid (α-hydroxyphenylacetic acid) in 70% yield [1003].

Refluxing of 9-fluorenone-1-carboxylic acid with zinc dust and copper sulfate in aqueous potassium hydroxide for 2.5 hours afforded 9-fluorenol-1-carboxylic acid in 94% yield [1004]. Reduction with *sodium borohydride* in aqueous methanol at 0–25° converted 5-ketopiperidine-2-carboxylic acid to *trans*-5-hydroxypiperidine-2-carboxylic acid in 54–61% yield [1005]. On the other hand, reduction of N-benzyloxycarbonyl-5-ketopiperidine-2-carboxylic acid gave 89% yield of N-benzyloxycarbonyl-*cis*-5-hydroxypiperidine-2-carboxylic acid under the same conditions [1005].

(140)

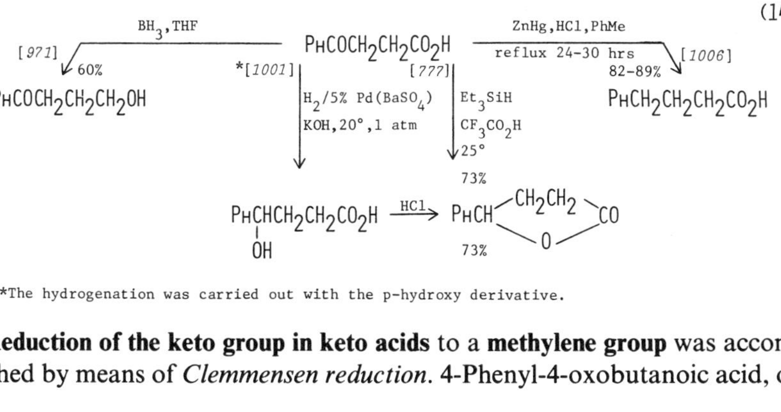

*The hydrogenation was carried out with the p-hydroxy derivative.

Reduction of the keto group in keto acids to a **methylene group** was accomplished by means of *Clemmensen reduction.* 4-Phenyl-4-oxobutanoic acid, on refluxing with amalgamated zinc, hydrochloric acid and toluene, afforded 82–89% yield of 4-phenylbutanoic acid [*1006*]. *Wolff–Kizhner reduction* or its modifications were used for the conversion of 6-oxoundecanedioic acid to undecanedioic acid (yield 87-93%) [*1007*], and for the transformation of *p*-phenoxy-4-phenyl-4-oxobutanoic acid to *p*-phenoxy-4-phenylbutanoic acid (yield 95-96%) [*281*]. *o*-Benzoylbenzoic acid was reduced to diphenylmethane-*o*-carboxylic acid in 85% yield by refluxing for 120 hours with ethanolic aqueous *hydriodic acid* and phosphorus [*228*], and *p*-benzoylbenzoic acid gave *p*-benzylbenzoic acid on reduction with *triethylsilane* and trifluoroacetic acid in 50% yield (50% recovered) [*777*]. The latter reagent reduced only keto acids with the keto group adjoining the benzene ring, and not without exceptions [*777*].

REDUCTION OF ACYL CHLORIDES

Reduction of acyl chlorides is exceptionally easy. Depending on the reagents and reaction conditions it can lead to aldehydes or to alcohols. *Catalytic hydrogenation* to *aldehydes* can be achieved by simply passing hydrogen through a solution of an acyl chloride in the presence of a catalyst. Since the product of the reaction of the acyl chloride with one molecule of hydrogen – the aldehyde – is further reducible to the stage of alcohol, a very inefficient catalyst must be used which catalyzes only the addition of hydrogen to the acyl chloride but does not catalyze the reaction of hydrogen with the aldehyde. Such a catalyst was made by deactivating palladium on barium sulfate by sulfur compounds like the so-called quinoline-*S* prepared by boiling quinoline with sulfur [*35*]. Even better results were obtained by deactivating platinum oxide with thiourea [*1008*]. The reaction referred to as *Rosenmund reduction* [*35*] is carried out by bubbling hydrogen through a solution of an acyl chloride in boiling toluene or xylene containing deactivated palladium on barium sulfate. Hydrogen chloride produced by the reaction is removed from the reaction mixture by a current of hydrogen which is then passed through a scrubber containing sodium or potassium hydroxide. In this way the progress of the reaction can be followed by determination of the amount of hydrogen chloride absorbed in the scrubber [*35*]. Reduction yields range from 50% to 97% [*1008, 1009*]; 2,4,6-trimethylbenzoyl chloride was converted to 2,4,6-trimethylbenzaldehyde in 70–80% yield [*1010*].

A variation of the Rosenmund reduction is heating of an acyl chloride at 50° with an equivalent of *triethylsilane* in the presence of 10% palladium on charcoal. Yields of aldehydes obtained by this method ranged from 45% to 75% [*80*].

Disadvantages of the Rosenmund reduction are high temperature, sometimes necessary to complete the reaction, and long reaction times. In this respect, reduction with complex hydrides offers considerable improvement. Again special, not too efficient reagents must be used; otherwise the reduction proceeds further and gives alcohols (p. 145). One of the most suitable complex hydrides proved to be *lithium tris(tert-butoxy)aluminum hydride* which, because of its bulkiness, does not react at low temperatures with the aldehydes formed by the reduction. The highest yields of aldehydes (52-85%) were obtained if the reduction was performed in diglyme at $-75°$ to $-78°$ for 1 hour [*96, 1011*] (*Procedure 16*, p. 208).

Another hydride for the preparation of aldehydes from acyl chlorides is obtained by treatment of a mixture of cuprous chloride and triphenylphosphine, trimethyl phosphite or triisopropyl phosphite in chloroform with an ethanolic solution of sodium borohydride. Such reagents reduce acyl chlorides to aldehydes in acetone solutions at room temperature in 15-90 minutes in yields ranging from 57% to 83% [*115*].

Both above mentioned complex hydrides have been successfully used for the **preparation of unsaturated aldehydes from unsaturated acyl chlorides** (yields 48-71%) [*1011*] and for the synthesis of *p*-nitrobenzaldehyde from *p*-nitrobenzoyl chloride [*115, 1011*], a reduction which could hardly be achieved by applying catalytic hydrogenation.

(141)

$$\text{PhCOCL} \xrightarrow{\hspace{4cm}} \text{PhCHO}$$

[*1009*]	H_2/5% Pd(BaSO$_4$),xylene,reflux	97%
[*1009*]	H_2/Ni,xylene,reflux 4 hrs	95%
[*1008*]	H_2/PtO$_2$,CS(NH$_2$)$_2$,PhMe,reflux 6-12 hrs	96%
[*80*]	SiHEt$_3$/10% Pd(C),60-80°	31-70%
[*1011*]	LiAlH(OCMe$_3$)$_3$,diglyme,-78°,1 hr; 25°,1 hr	81%
[*115*]	(Ph$_3$P)$_2$Cu⟨BH$_4$⟩,Me$_2$CO,25°,1 hr	83%
[*115*]	([MeO]$_3$P)$_2$Cu⟨BH$_4$⟩,Me$_2$CO,25°,15 min	82%

α,β-Unsaturated acyl chlorides are also converted to α,β-unsaturated aldehydes in 94-100% yields by treatment with *triethyl phosphite* and subsequent reduction of the diethyl acylphosphonate with *sodium borohydride* at room temperature followed by alkaline hydrolysis [*1012*].

Complete reduction of acyl chlorides to primary alcohols is not nearly as important as the reduction to aldehydes since alcohols are readily obtained by reduction of more accessible compounds such as aldehydes, free carboxylic acids or their esters [*83, 968*]. Because aldehydes are the primary products of the reduction of acyl chlorides strong reducing agents convert acyl chlorides directly to alcohols.

Catalytic hydrogenation is hardly ever used for this purpose since the reaction by-product – hydrogen chloride – poses some inconveniences in the experimental procedures. Most transformations of acyl chlorides to alcohols are effected by hydrides or complex hydrides. Addition of acyl chlorides to ethereal solutions of *lithium aluminum hydride* under gentle refluxing produced alcohols from aliphatic, aromatic and unsaturated acyl chlorides in 72-99% yields [*83*]. The reaction is suitable even for the preparation of halogenated alcohols. Dichloroacetyl chloride was converted to dichloroethanol in 64-65% yield [*1013*]. *Sodium aluminum hydride* converted acyl chlorides to alcohols in 94-99% yields [*88*], *sodium bis(2-methoxyethoxy)aluminum hydride* in 52-99% yields [*969*]. For the reduction of α,β-*unsaturated acyl* chlorides a reverse technique is used [*544, 969*]. Reduction of acyl chlorides to alcohols can also be accomplished using *alanes* generated *in situ* from lithium aluminum hydride and aluminum chloride in ethereal solutions. These reagents are especially suitable for reduction of halogenated acyl chlorides which, in reductions with lithium aluminum hydride, sometimes yield halogen-free alcohols [*993*]. Reductions of 3-bromopropanoyl chloride with alane gave higher yields (76-90%) than reduction with lithium aluminum hydride (46-87%) even when a normal (not reverse) technique was used [*787, 993*].

(142)

$$PhCOCl \longrightarrow PhCH_2OH$$

[*83*]	LiAlH$_4$,Et$_2$O,reflux	72%
[*751*]	NaBH$_4$,dioxane,100°	76%
[*1014*]	NaBH$_4$/Al$_2$O$_3$,25°,2-4 hrs	90%
[*99*]	NaBH(OMe)$_3$,Et$_2$O,reflux 4 hrs	66%
[*771*]	Bu$_4$NBH$_4$,CH$_2$Cl$_2$,-78°,15 min	97%

High yields (76-81%) of alcohols are also obtained by adding solutions of acyl chlorides in anhydrous dioxane or diethyl carbitol to a suspension of *sodium borohydride* in dioxane and brief heating of the mixtures on the steam bath [*751*], by stirring solutions of acyl chlorides in ether for 2-4 hours at room temperature with aluminum oxide (activity I) impregnated with a 50% aqueous solution of sodium borohydride (Alox) (yields 80-90%) [*1014*], by refluxing acyl chlorides with ether solutions of *sodium trimethoxyborohydride* [*99*], or by treatment of acyl chlorides in dichloromethane solutions with *tetrabutylammonium borohydride* at −78° [*771*]. A 94% yield of neopentyl alcohol was achieved by the reaction of trimethylacetyl chloride with *tert-butylmagnesium chloride* [*324*].

REDUCTIONS OF ACID ANHYDRIDES

Reductions of anhydrides of monocarboxylic acids to alcohols are very rare but can be accomplished by complex hydrides [*83, 99*]. More frequent are **reductions of cyclic anhydrides** of dicarboxylic acids, which give lactones. Such reductions were carried out by catalytic hydrogenation, by complex hydrides and by metals.

Hydrogenation of phthalic anhydride over copper chromite afforded 82.5% yield of the lactone, phthalide, and 9.8% of *o*-toluic acid resulting from hydrogenolysis of a carbon–oxygen bond [*1015*]. Homogeneous hydrogenation of α,α-dimethylsuccinic anhydride over tris(triphenylphosphine)rhodium chloride gave 65% of α,α-dimethyl- and 7% of β,β-dimethylbutyrolactone [*1016*].

Powerful complex hydrides like *lithium aluminum hydride* in refluxing ether [*83*] or refluxing tetrahydrofuran [*1017*] reduce cyclic anhydrides to diols. Phthalic anhydride was thus transformed to phthalyl alcohol (*o*-hydroxymethylbenzyl alcohol) in 87% yield [*83*]. Similar yields of phthalyl alcohol were obtained from phthalic anhydride and *sodium bis(2-methoxyethoxy)-aluminum hydride* [*544, 969*].

Under controlled conditions, especially avoiding an excess of lithium aluminum hydride and performing the reaction at $-55°$, cyclic anhydrides were converted to lactones in high yields. Both *cis*- and *trans*-1-methyl-1,2,3,6-tetrahydrophthalic anhydride and *cis*- and *trans*-hexahydrophthalic anhydride were reduced to lactones at the carbonyl group adjoining the methyl-carrying carbon in yields of 75–88.6% [*1017*]. Similar results, including the stereospecificity (where applicable), were obtained by reduction with *sodium borohydride* in refluxing tetrahydrofuran or dimethylformamide at 0–25° in yields of 51–97% [*1018, 1019*].

Other reagents used for the preparation of lactones from acid anhydrides are *lithium borohydride* [*1019*], *lithium triethylborohydride* (Superhydride*) [*1019*] and *lithium tris(sec-butyl)borohydride* (L-Selectride*) [*1019*]. Of the three complex borohydrides the last one is most stereoselective in the reduction of 3-methylphthalic anhydride, 3-methoxyphthalic anhydride, and 1-methoxynaphthalene-2,3-dicarboxylic anhydride. It reduces the less sterically hindered carbonyl group with 85–90% stereoselectivity and is 83–91% yield [*1019*].

Results of reductions of cyclic anhydrides to lactones with *sodium amalgam* or with *zinc* are inferior to those achieved by complex hydrides [*1020*]. However, 95.6% yield of phthalide was obtained by reduction of phthalimide with zinc in sodium hydroxide [*1021*].

(143)

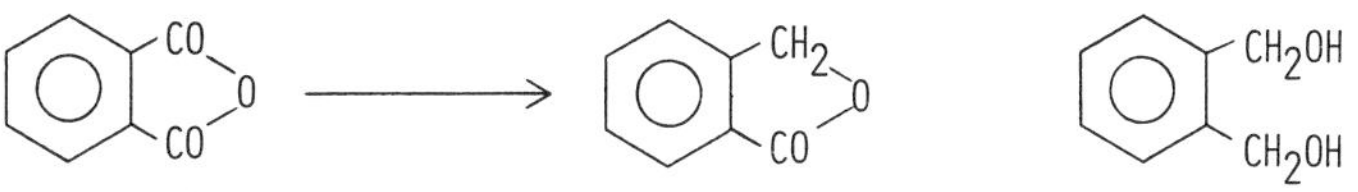

[*1015*] $H_2/CuCr_2O_4,C_6H_6,260°,215$ atm	82.5%	
[*83*] $LiAlH_4,Et_2O$,reflux		87%
[*969*] $NaAlH_2(OCH_2CH_2OMe)_2,C_6H_6,80°,1.5$ hr		88.5%
[*1018*] $NaBH_4$,DMF,0°–25°,1 hr	97%	
[*100*] $LiBHEt_3$,THF,0°,1 hr	76%	
[*1020*] Zn,AcOH	30–35%	

REDUCTION OF ESTERS AND LACTONES OF CARBOXYLIC ACIDS

Reduction of esters of carboxylic acids is complex and takes place in stages and by different routes. In the first stage hydrogen adds across the carbonyl double bond and generates a hemiacetal (route a), or else it can hydrogenolyze

*Trade name of the Aldrich Chemical Company.

the alkyl–oxygen bond and form an acid and a hydrocarbon (route b). The latter reaction is very rare and is mainly limited to esters of tertiary and benzyl-type alcohols. The hemiacetal is further reduced so that hydrogen replaces either the hydroxylic group (route c) and gives an ether, or else replaces the alkoxy group and gives an alcohol (route d). Alternatively the hemiacetal may eliminate one molecule of alcohol and afford an aldehyde (route e) which is subsequently reduced to the same alcohol which was formed by route d. The hemiacetals can be isolated in esters of carboxylic acids containing electron-withdrawing substituents. One electron reduction achieved by metals leads to acyloins (route f). The difference in the reaction pathways depends mainly on the reducing agents.

(144)

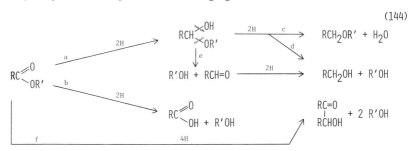

By the use of aluminum hydrides and complex hydrides **esters can be reduced to aldehydes** in respectable yields. The reductions are usually carried out at subzero temperatures, frequently under cooling with dry ice. *Lithium aluminum hydride* is too powerful and even at −78° reduced methyl hexanoate in tetrahydrofuran not only to hexanal (49%) but also to 1-hexanol (22%) [*1022*]. *Potassium aluminum hydride* in diglyme at −15° produced 54% yield of butanal from ethyl butyrate [*1022*], and *sodium aluminum hydride* in tetrahydrofuran at −45° to −60° afforded 80% yield of 3-phenylpropanal from methyl hydrocinnamate [*1022*]. *Diisobutylaluminum hydride* in ether, hexane or toluene at −70°, and *sodium diisobutylaluminum hydride* in ether at −70° were also used and gave, respectively, 50–88% and 60–80% yields of aldehydes from methyl and ethyl esters of aliphatic as well as aromatic carboxylic acids. *Aromatic esters* require lower temperatures and longer times, and give 10–15% lower yields than the aliphatic ones [*1022, 1023*]. Diisobutylaluminum hydride is frequently used for reduction of esters to aldehydes in syntheses of natural products [*1024*]. Diethyl octafluoroadipate in tetrahydrofuran was reduced quantitatively with *sodium bis(2-methoxyethoxy)aluminum hydride* in toluene at −70° to −50° to the bis-hemiacetal, 2,2,3,3,4,4,5,5-octafluoro-1,6-diethoxy-1,6-hexanediol, which on distillation with phosphorus pentoxide gave a 64% yield of octafluorohexanediol [*68, 1025*]. With less powerful hydrides such as *lithium tris(tert-butoxy)aluminum hydride* or *bis(N-methylpiperazino)aluminum hydride* and similar amino hydrides the reductions were carried out at 0°, at room temperature and even at higher temperatures. The former reagent converted phenyl esters to aldehydes in tetrahydrofuran at 0° in yields of 33–77% [1026]; the latter hydride reduced alkyl esters in tetrahydrofuran at 25° or on refluxing in 56–86% yields [1027].

(145)

$$C_5H_{11}CO_2ME \xrightarrow{\hspace{5cm}} C_5H_{11}CHO$$

[1022] LiAlH$_4$,THF,-78°	49%*
[1022] NaAlH$_4$,THF,-60° to -45°,3 hrs	85%
[1023] NaAlH$_2$(CH$_2$CHMe$_2$)$_2$,Et$_2$O,-70°	72%
[1023] AlH(CH$_2$CHMe$_2$)$_2$,PhMe or C$_6$H$_{14}$,0.5-1 hr	85%
[1027] AlH(N̄ NMe)$_2$,THF,reflux 6 hrs**	78%

*Hexanol was formed in 22% yield. **Ethyl ester was reduced.

Reduction of lactones leads to cyclic hemiacetals of aldehydes. With a sto-ichiometric amount of *lithium aluminum hydride* in tetrahydrofuran at −10° to −15° and using the inverse technique, γ-valerolactone was converted in 58% yield to 2-hydroxy-5-methyltetrahydrofuran, and α-methyl-δ-caprolac-tone in 64.5–84% yield to 3,6-dimethyl-2-hydroxytetrahydropyran [1028]. Also *diisobutylaluminum hydride* in tetrahydrofuran solutions at subzero tem-peratures afforded high yields of lactols from lactones [1024].

Reduction of lactones to aldehyde-hemiacetals is of the utmost importance in the realm of saccharides. The old method of converting aldonolactones to aldoses by means of sodium amalgam was superseded by the use of *sodium borohydride*, which usually gives good yields, is easy to handle and is very suitable because of its solubility in water, the best solvent for sugars [1029]. However, the yields of sodium borohydride reductions are not always high [1030], and yields of *sodium amalgam* reductions can be greatly improved if the reductions are carried out at pH 3–3.5 with 2.5–3 g-atoms of sodium in the form of a 2.5% sodium amalgam with particle size of 4–8 mesh. Yields of 50–84% of aldoses were thus obtained [1031].

The **reduction of esters to ethers** is of limited practical use. Good results were obtained only with esters of tertiary alcohols using 2 mol of *borane* generated from sodium borohydride and an enormous excess (30 mol) of boron trifluoride etherate in tetrahydrofuran and diglyme. Butyl, *sec*-butyl and *tert*-butyl cholanates were converted to the corresponding ethers, 24-butoxy-, *sec*-butoxy- and *tert*-butoxy-5β-cholanes, in respective yields of 7%, 41% and 76%, and *tert*-butyl 5α-pregnane-20-S-carboxylate to 20-S-*tert*-butoxymethylene-5α-pregnane in 76% yield [1032].

[1032]   (146)

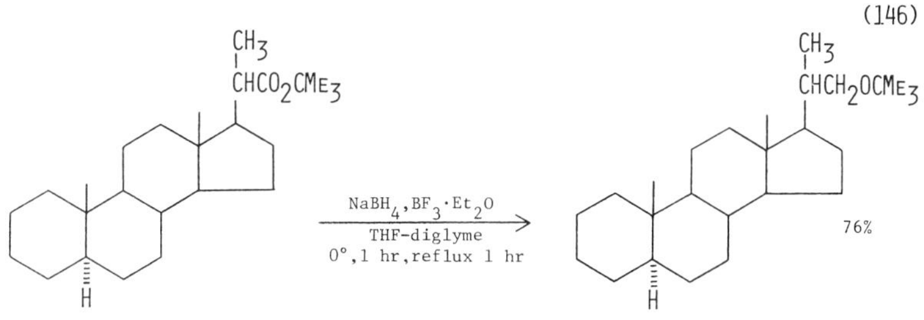

Reduction of esters by *trichlorosilane* in tetrahydrofuran in the presence of *tert*-butyl peroxide and under ultraviolet irradiation gave predominantly ethers from esters of primary alcohols, while esters of tertiary alcohols were cleaved to acids and hydrocarbons. Esters of secondary alcohols gave mixtures of ethers and acids/hydrocarbons in varying ratios. 1-Adamantyl trimethylacetate, for example, afforded 50–100% yields of mixtures containing 2–42% of 1-adamantyl neopentyl ether and 58–98% of adamantane and trimethylacetic acid [1033].

A more useful way of reducing esters to ethers is a two-step procedure applied to the **reduction of lactones to cyclic ethers.** First the lactone is treated with *diisobutylaluminum hydride* in toluene at −78°, and the product – a lactol – is subjected to the action of triethylsilane and boron trifluoride etherate at −20° to −70°. γ-Phenyl-γ-butyrolactone was thus transformed to 2-phenyltetrahydrofuran in 75% yield, and δ-lactone of 3-methyl-5-phenyl-5-hydroxy-2-pentenoic acid to 4-methyl-2-phenyl-2,3-dihydropyran in 72% yield [1034].

Cleavage of esters to acids and hydrocarbons mentioned above was achieved not only with hydrides but also by *catalytic hydrogenation* and reduction with metals. For example the acetate of mandelic acid was converted to mandelic acid and acetic acid by hydrogenation at 20° and 1 atm over palladium on barium sulfate in ethanol in the presence of triethylamine in 10 minutes [1035], and α,α-diphenylphthalide was reduced by refluxing for 5 hours with *zinc* in formic acid to α,α-diphenyl-*o*-toluic acid in 92% yield [1036]. Such reductions are of immense importance in esters of benzyl-type alcohols where the yields of the acids are almost quantitative.

The most common hydrogenolysis of benzyl esters is accomplished by *catalytic hydrogenation over palladium catalysts.* This reaction is very useful since it removes benzylic protective groups under very gentle conditions (room temperature, atmospheric pressure) without using media which could damage sensitive functions in the molecule. Hydrogenolysis of benzyl esters is very easy and takes preference to saturation of double bonds, to reduction of carbonyls in amides, and takes place even in the presence of divalent sulfur in the molecule [1037]. Dibenzyl 4-methoxycarbonyl-3-indolylmethylacetamidomalonate over palladium on charcoal in methanol at 20° and 1 atm was debenzylated to the free dicarboxylic acid in 1.5 hours in 75% yield [1038]. Dibenzyl esters of acylmalonic acids are hydrogenolyzed over 10% palladium on carbon or strontium carbonate at 30°. Under these conditions double decarboxylation takes place, giving ketones in 60–91% yields [1039].

[1038]   (147)

$$\text{indolyl-}CO_2Me,\ CH_2C(CO_2CH_2Ph)_2,\ NHAc \xrightarrow[20°,1\ atm,1.5\ hr]{H_2/Pd(C),MeOH} \text{indolyl-}CO_2Me,\ CH_2C(CO_2H)_2,\ NHAc\quad 75\%$$

[1039] (148)

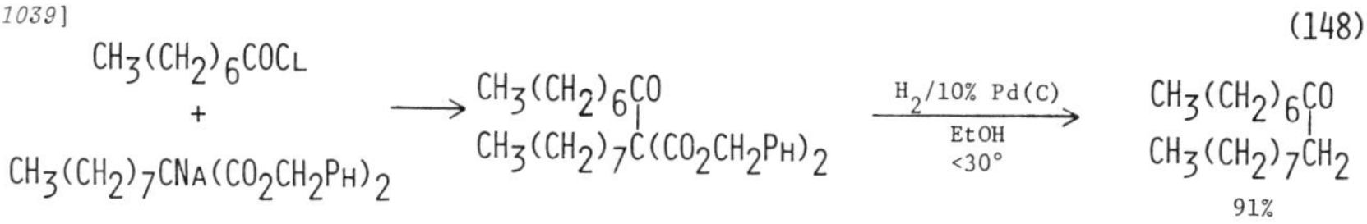

$$CH_3(CH_2)_6COCL$$
$$+$$
$$CH_3(CH_2)_7CNa(CO_2CH_2PH)_2$$
$$\longrightarrow$$
$$CH_3(CH_2)_6CO$$
$$CH_3(CH_2)_7C(CO_2CH_2PH)_2$$
$$\xrightarrow[<30°]{H_2/10\% \ Pd(C) \atop EtOH}$$
$$CH_3(CH_2)_6CO$$
$$CH_3(CH_2)_7CH_2$$
91%

Catalytic hydrogenolysis is frequently used for **cleavage of benzyl alkyl carbonates or carbamates**. The free-acid resulting from the hydrogenolysis – alkylcarbonic or alkylcarbamic acid – spontaneously decarboxylates to a hydroxy or amino compound. Many examples are found in the field of amino acids where benzyloxycarbonyl and *p*-nitrobenzyloxycarbonyl groups temporarily protect amino groups [1040, 1041, 1042, 1043, 1044]. Hydrogenolysis is very easy and takes place in preference to saturation of double bonds [728], prior to reduction of the nitro group [1042, 1044], and even in the presence of divalent sulfur in the molecule [599, 728]. However, the azido group is hydrogenated preferentially to the hydrogenolysis of the benzyl ester [599]. Cyclohexene and cyclohexadiene can be used as hydrogen donors in the catalytic hydrogen transfer over palladium on charcoal [77, 1045] (*Procedure 10*, p. 206).

(149)

$$P_HCH_2OCOOR \xrightarrow{H_2/Pd} P_HCH_3 + HOCOOR \longrightarrow HOR + CO_2$$

$$P_HCH_2OCONHR \xrightarrow{H_2/Pd} P_HCH_3 + HOCONHR \longrightarrow H_2NR + CO_2$$

Cleavage of alkyl–oxygen bonds of other than benzyl-type esters took place in esters of α-keto alcohols. 3β,17α-Diacetoxyallopregnan-20-one gave 3β-acetoxyallopregnan-20-one on refluxing with zinc in acetic acid (yield 89%) [918], and a lactone of *trans*-2,9-diketo-la-carboxymethyl-1-hydroxy-8-methyl-1,1a,2,3,4,4a,-hexahydrofluorene was hydrogenolyzed to the free acid with *chromous chloride* in acetone at room temperature (yield 65%) [1046]. Also the vinylog of an ester of an α-keto alcohol, namely Δ⁴-cholesten-6β-ol-3-one acetate, underwent alkyl–oxygen hydrogenolysis by refluxing with zinc dust in acetic acid and gave cholest-4-en-3-one in 67% yield [1047].

[1046] (150)

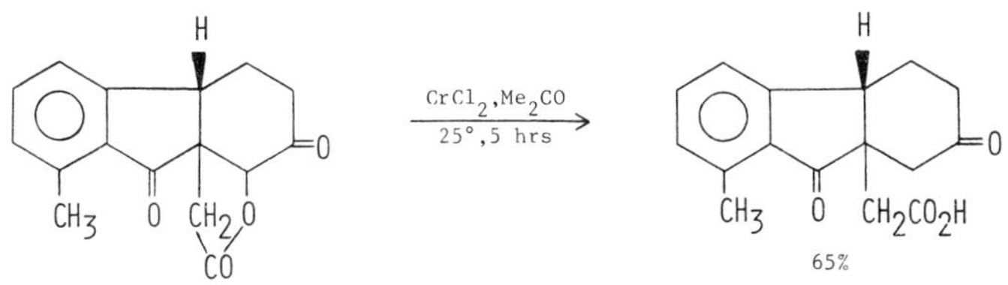

Reactions reminiscent of pinacol reduction take place if esters are treated with sodium in aprotic solvents. The initially formed radical anion dimerizes and ultimately forms an α-hydroxy ketone, an acyloin. Such **'acyloin conden-**

sation' of esters is especially useful with esters of α,ω-dicarboxylic acids of at least six carbons in the chain for cyclic acyloins are formed in very good yields. The best yields are obtained if the reduction is carried out with an exactly stoichiometric amount of sodium (2 g-atoms per molecule of ester) under an inert gas in boiling toluene or xylene which dissolves the reaction intermediates and keeps sodium in a molten state. Open-chain acyloins of 8–18 carbons were thus synthesized in 80–90% yields [*1048*], and cyclic acyloins of 9, 10, 12, 14 and 20 carbons in 30, 43, 65, 47 and 96% yields [*1049, 1050, 1051*] (*Procedure 27*, p. 211).

(151)

$$
\begin{array}{c}
\underset{\substack{| \\ RC=O \\ | \\ OR'}}{\overset{OR'}{\underset{RC=O}{}}}
\xrightarrow{\ 2\ Na\ }
\underset{\substack{| \\ RC-O \\ | \\ OR'}}{\overset{OR'}{\underset{RC-O}{}}} 2\ Na^{\oplus}
\xrightarrow{\ -2\ R'ONa\ }
\underset{\substack{| \\ RC=O}}{\overset{RC=O}{}}
\xrightarrow{\ 2\ Na\ }
\underset{\substack{\| \\ RC-O}}{\overset{RC-O^{\ominus}}{}} 2\ Na^{\oplus}
\end{array}
$$

[*1048*]

$$
\underset{\substack{\| \\ RC-O}}{\overset{RC-O^{\ominus}}{}} 2\ Na^{\oplus}
\xrightarrow[-2\ NaOH]{\ 2\ H_2O\ }
\underset{\substack{\| \\ RC-OH}}{\overset{RC-OH}{}}
\ \rightleftharpoons\
\underset{\substack{| \\ RCH-OH}}{\overset{RC=O}{}}
$$

[*1050*]
$$
\underset{\substack{| \\ (CH_2)_{18} \\ | \\ CO_2ME}}{\overset{CO_2ME}{}}
\xrightarrow[110^{\circ},\,3\ hrs]{\ 4\ Na,\,xylene\ }
(CH_2)_{18}\!\!\overbrace{\underset{CHOH}{\overset{CO}{}}}\quad 96\%
$$

By far the most frequent **reduction of esters** is their conversion **to alcohols.** The reaction is important not only in the laboratory but also on an industrial scale where it is used mainly for hydrogenolysis of fats to fatty alcohols and glycerol. Lactones are reduced to diols.

The reduction of esters to alcohols has an interesting history. The oldest reduction – the *Bouveault–Blanc reaction* [*1052*] – carried out by adding sodium into a solution of an ester in ethanol, did not give very high yields, possibly because of side reactions like Claisen condensation. With some modifications good yields (65–75%) were obtained with esters of monocarboxylic acids [*1053*], dicarboxylic acids (73–75%) [*1054*], and even unsaturated acids containing non-conjugated double bonds (49–51%) [*1055*]. The main improvement was use of an inert solvent such as toluene or xylene and of a secondary alcohol which is acidic enough to decompose sodium-containing intermediates but does not react too rapidly with sodium: the best of all is methylisobutylcarbinol (4-methyl-2-pentanol). Under these conditions the precise theoretical amounts of sodium and the alcohol may be used, which prevents side reactions such as acyloin and Claisen condensations. The reaction is best carried out by adding a mixture of the ester, alcohol and toluene or xylene into a stirred refluxing mixture of sodium and xylene at a rate sufficient to maintain reflux. Yields obtained in this modification are close to theoretical, or at least 80% [*135*]. The amount of sodium and the alcohol is given by the following stoichiometric equation accounting for the individual reaction steps:

[122]   (152)

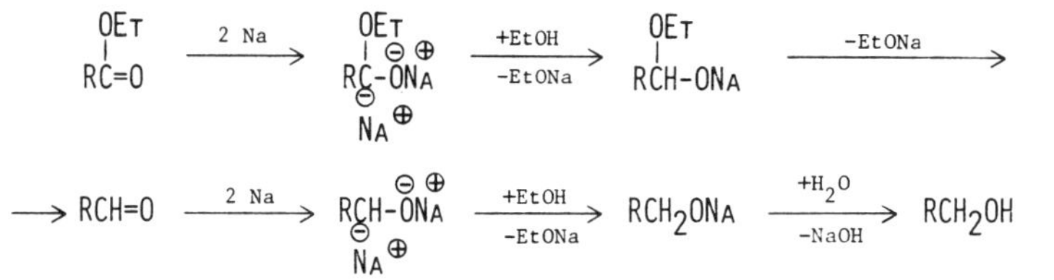

$$RC{=}O \text{ (OET)} \xrightarrow{2 \text{ Na}} R\overset{\ominus}{\underset{\underset{Na^{\oplus}}{\ominus}}{C}}{-}ONa \text{ (OET)} \xrightarrow[-EtONa]{+EtOH} RCH{-}ONa \text{ (OET)} \xrightarrow{-EtONa}$$

$$\longrightarrow RCH{=}O \xrightarrow{2 \text{ Na}} R\overset{\ominus}{\underset{\underset{Na^{\oplus}}{\ominus}}{C}}H{-}ONa \xrightarrow[-EtONa]{+EtOH} RCH_2ONa \xrightarrow[-NaOH]{+H_2O} RCH_2OH$$

$$RCO_2Et + 4 \text{ Na} + 2 \text{ EtOH} + H_2O = RCH_2OH + 3 \text{ EtONa} + \text{NaOH}$$

The reduction as described is applicable to small as well as large scale operations provided safety precautions necessary for handling molten sodium and sodium dispersion are observed.

Catalytic hydrogenation of esters to alcohols is rather difficult, therefore it is not surprising that feasible ways of hydrogenolyzing esters were developed relatively recently. The main breakthrough was the discovery of special catalysts based on oxides of copper, zinc, chromium [50, 51, 52] and other metals [42]. Such catalysts are especially suited for hydrogenolysis of carbon-oxygen bonds. They are fairly resistant to catalytic poisons but require high temperatures (100–300°) and high pressures (140–350 atm) [620, 1056, 1057, 1058]. Yields of the alcohols are high to quantitative [620, 1057]. Side reactions include hydrogenolysis of carbon-oxygen bonds in diols resulting from hydrogenation of hydroxy or keto esters, hydrogenolysis of glycerol to propylene glycol during the hydrogenation of fats to fatty alcohols, and hydrogenolysis of some aromatic alcohols to hydrocarbons or basic heterocycles [428]. Unsaturated esters are usually converted to saturated alcohols on the *copper chromite* catalysts, but unsaturated alcohols are obtained in 37–65% yield if *zinc chromite* is used at 280–300° and 200 atm [52]. Benzene rings resist hydrogenation but pyridine rings are hydrogenated to piperidine rings [620]. Also pyrrole nuclei are partly saturated to pyrrolidines with the con-

(153)

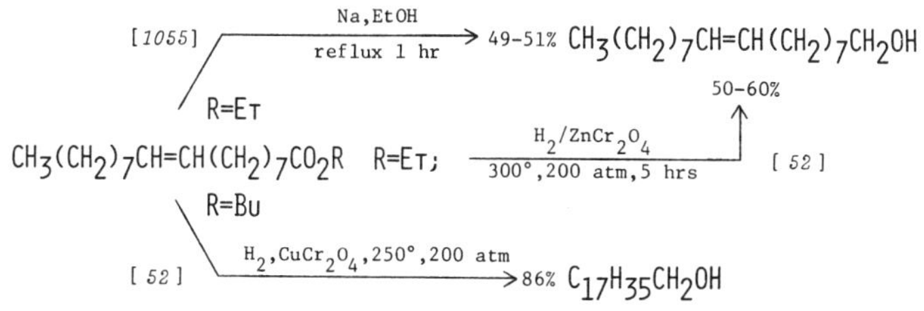

$$[1055] \nearrow \xrightarrow[\text{reflux 1 hr}]{\text{Na,EtOH}} 49\text{--}51\% \ CH_3(CH_2)_7CH{=}CH(CH_2)_7CH_2OH$$

$$CH_3(CH_2)_7CH{=}CH(CH_2)_7CO_2R \quad R{=}Et; \quad \begin{array}{c} 50\text{--}60\% \\ \uparrow \\ \xrightarrow[300°,200 \text{ atm},5 \text{ hrs}]{H_2/ZnCr_2O_4} \ [52] \end{array}$$

$$[52] \searrow \xrightarrow[]{H_2,CuCr_2O_4,250°,200 \text{ atm}} 86\% \ C_{17}H_{35}CH_2OH$$

where R=Et, R=Bu

(154)

[428] $H_2/CuCr_2O_4$
220°, 200–300 atm   55%   8%   25%
18 min

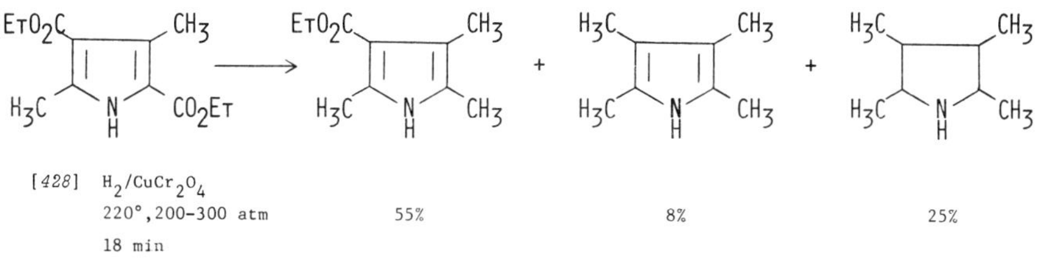

comitant hydrogenolysis of the intermediate alcohols. Esters of alkyl pyr-rolecarboxylates are converted to alkylpyrroles and alkylpyrrolidines [428].

Hydrogenation temperatures can be lowered down to 125–150° when a large excess of the catalyst is used. Such a modification is especially useful in reductions of hydroxy and keto esters since the hydrogenolysis of diols or triols to alcohols is considerably lower and yields of the products range from 60% to 80%. The amounts of catalysts used in such experiments were up to 1.5 times the weight of the ester under the pressure of 350 atm [65].

[65]   (155)

$$\text{PhCHCH}_2\text{OH} \underset{\substack{\text{EtOH, 0.2 hr}}}{\overset{\substack{\text{H}_2/\text{CuCr}_2\text{O}_4 \quad *}{125°, 350 \text{ atm}}}{\longleftarrow}} \text{PhCHCO}_2\text{Et} \underset{\substack{\text{EtOH, 1.3 hr}}}{\overset{\substack{\text{H}_2/\text{Raney Ni} \quad *}{100°, 350 \text{ atm}}}{\longrightarrow}}$$

PhCHCH$_2$OH   OH 80%

PhCHCO$_2$Et   OH

$$\longrightarrow c\text{-C}_6\text{H}_{11}\text{CHCH}_2\text{OH}$$
OH 32–53%

$$\longrightarrow c\text{-C}_6\text{H}_{11}\text{CH}_2\text{CO}_2\text{Et}$$
16–24%

$$\longrightarrow c\text{-C}_6\text{H}_{11}\text{CH}_2\text{CH}_2\text{OH}$$
12–24%

*1.5 Parts per 1 part of the ester.

With the same excess of catalysts hydrogenations of the esters over Raney nickel could be carried out at temperatures as low as 25–125° at 350 atm with comparable results (80% yields). However, benzene rings were saturated under these conditions [65]. In addition to nickel and copper, zinc and chromium oxides, rhenium obtained by reduction of rhenium heptoxide also catalyzes hydrogenation of esters to alcohols at 150–250° and 167–340 atm in 35–100% yields [42].

Both older methods for the reduction of esters to alcohols, catalytic hydro-genation and reduction with sodium, have given way to reductions with *hydrides* and *complex hydrides* which have revolutionized the laboratory preparation of alcohols from esters.

Countless reductions of esters to alcohols have been accomplished using *lithium aluminum hydride*. One half of a mol of this hydride is needed for reduction of 1 mol of the ester. Ester or its solution in ether is added to a solution of lithium aluminum hydride in ether. The heat of reaction brings the mixture to boiling. The reaction mixture is decomposed by ice-water and acidified with mineral acid to dissolve lithium and aluminum salts. Less frequently sodium hydroxide is used for this purpose. Yields of alcohols are frequently quantitative [83, 1059]. Lactones afford glycols (diols) [576].

Esters are also reduced by *sodium aluminum hydride* (yields 95–97%) [88] and by *lithium trimethoxyaluminum hydride* (2 mol per mol of the ester) [94] but not by *lithium tris(tert-butoxy)aluminum hydride* [96]. Another complex hydride, *sodium bis(2-methoxyethoxy)aluminum hydride*, reduces esters in ben-zene or toluene solutions (1.1–1.2 mol per ester group) at 80° in 15–90 minutes in 66–98% yields [969]. *Magnesium aluminum hydride* (in the form of its tetrakistetrahydrofuranate) reduced methyl benzoate to benzyl alcohol in 58% yield on refluxing for 2 hours in tetrahydrofuran [89].

Alane, formed *in situ* from lithium aluminum hydride and aluminum chloride, reduces esters (1.3 mol per mol of ester) in tetrahydrofuran at 0° in 15–30 minutes in 83–100% yields [*994*]. Because it does not reduce some other functions it is well suited for selective reductions [*993, 994*] (p. 159).

Other reagents used for reduction are *boranes* and complex borohydrides. *Lithium borohydride* whose reducing power lies between that of lithium aluminum hydride and that of sodium borohydride reacts with esters sluggishly and requires refluxing for several hours in ether or tetrahydrofuran (in which it is more soluble) [*750*]. The reduction of esters with lithium borohydride is strongly catalyzed by boranes such as B-methoxy-9-borabicyclo[3.3.1]nonane and some other complex lithium borohydrides such as lithium triethylborohydride and lithium 9-borabicyclo[3.3.1]nonane. Addition of 10 mol% of such hydrides shortens the time necessary for complete reduction of esters in ether or tetrahydrofuran from 8 hours to 0.5–1 hour [*1060*].

Sodium borohydride reduces esters only under special circumstances. An excess of up to 16 mol of sodium borohydride refluxed with 1 mol of the ester in methanol for 1–2 hours produces 72–93% of alcohols from aliphatic, aromatic, heterocyclic and α,β-unsaturated esters. The conjugated double bonds, especially if also conjugated with aromatic rings, are saturated in many instances [*1061*]. In polyethylene glycols reductions of aliphatic, aromatic and α,β-unsaturated esters are accomplished by heating at 65° for 10 hours with 3 mol of sodium borohydride per mol of the ester [*1062*]. Tetrabutylammonium borohydride is unsuitable for the reduction of esters [*771*]. On the other hand, *calcium borohydride* in tetrahydrofuran reduces esters at a rate higher than sodium borohydride and lower than lithium borohydride, and in alcoholic solvents even higher than lithium borohydride. When esters are treated with stoichiometric amounts (0.5 mol) of lithium borohydride in ether or tetrahydrofuran, or calcium borohydride in tetrahydrofuran, and the mixtures after addition of toluene are heated to 100° while the more volatile solvents distill off, 70–96% yields of alcohols are usually obtained in 0.5–2 hours [*1063*]. *Sodium trimethoxyborohydride* was also used for reduction of esters to alcohols at higher temperatures but in poor yields [*99*].

The efficient *lithium triethylborohydride* converts esters to alcohols in 94–100% yields in tetrahydrofuran at 25° in a few minutes when 2 mol of the hydride per mol of the ester are used [*100*].

Borane prepared by adding aluminum chloride to a solution of sodium borohydride in diethylene glycol dimethyl ether (diglyme) reduced aliphatic and aromatic esters to alcohols in quantitative yields in 3 hours at 25° using a 100% excess, or in 1 hour at 75° using a 25% excess of lithium borohydride over 2 mol of the hydride per mol of the ester [*738*] (*Procedure 20*, p. 209).

A solution of borane in tetrahydrofuran reduces esters at room temperature only slowly [*977*]. Under such conditions free carboxylic groups of acids are reduced preferentially: monoethyl ester of adipic acid treated with 1 mol of borane in tetrahydrofuran at −18° to 25° gave 88% yield of ethyl 6-hydroxyhexanoate [*971*]. Borane-dimethyl sulfide in tetrahydrofuran was used for reduction of esters to alcohols on refluxing for 0.5–16 hours and allowing the dimethyl sulfide to distill off (yields 73–97%) [*1064*].

(156)

PhCO$_2$Et $\longrightarrow$ PhCH$_2$OH

[*83*]	LiAlH$_4$,Et$_2$O,reflux	90%
[*969*]	NaAlH$_2$(OCH$_2$CH$_2$OMe)$_2$,C$_6$H$_6$,80°,0.5 hr	91%
[*1063*]	LiBH$_4$,THF,PhMe,100°,1 hr	87%
[*1060*]	LiBH$_4$,MeOB⌂ ,Et$_2$O,reflux 2 hrs	81%
[*1061*]	10 NaBH$_4$,MeOH,reflux 1–2 hrs*	64%
[*1063*]	Ca(BH$_4$)$_2$,THF,PhMe,100°,1.5 hr	87%
[*1064*]	BH$_3$·Me$_2$S,THF,reflux 4 hrs	90%

*Methyl ester.

Other hydrides used for the conversion of esters to alcohols are *magnesium aluminum hydride* in tetrahydrofuran [*89, 577*] and *magnesium bromohydride* prepared by decomposition of ethylmagnesium bromide at 235° for 2.5 hours at 0.5 mm [*1065*]. They do not offer special advantages (the latter giving only 35% yield of benzyl alcohols from ethyl benzoate).

Reduction of Unsaturated Esters

The reduction of unsaturated esters encompasses the reduction of esters containing double and triple bonds, usually in α,β positions, and/or aromatic rings. Such esters may be converted to less saturated or completely saturated esters, to unsaturated or less unsaturated alcohols or to saturated alcohols. Presence of aromatic rings in conjugation with the multiple bonds facilitates saturation of the bonds. Aromatic rings are hydrogenated only after the saturation of the double bonds.

Catalytic hydrogenation of **olefinic esters** under mild conditions affords **saturated esters** whether or not the double bonds are conjugated. Methyl 6-methoxy-1,2,3,4-tetrahydro-1-naphthylidenecrotonate absorbed 1 mol of hydrogen in 8.5 minutes on hydrogenation over Raney nickel in ethanol and acetic acid at room temperature and atmospheric pressure and gave methyl γ-(6-methoxy-1,2,3,4-tetrahydronaphthylidene)butyrate in 96% yield. Here the α,β-conjugated double bond was saturated in preference to the γ,δ bond [*1066*].

Hydrogenation over nickel catalyst at high temperatures and pressures affects aromatic rings. Over Urushibara nickel at 106–150° and 54 atm, ethyl benzoate gave ethyl hexahydrobenzoate in 82% yield [*48*].

Methyl cinnamate was reduced quantitatively to methyl 3-phenylpropanoate by hydrogen over colloidal palladium at room temperature and atmospheric pressure [*1067*]; ethyl cinnamate was reduced to ethyl 3-phenylpropanoate over tris(triphenylphosphine)rhodium chloride in ethanol at 40–60° and 4–7 atm in 93% yield [*56*], and over copper chromite at 150° and 175 atm in 97% yield [*420*]. On the other hand, hydrogenation of ethyl cinnamate over

Raney nickel at 250° and 100–200 atm afforded 88% yield of ethyl 3-cyclo-hexylpropanoate [*1068*], and hydrogenation of the same compound over copper chromite at 250° and 220 atm gave 83% yield of 3-phenylpropanol [*1057*] (p. 158).

Sodium borohydride under normal conditions does not reduce ester groups but reduces (in 59–74% yields) double bonds in dialkyl alkylidenemalonates (and cyanoacetates) where the double bonds are cross-conjugated with two carboxylic groups [*1069, 1070*]. A high yield (80%) was also obtained in the reduction of ethyl atropate (ethyl α-phenylacrylate), where the double bond is conjugated with one carboxyl and one phenyl group [*1070*]. Ethyl acrylate gave, under the same conditions, only a 25% yield (possibly because of polymerization). The reduction was carried out by allowing equimolar quantities of sodium borohydride and the ester in ethanol or isopropyl alcohol to react first at 0–5° for 1 hour and then at room temperature for 1–2 hours [*1070*].

Conjugated double bonds were also reduced in methyl 3-methyl-2-buten-oate with *tributylstannane* on irradiation with ultraviolet light at 70° (yield of methyl isovalerate was 90%) [*1071*], and in diethyl maleate and diethyl fumarate which afforded diethyl succinate in respective yields of 95% and 88% on treatment with *chromous sulfate* in dimethylformamide at room temperature [*974, 1072*].

Unsaturated alcohols can be obtained from unsaturated esters under special precautions. *Catalytic hydrogenation* over zinc chromite converted butyl oleate to oleyl alcohol (9-octadecenol) in 63–65% yield even under very energetic conditions (283–300°, 200 atm) [*52*].

Much more conveniently, even α,β-unsaturated esters can be transformed into α,β-unsaturated alcohols by very careful treatment with *lithium aluminum hydride* [*1073*], *sodium bis(2-methoxyethoxy)aluminum hydride* [*544*] or *diisobutylalane* [*1151*] (*Procedure 18*, p. 208). An excess of the reducing agent must be avoided. Therefore the 'inverse technique' (addition of the hydride to the ester) is used and the reaction is usually carried out at low temperature. In hydrocarbons as solvents the reduction does not proceed further even at elevated temperatures. Methyl cinnamate was converted to cinnamyl alcohol in 73% yield when an equimolar amount of the ester was added to a suspension of lithium aluminum hydride in benzene and the mixture was heated at 59–60° for 14.5 hours [*1073*]. Ethyl cinnamate gave 75.5% yield of cinnamyl alcohol on 'inverse' treatment with 1.1 mol of sodium bis(2-methoxy-ethoxy)aluminum hydride at 15–20° for 45 minutes [*544*].

Complete reduction of unsaturated esters to **saturated alcohols** takes place when the esters are hydrogenated over Raney nickel at 50° and 150–200 atm [*44*] or over copper chromite at temperatures of 250–300° and pressures of 300–350 atm [*52, 1056*] (p. 153). In contrast to most examples in the literature the reduction of ethyl oleate was achieved at atmospheric pressure and 270–280° over copper chromite, giving 80–90% yield of octadecanol [*1074*]. α,β-Unsaturated lactones are reduced to saturated ethers or alcohols, depending on the reaction conditions. The results are demonstrated by the hydrogenation of coumarin, the δ-lactone of o-hydroxy-cis-cinnamic acid [1075].

[1075]   (157)

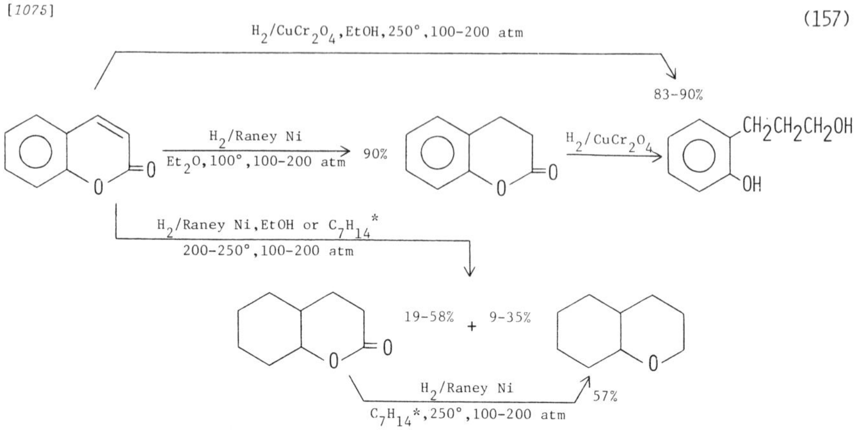

*Methylcyclohexane

α,β-Unsaturated esters, especially those in which the double bond is also conjugated with an aromatic ring, are more comfortably converted to saturated alcohols on treatment with an excess of *lithium aluminum hydride* [1076], with *potassium borohydride* in the presence of lithium chloride [1077], or with *sodium bis(2-methoxyethoxy)aluminum hydride* [544].

Sodium borohydride, even in a very large excess, reduced methyl 2-nonenoate and methyl cinnamate incompletely to mixtures of saturated esters, unsaturated alcohols and saturated alcohols [1061]. On the other hand, α,β-unsaturated esters of pyridine and pyrimidine series were converted predominantly and even exclusively to saturated alcohols. Methyl 3-(γ-pyridyl)acrylate gave, on refluxing for 1–2 hours in methanol with 10 mol of sodium hydride per mol of the ester, 67% of 3-(γ-pyridyl)propanol and 6% of 3-(γ-pyridyl)-2-propenol; methyl 3-(6-pyrimidyl)acrylate gave 77% of pure 3-(6-pyrimidyl)propanol [1061].

Complete reduction of α,β-unsaturated esters to saturated alcohols was also accomplished by *sodium* in alcohol [1078].

(158)

$PhCH{=}CHCO_2R$	$\longrightarrow$	$PhCH_2CH_2CO_2R$	$PhCH{=}CHCH_2OH$	$PhCH_2CH_2CH_2OH$
[56] $^{**}H_2/(Ph_3P)_3RhCl,EtOH,40{-}60°,4{-}7$ atm		93%		
[420] $^{**}H_2/CuCr_2O_4,150°,175$ atm		97%		
[1061] ^{*10} NaBH$_4$,MeOH,reflux 1–2 hrs		29%	38%	33%
[1073] ^{*1} LiAlH$_4$,C$_6$H$_6$,60°,14.5 hrs			73%	
[1077] **KBH$_4$,LiCl,THF,reflux 4–5 hrs			10%	66%
[534] $^{**}1.1$ NaAlH$_2$(OCH$_2$CH$_2$OMe)$_2$,C$_6$H$_6$, 15–20°,45 min			75.5%	
[534] $^{**}2.4$ NaAlH$_2$(OCH$_2$CH$_2$OMe)$_2$,C$_6$H$_6$, 15–20°,2 hrs				44.5%
[1078] **Na,EtOH,reflux				+

*R=Me **R=Et

**Esters containing conjugated triple bonds may be reduced to acetylenic and
ethylenic esters** by controlled *catalytic hydrogenation* (p. 45) and to **acetylenic
or ethylenic alcohols** by *lithium aluminum hydride*. Methyl phenylpropiolate
treated with 0.5 mol of lithium aluminum hydride in ether at −78° for 5
minutes afforded phenylpropargyl alcohol in 96% yield while addition of the
same ester to 0.75 mol of lithium aluminum hydride in ether at 20° gave, after
15 minutes, 83% yield of *trans*-cinnamyl alcohol [*1079*]. Treatment of
dimethyl acetylenedicarboxylate with triphenylphosphine and deuterium oxide
in tetrahydrofuran gave dimethyl dideuterofumarate in 70% yield [*1080*].

Reduction of Substitution Derivatives of Esters

Reductions of halogen, nitro, diazo and azido derivatives of esters resemble
closely those of the corresponding derivatives of carboxylic acids (p. 141).

Catalytic hydrogenation of **halogenated esters** may yield halogen-free esters.
While the reduction of ethyl 2-chloro-4-nitrobenzoate over 5% palladium on
barium sulfate at room temperature and atmospheric pressure in ethyl acetate
gave a quantitative yield of ethyl 2-chloro-4-aminobenzoate after 3 hours,
after 5 hours in isopropyl alcohol 100% of ethyl 4-aminobenzoate was
obtained [*1081*]. Tetraacetate of 6-iodo-6-deoxy-β-glucopyranose, on treatment with hydrogen and Raney nickel in methanol containing triethylamine,
gave 52% of tetraacetate of 6-deoxy-β-glucose [*1082*].

Use of *lithium aluminum hydride* in reducing halogenated esters to halogenated alcohols calls for certain precautions since halogens were hydrogenolyzed occasionally, and not even fluorine was immune [*68, 92*]. More reliable
hydrides which reduce the ester group without endangering the halogen are
alane prepared *in situ* from lithium aluminum hydride and aluminum chloride
[*993*] or sulfuric acid [*994*] and *calcium borohydride* obtained by adding
anhydrous calcium chloride to a mixture of sodium borohydride and tetrahydrofuran at 0° [*92*]. The former reagent reduced ethyl 3-bromopropanoate
to 3-bromopropanol in better yields than lithium aluminum hydride even
when inverse technique was used (72–90% as opposed to 4–52%) [*993*]. Calcium borohydride converted ethyl 2-fluoro-3-hydroxypropanoate to 2-
fluoro-1,3-propanediol in 70% yield while lithium borohydride gave only a
45% yield and lithium aluminum hydride hydrogenolyzed fluorine [*92*].

Reduction of nitro esters to amino esters is easy, since most of the reagents
used for the reduction of a nitro group to an amino group do not attack the
ester group. Ethyl 2-chloro-4-nitrobenzoate was quantitatively *hydrogenated*
to either ethyl 2-chloro-4-aminobenzoate or 4-aminobenzoate over palladium
[*1081*].

Iron and acetic or dilute hydrochloric acid can be safely used for the
reduction of nitro group to an amino group in nitro esters. The problem
arises when a **nitro ester** is to be **reduced to a nitro alcohol.** Nitro groups are
not inert toward the best reagents for the reduction of esters to alcohols,
complex hydrides. However the rate of reduction of a nitro group by lithium

aluminum hydride is considerably lower than that of the ester group. Consequently, when a reduction is carried out at low enough temperatures (− 30 to −60°) both aliphatic and aromatic nitro esters can be reduced to nitro alcohols [1083]. On treatment with a 0.5 mol quantity of *lithium aluminum hydride* in ether at −35° methyl 4-nitropentanoate gave 61% of 4-nitropentanol. Even better yields of nitro alcohols were obtained on treatment of nitro esters with *alane* or *borane*. Alane (2 mol) in tetrahydrofuran converted methyl 4-nitropentanoate to 4-nitropentanol at − 10° in 80% yield [994], and ethyl *p*-nitrobenzoate to *p*-nitrobenzyl alcohol in 68% yield [994]. An even better reagent for this purpose proved to be borane, which was prepared from sodium borohydride and aluminum chloride in diglyme and which reduced ethyl *p*-nitrobenzoate to *p*-nitrobenzyl alcohol at 50–75° in 77% yield [738].

Diazo esters were reduced to ester hydrazones, ester hydrazines, or amino esters. Reduction of ethyl diazoacetate with aqueous *potassium sulfite* at 20–30° gave a derivative of ethyl glyoxylate hydrazone which was hydrolyzed to ethyl glyoxylate [1084]. *Sodium amalgam* reduced the same diazo ester to ethyl hydrazinoacetate in 90% yield [1085]. *Catalytic hydrogenation* over platinum oxide at room temperature converted ethyl α-diazoacetoacetate in ethanolic sulfuric acid to 2-amino-3-hydroxybutanoic acid (yield 39%) by reducing simultaneously the diazo group to an amino group and the keto group to hydroxyl [907].

(159)

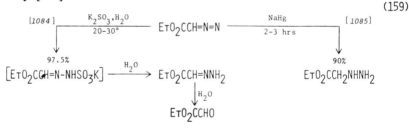

Azido esters (triazo esters) are reduced to amino esters very easily. *tert*-Butyl azidoacetate *hydrogenated* over 5% palladium on carbon in methanol at room temperature and at atmospheric pressure afforded 75–85% yield of glycine *tert*-butyl ester [1086]. Unsaturated azido esters were reduced to unsaturated amino esters by *aluminum amalgam* in ether at room temperature in 71–86% yields [149]. Ethyl α-azidocrotonate gave 75.6% yield of ethyl α-aminocrotonate, and ethyl α-azidocinnamate gave ethyl α-aminocinnamate in 70.9% yield [149].

Reduction of Hydroxy Esters, Amino Esters, Keto Esters, Oximino Esters and Ester Acids

Hydroxy esters are reduced at the ester group by energetic *catalytic hydrogenation* over nickel or copper chromite at high temperatures and high pressures [420, 1087]. As a consequence secondary reactions ensue, such as the hydrogenolysis of hydroxyl groups and even carbon–carbon bond cleavage. Especially easy is the replacement by hydrogen of benzylic hydroxyl. Ethyl mandelate, for example, on hydrogenation over copper chromite at 250° and 100–170 atm gave 70% yield of 2-phenylethanol and 13% of ethylbenzene [1087]. But even nonactivated hydroxyls in aliphatic hydroxy esters are frequently hydrogenolyzed during the catalytic hydrogenation over nickel or copper chromite [420].

More reliable therefore is reduction with *lithium aluminum hydride, lithium borohydride* or *calcium borohydride*, which do not hydrogenolyze hydroxy groups [92].

Reduction of amino esters to amino alcohols was accomplished by hydrogenation over Raney nickel at 50° and 150–200 atm with yields up to 98% [44].

In **keto esters** reduction can affect the **keto group** and convert it to an alcoholic group or to a methylene group, or else it can hydrogenolyze the **ester group,** either to an acid or to an alcohol. The latter reduction may be accompanied by the reduction of the keto group at the same time.

Catalytic hydrogenation transforms keto esters to hydroxy esters under very gentle conditions. In cyclic ketones products of different configuration may result. Ethyl 3,3-dimethylcyclohexanone-2-carboxylate on hydrogenation over platinum oxide in acetic acid gave 96.3% yield of *cis*, and over Raney nickel in methanol gave 97% yield of *trans* ethyl 3,3-dimethyl-cyclohexanol-2-carboxylate, both at room temperature and atmospheric pressure [847].

Different stereoisomeric cyclohexanols resulted also from reduction of ethyl 4-*tert*-butyl-2-methylcyclohexanone-2-carboxylate with *sodium borohydride* (equatorial hydroxyl) or with *aluminum isopropoxide* (axial hydroxyl) [1088].

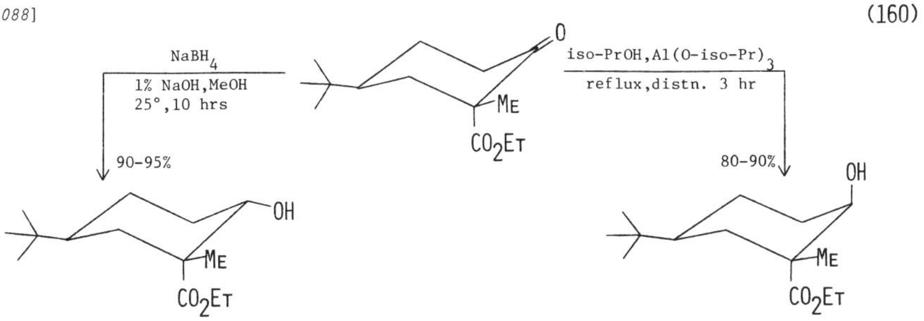

Metal amalgams may be used for reduction of the keto groups in keto esters provided the medium does not cause hydrolysis of the ester. Because of that *aluminum amalgam* in ether is preferable to *sodium amalgam* in aqueous solutions. Diethyl oxalacetate was reduced to diethyl malate by sodium amalgam in 50% yield and with aluminum amalgam in 80% yield [148].

Stereospecific reduction of α- and β-keto esters to optically pure hydroxy esters was achieved by *biochemical reduction* in moderate to good yields. *Saccharomyces cerevisiae* converted methyl 2-keto-2-phenylacetate to methyl

(−)-(R)-2-hydroxy-2-phenylacetate (mandelate) in 59% yield, ethyl 2-keto-2-(α-thienyl)acetate to ethyl (−)-(R)-2-hydroxy-2-(α-thienyl)acetate in 49% yield, and ethyl benzoylacetate to ethyl (−)(S)-3-hydroxy-3-phenylpropanoate in 70% yield. Ethyl acetoacetate gave also optically pure ethyl (+)-(S)-3-hydroxybutanoate (yield 80%) whereas the optical yield of ethyl (+)-lactate from ethyl pyruvate was only 92% [1089].

Reductions of **keto esters to esters** are not very frequent. Both Clemmensen and Wolff–Kizhner reductions can hardly be used. The best way is **desulfurization of thioketals with Raney nickel** (p. 130). Thus ethyl acetoacetate was reduced to ethyl butyrate in 70% yield, methyl benzoylformate (phenylglyoxylate) to methyl phenylacetate in 79% yield, and other keto esters gave equally high yields (74–77%) [823].

Another reagent capable of accomplishing the reduction of keto esters to esters is *sodium cyanoborohydride*, which converted isopropylidene acylmalonates to isopropylidene alkylmalonates in 50–85% yields [1090].

[1090]   (161)

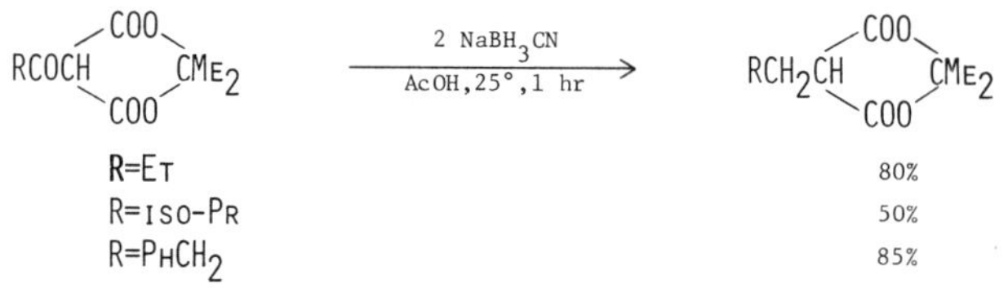

$$RCOCH(COO)(COO)CMe_2 \xrightarrow[\text{AcOH,}25°,\text{1 hr}]{\text{2 NaBH}_3\text{CN}} RCH_2CH(COO)(COO)CMe_2$$

R=Et

R=iso-Pr

R=PhCH_2

80%

50%

85%

In an interesting reaction, β-santonin was reduced with *lithium* in liquid ammonia so that the **lactone was hydrogenolyzed to an acid** and one of the double bonds conjugated with the carbonyl was reduced. The other double bond as well as the keto group did not undergo reduction [1091].

[1091]   (162)

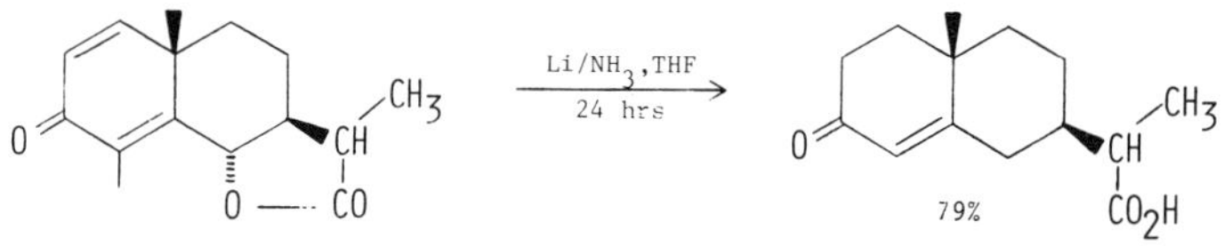

$$\xrightarrow[\text{24 hrs}]{\text{Li/NH}_3\text{,THF}}$$ 79%

The reason why the carbonyl group in β-santonin remained intact may be that, after the reduction of the less hindered double bond, the ketone was enolized by lithium amide and was thus protected from further reduction. Indeed, treatment of ethyl 1-methyl-2-cyclopentanone-1-carboxylate with lithium diisopropylamide in tetrahydrofuran at −78° enolized the ketone and prevented its reduction with lithium aluminum hydride and with diisobutyl-alane (DIBAL®). Reduction by these two reagents in tetrahydrofuran at −78° to −40° or −78° to −20°, respectively, afforded **keto alcohols from several keto esters** in 46–95% yields. Ketones whose enols are unstable failed to give keto alcohols [1092].

Another way of avoiding reduction of the keto group in a keto ester is protection by acetalization. Ketals are evidently not reduced by lithium aluminum hydride in ether at 0°. Consequently, the diethyl ketal of ethyl p-acetylbenzoate (prepared in 81% yield) was reduced with lithium aluminum hydride to diethylketal of p-acetylbenzyl alcohol (yield 93%). Hydrolysis with methanolic hydrochloric acid at 80° recovered the keto group with 95% yield (overall 71.5% based on the starting keto ester) [1093].

Reduction of oximino esters, i.e. oximes of keto esters, is very useful for the preparation of amino esters. Reductions are very selective since the oximes are easily reduced by *catalytic hydrogenation* over 10% palladium on carbon in ethanol (yield 78–82%) [1094], by *aluminum amalgam* in ether (yields 52–87%) [150, 1095], or by *zinc* dust in acetic acid (yield 77–78%). None of these reagents attacks the ester group. The last mentioned reaction gives an N-acetyl derivative [1096].

(163)

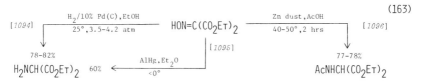

In ethyl 3-keto-2-oximino-3-phenylpropanoate *catalytic hydrogenation* over palladium on carbon reduced both the keto and oximino group, giving a 74% yield of ethyl ester of β-phenylserine (ethyl 2-amino-3-hydroxy-3-phenylpropionate). The reduction is stereospecific and only the *erythro* diastereomer was obtained, probably via a cyclic intermediate [1097]. Similarly, hydrogenation over Raney nickel at 25–30° and 1–3 atm converted ethyl α-oximimoacetoacetate quantitatively to ethyl 2-amino-3-hydroxybutanoate [45].

In dicarboxylic acids with one free and one esterified carboxyl group the free carboxyl only may be reduced with *hydrides* (alanes, boranes). The monoethyl ester or adipic acid was converted by an equimolar amount of *borane* in tetrahydrofuran at $-18°$ to 25° to ethyl 6-hydroxyhexanoate in 88% yield [971] (*Procedure 19*, p. 209).

Reduction with sodium in alcohols, on the other hand, reduces only the ester group and not the free carboxyl.

Reduction of Ortho Esters, Thio Esters and Dithio Esters

Ortho esters were reduced to acetals. Refluxing with 0.25 mol of *lithium aluminum hydride* in ether–benzene solution for 4 hours transformed 3-methylmercaptopropanoic acid trimethyl orthoester to 3-methylmercaptopropionaldehyde dimethylacetal in 97% yield [1098].

Thiol esters can be reduced to sulfides or aldehydes. *Alane* generated from lithium aluminum hydride and boron trifluoride etherate [1099] or aluminum chloride [1100] in ether reduced the carbonyl group to methylene and gave 37–93% yields of sulfides. Phenyl thiolbenzoate failed to give the sulfide, and phenyl thiolacetate gave only an 8% yield of ethyl phenyl sulfide [1100].

(164)

$$[1100] \xrightarrow[\text{Et}_2\text{O,reflux 2 hrs}]{1.4 \text{ LiAlH}_4, 1.2 \text{ AlCl}_3} \text{PHCOSR} \xrightarrow[\text{reflux 6 hrs}]{\text{Raney Ni,70\% EtOH}} [1101]$$

$$\downarrow 79\% \qquad R=\text{ME} \qquad R=\text{ET} \qquad \downarrow 62\%$$

$$\text{PHCH}_2\text{SME} \qquad\qquad\qquad\qquad \text{PHCHO}$$

Desulfurization by refluxing with Raney nickel in 70% ethanol for 6 hours converted thiol esters to aldehydes in 57–73% yields (exceptionally 22% yield) [1101] (*Procedure 6*, p. 205). Desulfurization of a dithioester, methyl dithiophenylacetate, by refluxing with Raney nickel in 80% ethanol for 1 hour afforded 65% yield of ethylbenzene [1102].

REDUCTION OF AMIDES, LACTAMS AND IMIDES

Reduction of amides may yield aldehydes, alcohols or amines. Which of these three classes is formed depends on the structure of the amide, on the reducing agent, and to a certain extent on reaction conditions.

From a simplified scheme of reduction of the amide function it can be seen that the first stage is formation of an intermediate with oxygen and nitrogen atoms linked to an sp^3 carbon. Such compounds tend to regenerate the original sp^2 system by elimination of ammonia or an amine. Thus an aldehyde is formed and may be isolated, or reduced to an alcohol. Alternatively the product is an amine resulting from direct hydrogenolysis of the sp^3 intermediate.

(165)

$$\text{RCONR}'_2 \xrightarrow{2 \text{ H}} \underset{\underset{\text{OH}}{|}}{\text{RCH-NR}'_2} \nearrow \xrightarrow{-\text{HNR}'_2} \text{RCH=O} \xrightarrow{2 \text{ H}} \text{RCH}_2\text{OH}$$

$$\searrow \xrightarrow{2 \text{ H}} \text{RCH}_2\text{NR}'_2 + \text{H}_2\text{O}$$

Reduction of amides to aldehydes was accomplished mainly by complex hydrides. Not every amide is suitable for reduction to aldehyde. Good yields were obtained only with some tertiary amides and *lithium aluminum hydride, lithium triethoxyaluminohydride* or *sodium bis(2-methoxyethoxy)aluminum hydride*. The nature of the substituents on nitrogen plays a key role. Amides derived from aromatic amines such as N-methylaniline [1103] and especially pyrrole, indole and carbazole were found most suitable for the preparation of aldehydes. By adding 0.25 mol of lithium aluminum hydride in ether to 1 mol of the amide in ethereal solution cooled to $-10°$ to $-15°$, 37–60% yields of benzaldehyde were obtained from the benzoyl derivatives of the above heterocycles [1104] and 68% yield from N-methylbenzanilide [1103]. Similarly 4,4,4-trifluorobutanol was prepared in 83% yield by reduction of N-(4,4,4-trifluorobutanoyl)carbazole in ether at $-10°$ [1105].

Even better results than those reached by reduction of the aromatic heterocyclic amides were reported from reductions of *N*-acylethyleneimines with 0.25–0.5mol of lithium aluminum hydride in ether at 0°: butanal, hexanal, 2,2-dimethylpropanal, 2-ethylhexanal and cyclopropanecarboxaldehyde were prepared in 67–88% yields [*1106*].

Better reagents than lithium aluminum hydride alone are its alkoxy derivatives, especially *di-* and *triethoxyaluminohydrides* prepared *in situ* from lithium aluminum hydride and ethanol in ethereal solutions. The best of all, lithium triethoxyaluminohydride, gave higher yields than its trimethoxy and tris(*tert*-butoxy) analogs. When an equimolar quantity of this reagent was added to an ethereal solution of a tertiary amide derived from dimethylamine, diethylamine, *N*-methylaniline, piperidine, pyrrolidine, aziridine or pyrrole, and the mixture was allowed to react at 0° for 1–1.5 hours aldehydes were isolated in 46–92% yields [*95, 1107*]. The reaction proved unsuccessful for the preparation of crotonaldehyde and cinnamaldehyde from the corresponding dimethyl amides [*95*].

(166)

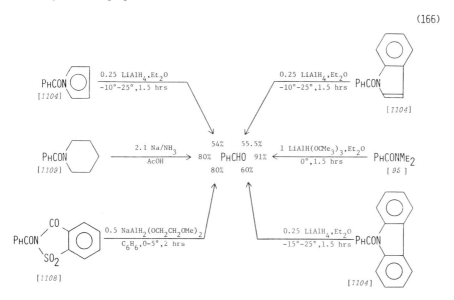

N-Acylsaccharins prepared by treatment of the sodium salt of saccharin with acyl chlorides were reduced by 0.5 molar amounts of *sodium bis(2-methoxyethoxy)aluminum hydride* in benzene at 0–5° to give 63–80% yields of aliphatic, aromatic and unsaturated aldehydes [*1108*]. Fair yields (45–58%) of some aliphatic aldehydes were obtained by *electrolytic reduction* of tertiary and even secondary amides in undivided cells fitted with platinum electrodes and filled with solutions of lithium chloride in methylamine. However, many secondary and especially primary amides gave 51–97% yields of alcohols under the same conditions [*130*].

Some tertiary amides were reduced in low yields by *sodium* in liquid ammonia and ethanol. Better results were reached when acetic acid (ammonium

acetate) was used as a proton source: benzaldehyde was isolated in yields up
to 80% from the reduction of *N*-benzoylpiperidine [*1109*].

Of the methods used for converting amides to aldehydes the one utilizing
lithium triethoxyaluminohydride is most universal, can be applied to many
types of amides and gives highest yields. In this way it parallels other methods
for the preparation of aldehydes from acids or their derivatives (p. 148).

While the reduction of amides to aldehydes competes successfully with
other synthetic routes leading to aldehydes, **reduction of amides to alcohols** is
only exceptionally used for preparative purposes. One such example is the
conversion of trifluoroacetamide to trifluoroethanol in 76.5% yield by *catalytic hydrogenation* over platinum oxide at 90° and 105 atm [*1110*].

Tertiary amides derived from pyrrole, indole and carbazole were hydrogenolyzed to alcohols and amines by refluxing in ether with a 75% excess
(0.88 mol) of *lithium aluminum hydride.* Benzoyl derivatives of the above
heterocycles afforded 80-92.5% yields of benzyl alcohol and 86-90% yields of
the amines [*1104*].

Good to excellent yields (62-90%) of alcohols were reported in reductions
of dimethyl and diethyl amides of benzoic acid and aliphatic acids by *lithium
triethylborohydride* (2.2 mol per mol of the amide) in tetrahydrofuran at room
temperature [*100, 1111*].

(167)

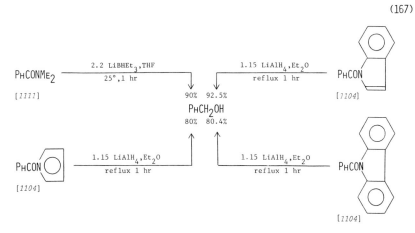

Usually alcohols accompany aldehydes in reductions with lithium aluminum hydride [*1104*] or *sodium bis(2-methoxyethoxy)aluminum hydride*
[*544*], or in hydrogenolytic cleavage of trifluoroacetylated amines [*1112*]. Thus
tert-butyl ester of *N*-(*N*-trifluoroacetylprolyl)leucine was cleaved on treatment with *sodium borohydride* in ethanol to *tert*-butyl ester of *N*-prolylleucine
(92% yield) and trifluoroethanol [*1112*]. During catalytic hydrogenations over
copper chromite, alcohols sometimes accompany amines that are the main
products [*1113*].

The predominant application of **reduction of amides** lies in their **conversion
to amines,** realized most often by hydrides and complex hydrides. Primary,

secondary and tertiary amines can be prepared from the corresponding amides in high yields. But a few other reductive methods are available for the same purpose.

Catalytic hydrogenation of amides does not take place with noble metal catalysts under gentle conditions but is accomplished over copper chromite at 210–250° and 100–300 atm using dioxane as a solvent [1113, 1114]. Because of the high temperatures needed for the reduction the yields of the major products are diminished by side reactions such as formation of secondary amines from primary amines produced from unsubstituted amides, and hydrogenolysis of secondary and tertiary amines. Alcohols are also formed to a small extent, and aromatic hydrocarbons were isolated where the benzylic amines were the primary reduction products [1113]. Salicylamide gave *o*-cresol as the sole product, lauramide afforded 48% yield of dodecylamine and 49% of didodecylamine, and lauranilide gave, in addition to 37% of *N*-dodecylaniline, 29% of aniline, 14% of dodecylamine, 5% of diphenylamine and 2% of didodecylamine [1113]. Complications inherent in the high-temperature catalytic hydrogenation do not occur during reduction of amides to amines by *lithium aluminum hydride* and other hydrides and complex hydrides. Reduction of a tertiary amide requires 0.5 mol of lithium aluminum hydride while secondary and primary amides require 0.75 and 1 mol of the hydride per mole of the amide, respectively. In practice a 25–30% excess over the stoichiometric amount was used for the disubstituted, and 200–250% excess for the monosubstituted amides. Refluxing for more than 1 hour was found necessary, sometimes for up to 34 hours [1104], Yields of amines are around 90% or even higher [1104, 1115, 1116]. Amides insoluble in ether were placed in a Soxhlet apparatus and extracted and flushed into the reaction flask containing the hydride [120].

Isolation of amines is best accomplished by adding a calculated amount of water (2*n* milliliters of water for *n* grams of lithium aluminum hydride followed by *n* milliliters of 15% sodium hydroxide and 5*n* milliliters of water). Other methods of decomposition can result in obstinate emulsions [1104].

High yields of amines have also been obtained by reduction of amides with an excess of *magnesium aluminum hydride* (yield 100%) [577], with *lithium trimethoxyaluminohydride* at 25° (yield 83%) [94] with *sodium bis(2-methoxyethoxy)aluminum hydride* at 80° (yield 84.5%) [544], with *alane* in tetrahydrofuran at 0–25° (isolated yields 46–93%) [994, 1117], with *sodium borohydride* and triethoxyoxonium fluoroborates at room temperature (yields 81–94%) [1121], with *sodium borohydride* in the presence of acetic or trifluoroacetic acid on refluxing (yields 20–92.5%) [1118], with *borane* in tetrahydrofuran on refluxing (isolated yields 79–84%) [1119], with borane-dimethyl sulfide complex (5 mol) in tetrahydrofuran on refluxing (isolated yields 37–89%) [1064], and by *electrolysis* in dilute sulfuric acid at 5° using a lead cathode (yields 63–76%) [1120].

In the reductions with borohydrides and boranes the isolation of products differs from that used in the reductions with lithium aluminum hydride, the

(168)

$$PhCONH_2 \longrightarrow PhCH_2NH_2$$

[577]	Mg(AlH$_4$)$_2$,THF	100%
[994]	2 AlH$_3$,THF,25°	82%
[1118]	5 NaBH$_4$,5 AcOH,dioxane,reflux 2 hrs	76%
[1119]	1.7 BH$_3$,THF,reflux 1 hr	87%
[984]	3 Na/NH$_3$,EtOH	74%

solvent usually being removed by evaporation prior to the final work-up [*1064, 1118, 1119, 1121*].

Sodium and *sodium amalgam* may be used for reduction of amides but the yields of amines are generally very low. Primary aromatic amides (benzamides) were reduced at the carbonyl function with 3.3 equivalents of sodium in liquid ammonia and ethanol while in *tert*-butyl alcohol reduction took place in the aromatic ring giving 1,4-dihydrobenzamides [*984*].

Reduction of lactams to amines resembles closely the reduction of amides except that *catalytic hydrogenation* is much easier and was accomplished even under mild conditions. α-Norlupinone (1-azabicyclo[4.4.0]-2-oxodecane) was converted quantitatively to norlupinane (1-azabicyclo[4.4.0]decane) over platinum oxide in 1.25% aqueous hydrochloric acid at room temperature and atmospheric pressure after 16 hours [*1122*]. Reduction of the same compound by *electrolysis* in 50% sulfuric acid over lead cathode gave 70% yield [*1122*].

Reduction of 5,5-dimethyl-2-pyrrolidone with 3 mol of *lithium aluminum hydride* by refluxing for 8 hours in tetrahydrofuran gave 2,2-dimethylpyrrolidine in 67–79% yields [*1123*]. Reduction of ε-caprolactam was accomplished by heating with *sodium bis(2-methoxyethoxy)aluminum hydride* [*544*], by successive treatment with triethyloxonium fluoroborate and *sodium borohydride* [*1121*], and by refluxing with *borane*–dimethyl sulfide complex [*1064*].

(169)

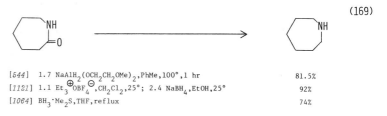

[544]	1.7 NaAlH$_2$(OCH$_2$CH$_2$OMe)$_2$,PhMe,100°,1 hr	81.5%
[1121]	1.1 Et$_3$$^{\oplus}OBF_4$$^{\ominus}$,CH$_2Cl_2$,25°; 2.4 NaBH$_4$,EtOH,25°	92%
[1064]	BH$_3$·Me$_2$S,THF,reflux	74%

α- and β-lactams may suffer ring cleavage and give amino alcohols on treatment with hydrides and complex hydrides.

Imides are reduced to lactams or amines. Succinimide was reduced *electrolytically* to pyrrolidone [*1124*] but was hydrogenolyzed by *sodium borohydride* in water to γ-hydroxybutyramide [*1125*].

N-Methylsuccinimide was converted to *N*-methylpyrrolidone in 92% yield on heating for 18 minutes at 100° with 3 equivalents of *sodium bis(2-methoxyethoxy)aluminum hydride* [*544*], and *N*-phenylsuccinimide was transformed into *N*-phenylpyrrolidone in 67% yield by *electroreduction* on a lead cathode in dilute sulfuric acid [*1120*].

(170)

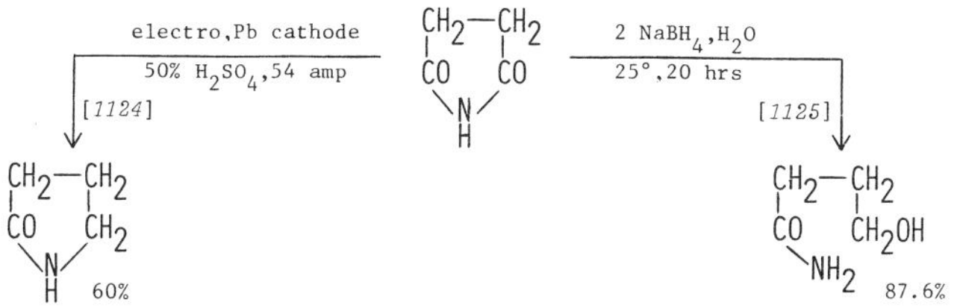

electro,Pb cathode
50% H_2SO_4,54 amp
[1124]

$CH_2{-}CH_2$
CO CO
N
H

2 $NaBH_4$,H_2O
25°,20 hrs
[1125]

$CH_2{-}CH_2$
CO CH_2
N
H 60%

$CH_2{-}CH_2$
CO CH_2OH
NH_2 87.6%

Borane generated from sodium borohydride and boron trifluoride etherate in diglyme reduced *N*-arylsubstituted succinimides to *N*-arylpyrrolidones in 42–72% yields [1126].

Glutarimides disubstituted in α or β positions were hydrogenolyzed by *sodium borohydride* in aqueous methanol mainly to disubstituted γ-hydroxyvaleramides [1125], but *N*-methylglutarimide was reduced *electrolytically* in sulfuric acid to a mixture of *N*-methylpiperidone (15–68%) and *N*-methylpiperidine (7–62%) [1127], and by *lithium aluminum hydride* in ether to *N*-methylpiperidine in 85% yield [1128].

Phthalimide was *hydrogenated catalytically* at 60–80° over palladium on barium sulfate in acetic acid containing an equimolar quantity of sulfuric or perchloric acid to phthalimidine [1129]. The same compound was produced in 76–80% yield by hydrogenation over nickel at 200° and 200–250 atm [43] and in 75% yield over copper chromite at 250° and 190 atm [1130]. Reduction with *lithium aluminum hydride*, on the other hand, reduced both carbonyls and gave isoindoline (yield 5%) [1130], also obtained by *electroreduction* on a lead cathode in sulfuric acid (yield 72%) [1130].

(171)

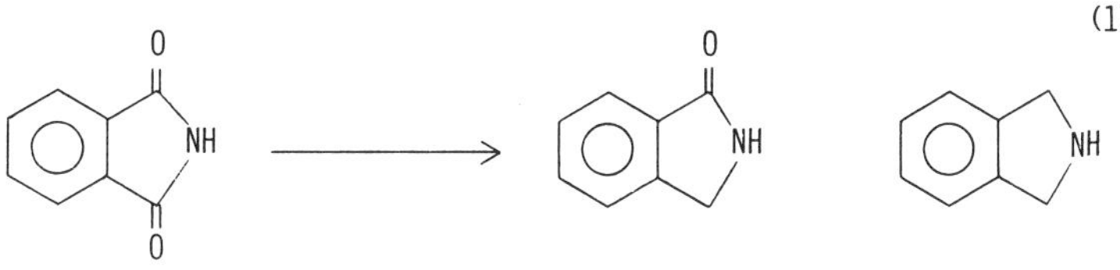

[43] H_2/Ni,C_6H_{11}Me,200°,200–250 atm   76–80%
[1130] H_2/$CuCr_2O_4$,dioxane,250°,190 atm   75%
[1130] electro,Pb cathode,4.5 amp,47°   72%
[1130] $LiAlH_4$,Et_2O 5%

Reduction of Amides and Lactams Containing Double Bonds and Reducible Functional Groups

Because of great differences in rates of hydrogenation, amides and lactams containing double bonds are hydrogenated easily to saturated amides or lactams. α-Acetamidocinnamic acid, on *hydrogenation* over platinum oxide in acetic acid at room temperature and 3 atm, gave 85–86% yield of α-acetamidohydrocinnamic acid [1131], and *N*-methyl-2-pyridone (1-methyl-2-oxo-1,2-dihydropyridine) under similar conditions gave 87.5% yield of *N*-methylpiperidone [435].

Reduction of α,β-unsaturated lactams, 5,6-dihydro-2-pyridones, with *lithium aluminum hydride, lithium alkoxyaluminum hydrides* and *alane* gave the corresponding piperidines. 5-Methyl-5,6-dihydro-2-pyridone (with no substituent on nitrogen) gave on reduction with lithium aluminum hydride in tetrahydrofuran only 9% yield of 2-methylpiperidine, but 1,6-dimethyl-5,6-dihydro-2-pyridone and 6-methyl-1-phenyl-5,6-dihydro-2-pyridone afforded 1,2-dimethylpiperidine and 2-methyl-1-phenylpiperidine in respective yields of 47% and 65% with an excess of lithium aluminum hydride, and 91% and 92% with *alane* generated from lithium aluminum hydride and aluminum chloride in ether. *Lithium mono-, di- and triethoxyaluminum hydrides* also gave satisfactory yields (45–84%) [*1132*].

Reductions of alkyl pyridones with *lithium aluminum hydride* or *alane* are very complex and their results depend on the position of the substituents and on the reducing reagent. Since the pyridones can be viewed as doubly unsaturated lactams with α,β- and γ,δ-conjugated double bonds, the products result from all possible additions of hydride ion: 1,2, 1,4 or 1,6. Consequently the products of reduction are alkylpiperidines and alkylpiperideines with double bonds in 3,4 or 4,5 positions [*449, 1133*].

(172)

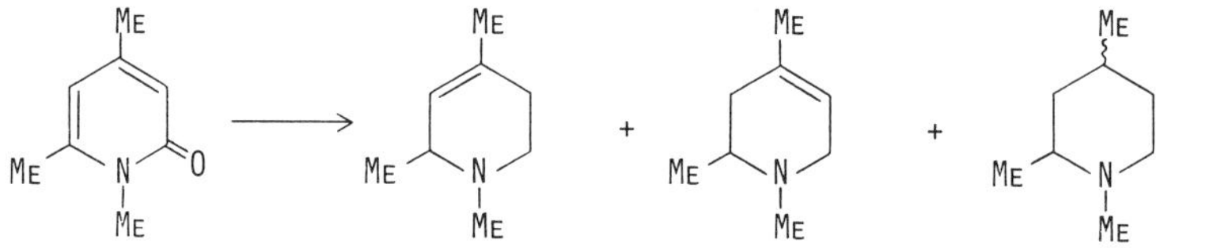

[*1133*] 2 LiAlH₄,THF, reflux 5 hrs	13.5%	2.5%	3.8% cis 5.6% trans
[*1133*] 1 LiAlH₄,1 AlCl₃, Et₂O,reflux 5 hrs	7.7%	11.8%	2% cis

Amides containing nitro groups are reduced to diamino compounds with *alane.* N,N-Dimethyl-p-nitrobenzamide, on reduction with *lithium aluminum hydride in the presence of sulfuric acid* in tetrahydrofuran, gave 98% yield of dimethyl-p-aminobenzylamine [*1117*].

Amides of keto acids were reduced to amides of hydroxy acids biochemically using *Saccharomyces cerevisiae* to give optically pure products [*1089*]. Refluxing with *lithium aluminum hydride* in ether for 6 hours reduced both the ketonic and the amidic carbonyl in N-methyl-5-phenyl-5-oxopentanamide and gave 82% yield of 5-methylamino-1-phenylpentanol [*1134*].

The **amidic group** in methyl N-acetyl-p-aminobenzoate was **reduced preferentially to an ester group** with *borane* in tetrahydrofuran (1.5–1.8 mol per mol of the amide), giving 66% yield of methyl p-N-ethylaminobenzoate. Similarly 1-benzyl-3-methoxycarbonyl-5-pyrrolidone afforded methyl 1-benzyl-3-pyrrolidinecarboxylate in 54% yield and 1,2-diethyl-5-ethoxycarbonyl-3-pyrazolidone gave ethyl 1,2-diethylpyrazolidine-3-carboxylate in 60% yield.

However, methyl hippurate (*N*-benzoylaminoacetate) under comparable conditions gave only 11% of methyl 2-benzylaminoacetate and 85% of 2-benzylaminoethanol [*1135*].

REDUCTIVE AMINATION WITH CARBOXYLIC ACIDS

When a compound containing a primary or secondary amino group (or which can generate such on reduction) is reduced in the presence of acids such as formic acid, acetic acid, trifluoroacetic acid, propionic acid, butyric acid, etc., the product is a **secondary or tertiary amine** containing one alkyl group with the same number of carbons as the acid used. When indole was heated with 4 mol of *sodium borohydride* in acetic acid at 50–60° the product was *N*-ethylindoline resulting from alkylation at the nitrogen atom. The same compound was obtained on treatment of indoline with sodium borohydride in acetic acid (yields 86–88%) [*457*]. Similar treatment of quinoline with 7 mol of sodium borohydride and acetic acid gave 65% yield of 1-ethyl-1,2,3,4-tetrahydroquinoline, and heating at 50–60° of isoquinoline with 7 mol of sodium borohydride and formic acid afforded 78% yield of 2-methyl-1,2,3,4-tetrahydroisoquinoline [*470*].

The reaction has broad applications and a large number of secondary and especially tertiary amines was prepared in isolated yields ranging from 60% to 84% [*1136*]. Although the mechanism of this reaction is not clear it is likely that the key step is reduction of the acid by borane, generated *in situ* from sodium borohydride and the acid, to an aldehyde which reacts with the amine as described in the section on reductive amination (p. 134–136).

[*1136*]   (173)

$$PHCH_2NHME \xrightarrow[\text{3 NaBH}_4, 50-55°]{\text{Me}_2\text{CHCO}_2\text{H}} \left[\begin{array}{c} PHCH_2NME \\ ME_2CHCHOH \end{array} \right] \longrightarrow \begin{array}{c} PHCH_2NME \\ ME_2CHCH_2 \end{array} \ 77\%$$

REDUCTION OF AMIDINES, THIOAMIDES, IMIDOYL CHLORIDES AND HYDRAZIDES

Several derivatives of carboxylic acids yield **aldehydes** on reduction and are frequently prepared just for this purpose.

Benzamidine, *N*-ethylbenzamidine and *N*-phenylbenzamidine afforded benzaldehyde on reduction with *sodium* in liquid ammonia with or without ethanol in yields of 54–100% [*1109*].

Thioamides were converted to aldehydes by cautious **desulfurization** with *Raney nickel* [*1137, 1138*] or by treatment with *iron* and acetic acid [*172*]. More intensive desulfurization with Raney nickel [*1139*], *electroreduction* [*172*], and reduction with *lithium aluminum hydride* [*1138*], with *sodium borohydride* [*1140*] or with *sodium cyanoborohydride* [*1140*] gave *amines* in good to excellent yields.

Imidoyl chlorides substituted at nitrogen (usually with an aryl group) are prepared from anilides and phosphorus pentachloride, and are **reduced to aldehydes via aldimines** (Schiff bases) [285]. Such intermediates were intercepted in the reduction with lithium tris(*tert*-butoxy)aluminum hydride. *N*-Cyclohexylimidoyl chlorides of propanoic, octanoic and benzoic acids were converted to imines on treatment with 120–300% excess of *lithium tris(tert-butoxy)aluminum hydride* in tetrahydrofuran at −78° to 25° (yields 60–85%) [1141]. Reduction of imidoyl chlorides to aldehydes was accomplished by *magnesium* [1142] but with better yields by *stannous chloride (Sonn–Müller reaction)* [182, 1143] and by *chromous chloride* [1144]. The latter reagent proved especially useful in the preparation of α,β-unsaturated aldehydes. Cinnamaldehyde was obtained from the imidoyl chloride of cinnamic acid in 80% yield [1144].

The **hydrazide** of 2,2-diphenyl-3-hydroxypropanoic acid was reduced with *lithium aluminum hydride* in *N*-ethylmorpholine at 100° to 3-amino-2,2-diphenylpropanol in 72.5% yield [1145]. Much more useful is reduction of *N*-arenesulfonylhydrazides* of acids to aldehydes (*McFadyen–Stevens reduction*) [284, 285] based on an alkali-catalyzed thermal decomposition according to Scheme 174.

(174)

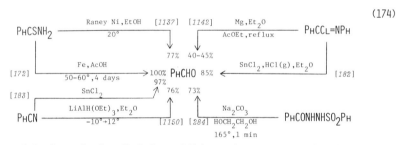

The original method called for addition of an excess of solid sodium carbonate to a solution of a benzenesulfonylhydrazide in ethylene glycol. The optimum temperature of decomposition was found to be around 160° and the optimum time around 1 minute. Yields of aromatic aldehydes ranged from 42% to 87%. The method failed with *p*-nitrobenzaldehyde, cinnamaldehyde and aliphatic aldehydes [284]. Later studies revealed that essential features of the reaction are the presence of a solid material in the reaction mixture and a short time of heating. A modification based on the addition of ground glass to the mixture gave higher yields of some aldehydes. Even *p*-nitrobenzaldehyde was obtained, albeit in only a 15% yield [1146]. The reaction is generally unsuitable for the synthesis of aliphatic aldehydes. Only a few exceptions were recorded such as preparation of cyclopropane carboxaldehyde (yield 16%) [1147], pivalaldehyde (2,2-dimethylpropanal) (yield of 40% after heating for only 30 seconds) [1148] and apocamphane-1-carboxaldehyde (yield 60%) [1148].

Reduction of *p*-toluenesulfonylhydrazides by *complex hydrides* **yields hydrocarbons.** The *N*-tosyl hydrazide of stearic acid gave a 50–60% yield of octadecane on reduction with lithium aluminum hydride [811].

REDUCTION OF NITRILES

Nitriles of carboxylic acids are an important source of primary amines which are produced by many reducing agents. However, a few reagents effect partial reduction to aldehydes [285].

Aromatic nitriles were converted to aldehydes in 50-95% yields on treatment with 1.3-1.7 mol of *sodium triethoxyaluminum hydride* in tetrahydrofuran at 20-65° for 0.5-3.5 hours [1149]. More universal reducing agents are *lithium trialkoxyaluminum hydrides*, which are applicable also to aliphatic nitriles. They are generated *in situ* from lithium aluminum hydride and an excess of ethyl acetate or butanol, respectively, are used in equimolar quantities in ethereal solutions at −10° to 12°, and produce aldehydes in isolated yields ranging from 55% to 84% [1150]. Reduction of nitriles was also accomplished with *diisobutylalane* but in a very low yield [1151].

The complex hydride reductions of nitriles to aldehydes compare favorably with the classical *Stephen reduction* which consists of treatment of a nitrile with anhydrous *stannous chloride* and gaseous hydrogen chloride in ether or diethylene glycol and applies to both aliphatic and aromatic nitriles [183, 285, 1152]. An advantage of the Stephen method is its applicability to polyfunctional compounds containing reducible groups such as carbonyl that is reduced by hydrides but not by stannous chloride [1153].

An alternative method for the conversion of aromatic nitriles to aldehydes is their heating with *Raney nickel and sodium hypophosphite* in water–acetic acid–pyridine (1:1:2) at 40-45° for 1-1.5 hours (yields 40-90%) [1154], or heating with Raney nickel and formic acid at 75-80° for 30 minutes (yields 35-83%) [1155], or even their refluxing for 30 minutes with Raney nickel alloy in 75% aqueous formic acid (yields 44-100%) [1156].

Catalytic hydrogenation **converts nitriles to amines** in good to excellent yields. Since the primary product – primary amines – may react with the starting nitriles, secondary and even tertiary amines are formed. Their amounts depend on the reaction conditions. Elevated temperatures favor formation of secondary and tertiary amines by condensation accompanied by elimination of ammonia. Such side reactions can be suppressed by carrying out the hydrogenations in the presence of ammonia, which affects the equilibrium of the condensation reaction (p. 135). Another way of preventing the formation of secondary and tertiary amines is to carry out the hydrogenation in acylating solvents such as acetic acid or acetic anhydride.

Excellent yields of primary amines were obtained by hydrogenation of nitriles over 5% rhodium on alumina in the presence of ammonia. The reaction was carried out at room temperature and at a pressure of 2-3 atm and was finished in 2 hours, giving 63-92.5% yields of primary amines. Under such conditions no hydrogenolysis of benzyl residues was noticed [1157]. Hydrogenation over Raney nickel at room temperature and atmospheric [1158] or slightly elevated pressures [45] also gave high yields of primary amines, especially in the presence of ammonia [1158]. Comparable results were achieved using nickel boride catalyst [*1159*]. In addition, high yields of primary amines were obtained in hydrogenations over Raney nickel at temperatures of 90-130° and pressures of 30-270 atm, especially in the presence of ammonia [*1160, 1161, 1162*]. Under such conditions Raney nickel gave even better yields of primary amines than rhodium or platinum on charcoal [*1162*]. Hydrogenation over nickel on infusorial earth at 115-150° and 100–175 atm afforded high proportions of secondary amines [*43*].

A reducing agent of choice for conversion of nitriles to primary amines without producing secondary or tertiary amines is *lithium aluminum hydride*. It is used in an equimolar ratio in ether at up to refluxing temperature and gives yields of 40-90% [*121, 576, 787*] (*Procedure 15*, p. 207). Modified lithium aluminum hydrides such as *lithium trimethoxyaluminum hydride* also gave high yields of primary amines [*94*].

Even better yields are obtained with *alane* produced *in situ* from lithium aluminum hydride and 0.5 mol of 100% sulfuric acid in tetrahydrofuran [*994*], or 1 mol of aluminum chloride in ether [*787*] (*Procedure 17*, p. 208). One or 1.3 mol of the alane is used per mole of the nitrile and the reduction is carried out at room temperature. Comparative experiments showed somewhat higher yields of amines than those obtained by lithium aluminum hydride alone [*787*].

(175)

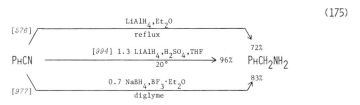

Isolation of the products is accomplished by decomposition with sodium potassium tartrate [*576*], by water and dilute sodium hydroxide [*121*], or by successive treatment with 6 N sulfuric acid followed by alkalization with potassium hydroxide [*787*].

In contrast to lithium aluminum hydride, borohydrides are not suitable for the reduction of nitriles. Lithium triethylborohydride gave erratic results [*100*], sodium trimethoxyborohydride reacted slowly even at higher temperature, and sodium borohydride reduced nitriles to amines only exceptionally. However, in the presence of 0.33 mol of aluminum chloride in diglyme *borane* is produced which gives very good yields of primary amines when used in a 25% excess at temperatures not exceeding 50°. Capronitrile gave a 65% yield of hexylamine and *p*-tolunitrile an 85% yield of *p*-methylbenzylamine [*738*]. Equally good results were obtained with borane made in diglyme from sodium borohydride and boron trifluoride etherate [*977*].

A long time before the hydrides were known, nitriles were reduced to primary amines in good yields by adding *sodium* into solutions of nitriles in boiling alcohols. Seven gram-atoms of sodium and at least 3 mol of butyl alcohol were used per mole of nitrile, giving 86% yield of pentylamine and

78% yield of hexylamine from the corresponding nitriles [*1163*]. The amount of sodium can be cut down to 4 gram-atoms if toluene or xylene is used as a diluent [*135*].

Reduction of Nitriles Containing Double Bonds and Other Reducible Functions

A nitrile group undergoes reductions very readily. It is therefore important to choose selective reagents for the reduction of nitriles containing reducible functions.

A **β,γ-unsaturated nitrile was converted to** the corresponding **unsaturated amine** by *catalytic hydrogenation*. Hydrogenation over Raney cobalt at 60° and 95 atm gave 90% yield of 2-(1-cyclohexenyl)ethylamine from 1-cyclohex-enylacetonitrile. The reduction of the nitrile group preceded even the saturation of the double bond [*1164*]. Reduction with *lithium aluminum hydride* in ether gave a 74% yield of the same product [*1164*]. Catalytic hydrogenation over other catalysts usually reduced the carbon–carbon double bond preferentially. Thus ethyl 2-ethoxycarbonylmethylcyclopentylidenecyanoacetate was reduced in 95% yields to ethyl 2-ethoxycarbonylmethylcyclopentyl-cyanoacetate over platinum oxide (*cis* isomer) and over palladium on alumina (*trans* isomer) [*1165*].

Since sodium borohydride usually does not reduce the nitrile function it may be used for **selective reductions of conjugated double bonds in α,β-un-saturated nitriles** in fair to good yields [*1069, 1070*]. In addition some special reagents were found effective for reducing carbon–carbon double bonds preferentially: *copper hydride* prepared from cuprous bromide and sodium *bis(2-methoxyethoxy)aluminum hydride* [*1166*], *magnesium* in methanol [*1167*], *zinc* and zinc chloride in ethanol or isopropyl alcohol [*1168*], and *triethylam-monium formate* in dimethyl formamide [*317*]. *Lithium aluminum hydride* reduced 1-cyanocyclohexene at −15° to cyclohexanecarboxaldehyde and under normal conditions to aminomethylcyclohexane, both in 60% yields [*117*].

In nitriles containing a nitro group the latter is reduced preferentially by *iron* and by *stannous chloride*. 2-Chloro-6-nitrobenzonitrile was reduced to 2-chloro-6-aminobenzonitrile with iron in methanol and hydrochloric acid in 89% yield [*1169*], and *o*-nitrobenzonitrile to *o*-aminobenzonitrile with stannous chloride and hydrochloric acid in 80% yield [*179*].

Nitriles of keto acids are reduced with *lithium aluminum hydride* at both functions. Benzoyl cyanide afforded 2-amino-1-phenylethanol in 86% yield [*1170*]. Selective reduction of the nitrile group in 87% yield without the reduction of the carbonyl was achieved by *stannous chloride* [*1153*]. **Oximes of keto nitriles** are reduced preferentially at the oximino group by *catalytic hydrogenation* in acetic anhydride over 5% platinum on carbon (yield 85%) [*1171*] or by *aluminum amalgam* (yield 78–82%) [*1172*].

REDUCTION OF ORGANOMETALLIC COMPOUNDS

Reduction of organometallic compounds is usually very easy. Hydrogenolysis of carbon–halogen bonds in organolithium compounds takes place even without a catalyst, just by treatment of alkyllithium compounds with *hydrogen* at room temperature and 0.2–7 atm. The ease of the formation of a hydrocarbon increases in the series $R_2Ca < RLi < RNa < RK < RRb < RCs$ [5]. Organometallic compounds of magnesium, zinc and lead require catalysts (*nickel or copper chromite*) and higher temperatures (160–200°) and pressures (125 atm) [1173].

Complex hydrides were used for reductions of organometallic compounds with good results. Trimethyllead chloride was reduced with *lithium aluminum hydride* in dimethyl ether at −78° to trimethylplumbane in 95% yield [1174], and 2-methoxycyclohexylmercury chloride with *sodium borohydride* in 0.5 N sodium hydroxide to methyl cyclohexyl ether in 86% yield [1175].

Reduction with *sodium dithionite* has been used for conversion of aryl *arsonic* and aryl *stibonic acids* to *arsenobenzenes* [260] and *stibinobenzenes* in good yields [261].

CORRELATION TABLES

The correlation tables on the following pages show how the main types of compounds are reduced to various products by different reducing agents. The tables list *only the reactions contained in this book* (with exceptional cases omitted). If a reagent for a certain conversion is not listed, that does not necessarily mean that it cannot accomplish a particular reduction. Reagents listed in parentheses apply only under special conditions or are used exceptionally.

Table 5   Reduction of alkanes, alkenes, alkadienes and alkynes

Starting compound	Product	Reducing agent	Page
$\geq$C—C$\leq$	$\geq$CH + HC$\leq$	H_2/Pt, H_2/Ni	39
		R_3SiH/CF_3CO_2H	39
`C=C<	>CH—CH<	H_2/Pt, Pd, Rh	6, 39–41
		H_2/Ru, Ni, Raney Ni	6, 39–41
		B_2H_6/$EtCO_2H$	41
		R_3SiH/CF_3CO_2H	41
		Electroreduction	41
		Na, Li	41
		N_2H_2	42
>C=C—C=C<	>CH—C=C—CH<	H_2/Pt, Ni_2B, P 1 Ni	42
		Na, Li, K/NH_3	42, 43
		N_2H_2	42
	>CHCHCHCH<	H_2/Pt	42
—C≡C—	$\overset{\diagdown}{_{H}}$C=C$\overset{\diagup}{_{H}}$	H_2/Pd deact.	6, 43, 44
		H_2/Raney Ni, H_2/Ni_2B	43, 44
		$Et_3N \cdot HCO_2H$/Pd	44
		BH_3,AlH(iso-Bu)$_2$	44
		$BH(CHMeCHMe_2)_2$	44
		MgH_2	44
		Electroreduction (Ni)	45
	$\overset{\diagdown}{_{H}}$C=C$\overset{H}{\diagdown}$	$LiAlH_4$	44
		LiAlHMe(iso-Bu)$_2$	44
		Electroreduction	45
		Na, Li, K/NH_3[a]	45
	—CH$_2$—CH$_2$—	H_2/PtO_2, Pd, Ni	6, 43, 46
		N_2H_2	43

[a] Mixtures of both stereoisomers are obtained from macrocyclic alkynes.

Table 6   Reduction of aromatic hydrocarbons

Starting compound	Product	Reducing agent	Page
		Electroreduction	48
		Na, Li, K, Ca/NH$_3$, ROH	48
		H$_2$/PtO$_2$, Pt, RhO$_2$, Rh	46–48
		H$_2$/Ni, Raney Ni	47
		N$_2$H$_4$ (N$_2$H$_2$)	48, 49
		H$_2$/Pt, Pd, Ni; C$_6$H$_8$/Pd	49
		Electroreduction	49
		LiAlH$_4$	49
		Na/NH$_3$,[a] AlHg[b]	49
		H$_2$/PtO$_2$, RhO$_2$–PtO$_2$, Rh	49
		Raney Ni	
		Na/MeNH$_2$[c]	49
		H$_2$/Raney Ni	49, 50
		Electroreduction	49, 50
		Na/MeOH	49, 50
		Zn	49
		CrSO$_4$	50
		H$_2$/Pd, Re$_2$S$_7$	49, 50
		Electroreduction	49, 50
		Na/EtOH	49, 50
		N$_2$H$_2$	50
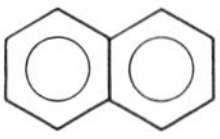	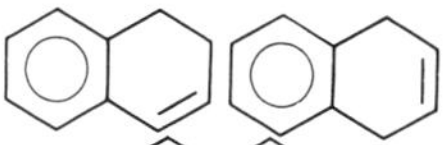	Na/NH$_3$, ROH	50, 51
		H$_2$/CuCr$_2$O$_4$	50
		Na/NH$_3$, ROH, Li, K	50, 51
	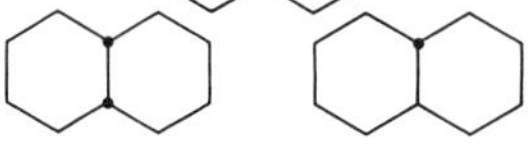	H$_2$/PtO$_2$, Pt, Ni	50
		LiAlH$_4$	51
	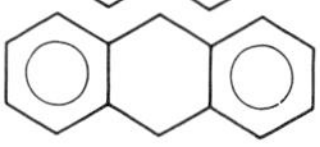	Na/EtOH, Li	51
	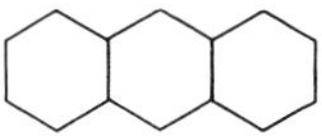	H$_2$/catalysts	51
		H$_2$/CuCr$_2$O$_4$	52, 53
		Na/ROH	52, 53
	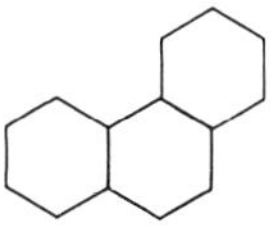	H$_2$/Raney Ni	52, 53

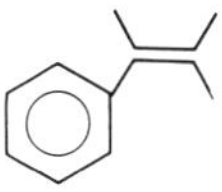

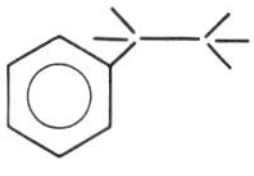

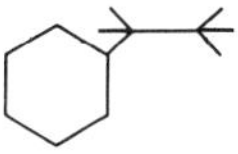

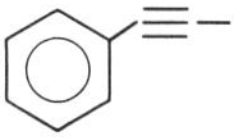

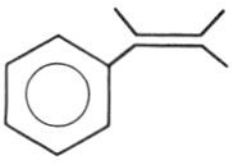

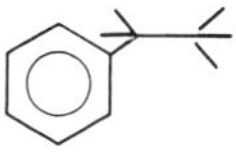

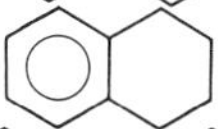

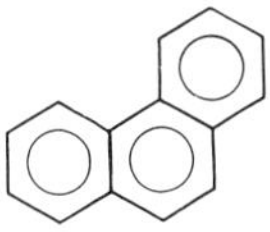

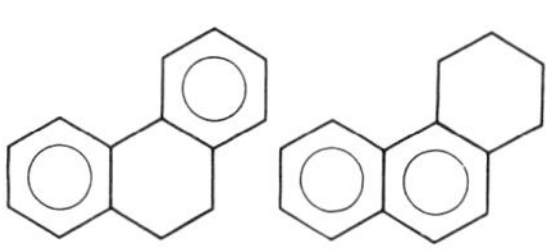

[a] Partial reduction of the ring may occur.
[b] Not general.
[c] Ethyl cyclohexene is also formed.
[d-g] More detailed partial reduction in Schemes 25, 26 and 27 (see pp. 51 and 52 respectively).

Table 7 **Reduction of heterocyclic aromatics**

Starting compound	Product	Reducing agent	Page
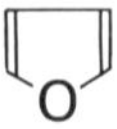 (furan)	 (tetrahydrofuran)	H_2/Pd, Raney Ni, Ni	53
(thiophene)	(tetrahydrothiophene)	H_2/Re$_2$S$_7$, Re$_2$Se$_7$	53
		H_2/CoS$_n$	53
	(2,3-dihydropyrrole)	Zn[a]	54
		PH$_4$I[a]	54
(pyrrole)	(pyrrolidine)	H_2/Pt, Raney Ni, Ni	53, 54
		H_2/CuCr$_2$O$_4$	53, 54
		Zn[a]	54
(pyridine)	(1,2,3,6-tetrahydropyridine)	Complex hydrides[b]	55, 56
		Electroreduction	55, 56
		Na/ROH	55
		(HCO$_2$H)	56
	(piperidine)	H_2/PtO$_2$, Pd, Rh, RuO$_2$	54, 55
		H_2/Raney Ni, CuCr$_2$O$_4$	54, 55
		Complex hydrides[b]	56
		Na/ROH, (HCO$_2$H)	55, 56
(indole)	(indoline)	H_2/PtO$_2$, H_2/CuCr$_2$O$_4$	56, 57
		NaBH$_4$, NaBH$_3$CN	56, 57
		BH$_3 \cdot$ NR$_3$	56, 57
		Na, Li/NH$_3$, Zn	56, 57
	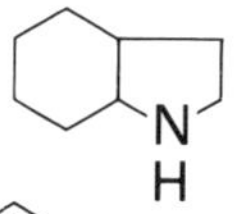 (octahydroindole)	H_2/Raney Ni	57
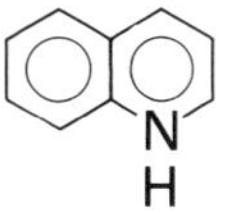 (quinoline)	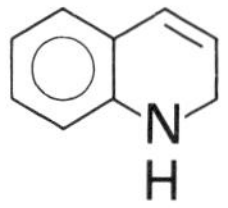 (1,2-dihydroquinoline)	Na/NH$_3$	58
	(1,2,3,4-tetrahydroquinoline)	H_2/Pt, Raney Ni	58
		H_2/Ni, CuCr$_2$O$_4$	58
		BH$_3 \cdot$ NR$_3$, NaBH$_3$CN	58
	(5,6,7,8-tetrahydroquinoline)	H_2/PtO$_2$, CF$_3$CO$_2$H	58
		H_2/Pd, Rh, CF$_3$CO$_2$H	58
	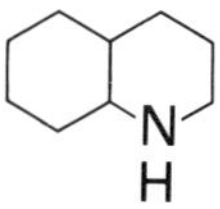 (decahydroquinoline)	H_2/Pt, AcOH	58
(isoquinoline)	(1,2-dihydroisoquinoline)	LiAlH$_4$	58
		NaBH$_3$CN	58
	(1,2,3,4-tetrahydroisoquinoline)	H_2/PtO$_2$, AcOH	58
		Na/NH$_3$, EtOH	58
	(5,6,7,8-tetrahydroisoquinoline)	H_2/PtO$_2$, CF$_3$CO$_2$H	58
	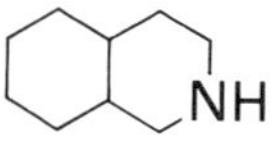 (decahydroisoquinoline)	H_2/PtO$_2$, AcOH, H$_2$SO$_4$	58

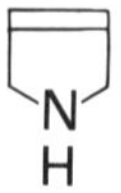

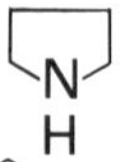

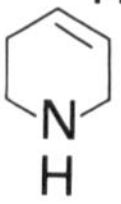

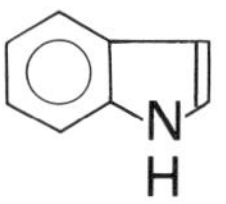

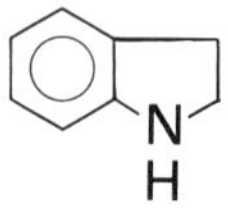

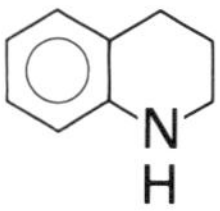

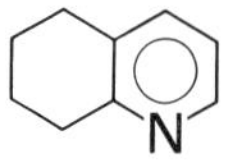

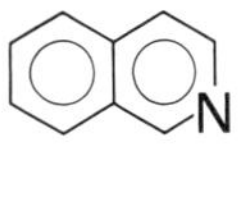

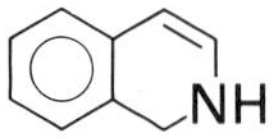

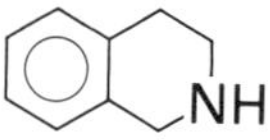

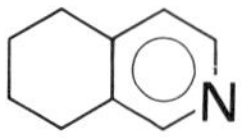

[a] In pyrrole derivatives.

[b] More detailed reduction of pyridine and its derivatives is given in Schemes 31 and 32 (pp. 55, 56).

Table 8 Reduction of halogen compounds

Starting compound (X = F, Cl, Br, I)	Product	Reducing agent	Page
$R\overset{\mid}{\underset{\mid}{C}}{-}X$	$R{-}\overset{\mid}{\underset{\mid}{C}}{-}H$	[a]H_2/Pt, Pd, Raney Ni/KOH	63
		$LiAlH_4$, $NaBH_4$, $NaAlH_4$, R_3SnH	63
		$LiAlH(OR)_3$, $LiBHEt_3$,	63
		$Mg,$[b]$Zn,$[b] $CrSO_4$[b]	64
$R\overset{\mid}{\underset{\mid}{C}}{-}X$ X	$R{-}\overset{\mid}{\underset{\mid}{C}}{-}H$ X	$LiAlH_4$, $NaAlH_2(OCH_2CH_2OMe)_2$	64
		Bu_3SnH, Electroreduction, Na/ROH	64
		AlHg, $CrSO_4$, Na_3AsO_3, Na_2SO_3	64
	$R{-}\overset{\mid}{\underset{\mid}{C}}{-}H$ H	H_2/Pt, Pd, Raney Ni, $LiAlH_4$	64
		$NaAlH_2(OCH_2CH_2OMe)_2$, Bu_3SnH	64
		Electroreduction, Na(ROH), $CrSO_4$	64
$-\overset{\mid}{\underset{\mid}{C}}X{-}\overset{\mid}{\underset{\mid}{C}}X{-}$	$-\overset{\mid}{C}{=}\overset{\mid}{C}{-}$	Mg,Zn	65
		R_3SnH	65
	$-\overset{\mid}{C}H{-}\overset{\mid}{C}H{-}$	H_2/Raney Ni, Li/tert-BuOH	65
$\rangle C{=}\overset{\mid}{C}{-}\overset{\mid}{\underset{\mid}{C}}{-}X$[c]	$\rangle CH{-}\overset{\mid}{C}H{-}\overset{\mid}{\underset{\mid}{C}}{-}X$	H_2/Pt, Pd, Rh	66
	$\rangle C{=}\overset{\mid}{C}{-}\overset{\mid}{\underset{\mid}{C}}{-}H$	$LiAlH_4$, Li/tert-BuOH, $CrSO_4$	66
	$\rangle CH{-}\overset{\mid}{C}H{-}\overset{\mid}{\underset{\mid}{C}}{-}H$	H_2/Raney Ni, KOH	66
$\rangle C{=}C\langle_X$	$\rangle C{=}C\langle_H$	$LiAlH_4$, $NaBH_4$, Li, Na/tert-BuOH	66, 67
		Zn[d]	
	$\rangle CH{-}C\langle^H_X$	H_2/Pt, Pd, Rh	66
	$\rangle CH{-}C\langle^H_H$	H_2/Raney Ni/KOH	66
$Ar{-}\overset{\mid}{\underset{\mid}{C}}{-}X$	$Ar{-}\overset{\mid}{\underset{\mid}{C}}{-}H$	H_2/Pd, KOH, H_2/Raney Ni, KOH	67
		$Et_3N{\cdot}HCO_2H$/Pd, $LiAlH_4$, BH_3	67
$Ar{-}X$	$Ar{-}H$	H_2/Pd, KOH, H_2/Raney Ni, KOH	67
		$LiAlH_4$, $NaAlH_2(OCH_2CH_2OMe)_2$	67, 68
		$NaAlH_4$, $NaBH_4$[d] Et_3SiH	67, 68
		Ph_3SnH, Na/NH_3, Mg, Zn	68

[a]Hydrogenolysis of fluorine and chlorine possible only under exceptionally energetic conditions; practically unfeasible.
[b]Only for X = Br, I.
[c]Similar reagents may be applied to acetylenic compounds.
[d]X = iodine only.

Table 9   Reduction of nitro compounds

Starting compound	Product	Reducing agent	Page
$>$CH$-$NO$_2$	$>$C$=$NOH	SnCl$_2$, TiCl$_3$, CrCl$_2$, Na$_2$S$_2$O$_3$	69, 70
	$>$CHNH$_2$	H$_2$/PtO$_2$, AlHg, Fe	69, 70
$-$CH$=$CHNO$_2$	$-$CH$_2$CH$_2$NO$_2$	H$_2$/Pd, H$_2$/(Ph$_3$P)$_3$RhCl, LiAlH$_4$, NaBH$_4$, Et$_3$N.HCO$_2$H	70, 71 70, 71
	$-$CH$_2$CH$=$NOH	H$_2$/Pd LiAlH$_4$ Fe	71 71 71
	$-$CH$_2$CH$_2$NHOH	LiAlH$_4$	71
	$-$CH$_2$CH$_2$NH$_2$	H$_2$/Pd Fe	71 71
ArNO$_2$	ArNHOH	H$_2$/Pd, Electroreduction, AlHg Zn/NH$_4$Cl	72 72
	ArN(O)$=$NAr	H$_2$/Pd, NaAlH$_2$(OCH$_2$CH$_2$OMe)$_2$ Na$_3$AsO$_3$, Sugars	72 72
	ArN$=$NAr	LiAlH$_4$, Mg(AlH$_4$)$_2$ NaAlH$_2$(OCH$_2$CH$_2$OMe)$_2$ Si/NaOH, Zn/NaOH	72 72 72
	ArNHNHAr	H$_2$/Pd,Zn/KOH	72
	ArNH$_2$	H$_2$/PtO$_2$, RhO$_2$-PtO$_2$,Pd H$_2$/Raney Ni, CuCr$_2$O$_4$, Re$_2$S$_7$ HCO$_2$H/Cu, N$_2$H$_4$/Ni Zn, Fe, SnCl$_2$, TiCl$_3$ FeSO$_4$, (H$_2$S, Na$_2$S$_2$O$_4$)	73 73 73 73 73
	Hydroaromatic amines	H$_2$/PtO$_2$, HCO$_2$H/Ni	74
Hal$-$ArNO$_2$	Hal$-$ArNH$_2$ ArNH$_2$	N$_2$H$_4$/Raney Ni, Fe, Na$_2$S$_2$O$_4$ N$_2$H$_4$/Pd, Zn, SnCl$_2$, NH$_4$SH	75 74

Table 10 Reduction of nitroso, diazonium and azido compounds

Starting compound	Product	Reducing agent	Page
ArNO	ArNHOH	Ph_3P, $(EtO)_3P$	75
	ArN=NAr	$LiAlH_4$	75
	$ArNH_2$	NH_4SH, $Na_2S_2O_4$	75
$\overset{\oplus \ominus}{ArN_2X}$	$ArNHNH_2$	Zn, $SnCl_2$, Na_2SO_3	76
	ArH	$NaBH_4$	75
		Na_2SnO_2	75
		H_3PO_2, $(H_2N)_3P$	75
		EtOH	75, 76
RN_3	RNH_2	H_2/Pt, H_2/Pd	76
		$LiAlH_4$, AlHg	76
		$TiCl_3$, VCl_2	76
		H_2S, $Na_2S_2O_4$	76
		HBr, $HS(CH_2)_3SH$	76
$-\overset{|}{C}=\overset{|}{C}-N_3$	$-\overset{|}{C}=\overset{|}{C}-NH_2$	AlHg, $HS(CH_2)_3SH$	76
$O_2N-\overset{\oplus \ominus}{\bigcirc}-N_2X$	$O_2N-\bigcirc$	$NaBH_4$	75
		Na_2SnO_2,	75
		H_3PO_2, EtOH	75
$O_2N-\bigcirc-N_3$	$O_2N-\bigcirc-NH_2$	H_2S, HBr	76
		$HS(CH_2)_3SH$	76
	 $H_2N-\bigcirc-NH_2$	H_2/Catalysts	
$-\overset{|}{C}-\overset{|}{C}-$	$>C=C<$	$LiAlH_4$	76
		$NaBH_4$	76
	$>\overset{C-C}{\underset{N}{}}<$	BH_3	76
Hal N_3	H	$LiAlHCl_3$	76

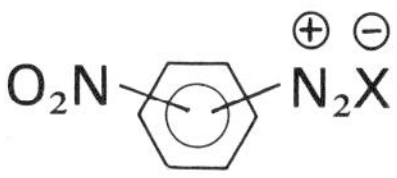

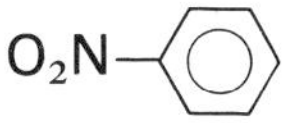

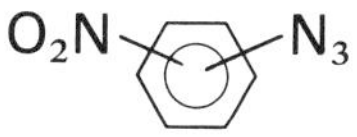

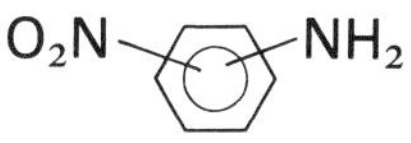

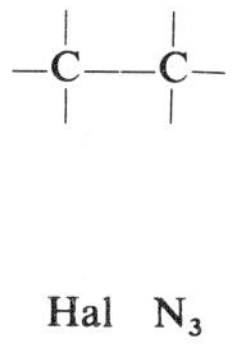

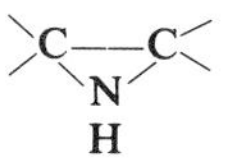

Table 11 Reduction of alcohols and phenols

Starting compound	Product	Reducing agent	Page
ROH	RH	H_2/V_2O_5, MoS_2, WS_2	77
		$LiAlH_4/AlCl_3$	77
		$R_3SiH/CF_3CO_2H^a$	77
		$+RN=C=NR$, H_2/Pd	77
RC—C—R (HO OH)	RCH—CHR (OH)	$H_2/CuCr_2O_4$	77
	RC=CR	$TiCl_3/K$, $(RO)_3P$	77
$>$C=C—C—OH	$>$CH—CH—C—OH	H_2/catalysts, $Na/NH_3{}^b$, $Li/EtNH_2{}^b$	77, 78
	$>$C=C—C—H	$LiAlH_4/AlCl_3$, Na/NH_3, Ph_3PI_2, Ph_3PHI,	77, 78
		$C_5H_5N{\cdot}SO_3 + LiAlH_4$	78
—C≡C—C—OH	$>$C=C—C—OH	H_2/Pd deact.	78
		$LiAlH_4$	78
		Zn, $CrSO_4$	
	—CH_2—CH_2—C—OH	H_2/catalyst	78
ArOH	Hydroaromatic $>$CHOH	H_2/PtO_2, RhO_2-PtO_2, Raney Ni	80
	ArH	$+RN=C=NR$, H_2/Pt;	78, 79
		$+BrCN+Et_2NH$, H_2/Pd	
		$LiAlH_4$ distn, Zn distn	79
		distn HI/AcOH^c	79
Ar—C—OH	Dihydro + tetrahydro^d cyclic alcohols	H_2/PtO_2, Raney Ni, $CuCr_2O_4$	80
	Ar—C—H	H_2/Pd, Raney Ni, Ni, $CuCr_2O_4$	79
		$AlHCl_2$	79
		$NaBH_4$, BH_3, Zn, HI	79
	Ar—C—C—Ar	$TiCl_3/LiAlH_4$	79
ArC=C—C—OH	ArCH—CH—C—OH	H_2/PtO_2, RhO_2—PtO_2, PdO,	80
		$LiAlH_4$	80
	ArC=C—C—H	$H_2/PtO_2/AcOH$	80
$>$C—C—OH (Hal)	$>$CH—C—OH	H_2/Raney Ni	81
	$>$C=C$<$	$TiCl_3/LiAlH_4{}^e$	81
$>$C—C—OH (NO_2)	$>$C—C—OH (NH_2)	H_2/Pt; Fe, Sn, Na_2S, $Na_2S_2O_4$	81

^{a}Only tertiary alcohols.
^{b}Truly exceptional; only in non-allylic unsaturated alcohols.
^{c}Only with naphthols and phenanthrols.
^{d}Only with naphthols, tetrahydro derivatives.
^{e}Bromohydrins only.

Table 12   Reduction of ethers

Starting compound	Product	Reducing agent	Page
 (epoxide, O on ring)	 H OH (diol/alcohol)	H_2/Raney Ni H_2/CuCr$_2$O$_4$ (LiAlH$_4$/AlCl$_3$)	81 81 81
$>C=\overset{\mid}{C}-OR$	$>C=\overset{\mid}{C}H$	Na, Li, K/NH$_3$	82
$>C=\overset{\mid}{C}-\overset{\mid}{C}-OR$	$>C=\overset{\mid}{C}-\overset{\mid}{C}-H$	Li/(CH$_2$NH$_2$)$_2$	82
ArOR	ArH	Na/EtOH	82
	Dihydro and tetrahydro[a] ethers	Li, Na/NH$_3$, ROH Na/EtOH	82 82
$Ar-\overset{\mid}{\underset{\mid}{C}}-OR$	$Ar-\overset{\mid}{\underset{\mid}{C}}-H$	H_2/Pd c-C$_6$H$_{10}$/Pd Na/NH$_3$ Na/BuOH	82 82 82 82

[a]In the naphthalene series.

Table 13   Reduction of epoxides, peroxides and ozonides

Starting compound	Product	Reducing agent	Page
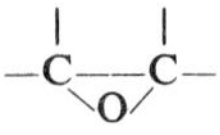 $-\overset{\mid}{C}\underset{O}{\diagdown}\overset{\mid}{C}-$	$-\overset{\mid}{C}=\overset{\mid}{C}-$	Zn, Zn/Cu TiCl$_3$/LiAlH$_4$, CrCl$_2$, FeCl$_3$/BuLi, (EtO)$_2$PONa/Te	83 83, 84 83 83, 84 83
	$-CH-\overset{\mid}{C}-OH$	H$_2$/Pt, LiAlH$_4$, LiAlH$_4$/AlCl$_3$, BH$_3$·BF$_3$, Li/(CH$_2$NH$_2$)$_2$	83, 84 83, 84 83 84
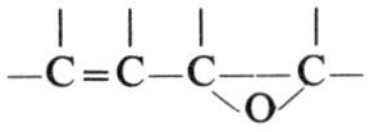 $-\overset{\mid}{C}=\overset{\mid}{C}-\overset{\mid}{C}\underset{O}{\diagdown}\overset{\mid}{C}-$	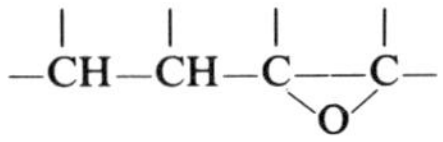 $-CH-CH-\overset{\mid}{C}\underset{O}{\diagdown}\overset{\mid}{C}-$	H$_2$/Catalysts	84
	$-CH-\overset{\mid}{C}=\overset{\mid}{C}-\overset{\mid}{C}-OH$	BH$_3$·THF	84
$-\overset{\mid}{C}-OOH$	$-\overset{\mid}{C}-OH$	H$_2$/PtO$_2$, Pd, Raney Ni, Zn, Na$_2$SO$_3$, R$_3$P, (RO)$_3$P	84, 85 84, 85 85
$\diagup\!\!\!\!\diagdown C=\overset{\mid}{C}-\overset{\mid}{C}-OOH$	$\diagup\!\!\!\!\diagdown C=\overset{\mid}{C}-\overset{\mid}{C}-OH$	H$_2$/Pd, Na$_2$SO$_3$, Ph$_3$P	84, 85
	$-CH-CH-\overset{\mid}{C}-OH$	H$_2$/Pd	84, 85
$-\overset{\mid}{C}-O-O-\overset{\mid}{C}-$	$-\overset{\mid}{C}-O-\overset{\mid}{C}-$	Et$_3$P, Ph$_3$P	85
$\diagup\!\!\!\!\diagdown C=\overset{\mid}{C}-\overset{\mid}{C}-OO-\overset{\mid}{C}-$	$\diagup\!\!\!\!\diagdown CH-CH-\overset{\mid}{C}-OO-\overset{\mid}{C}-$	H$_2$/Pd	85
	$\diagup\!\!\!\!\diagdown C=\overset{\mid}{C}-\overset{\mid}{C}-OH$	H$_2$/Pd	85
	$\diagup\!\!\!\!\diagdown CH-CH-\overset{\mid}{C}-OH$	H$_2$/Pd	85
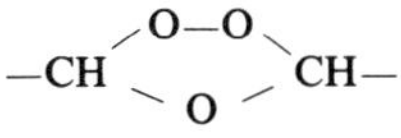 $-CH\overset{\diagup O-O\diagdown}{\underset{O}{\diagdown\!\!\!\diagup}}CH-$	$-CHO+OHC-$	H$_2$/Pd Zn/AcOH Me$_2$S	85 85, 86 86

Table 14   Reduction of sulfur compounds (except sulfur derivatives of aldehydes, ketones and acids)

Starting compound	Product	Reducing agent	Page
RSH	RH	Raney Ni, Ni_2B, Fe, $(EtO)_3P$	86
RSR	RH	Raney Ni, Ni_2B, $LiAlH_4/TiCl_3$ Na/NH_3[a] $(EtO)_3P$	86, 87 86, 87
RSSR	RH RSH RSR	Raney Ni, Ni_2B $LiAlH_4$, $LiBHEt_3$, $NaBH_4$ $(Et_2N)_3P$	86, 87 87 87
RSOR	RH RSR	AlHg H_2/Pd, $SnCl_2$, $TiCl_3$, $MoCl_3$ VCl_2, $CrCl_2$, $NaI/(CF_3CO)_2O$	88 88 88
RSO_2R'	RH RSR' RSO_2H	Raney Ni, $LiAlH_4$, $AlH(iso\text{-}Bu)_2$, $Zn/AcOH$ $LiAlH_4$, Na/NH_3, NaHg, AlHg	88 88, 89 89
RSCN	RSH	$LiAlH_4$	87
RSCl	RSH	$LiAlH_4$	89
RSO_2H	RSSR	$LiAlH_4$, NaH_2PO_2[b], $EtOPH_2(O)$[b]	89
RSOCl	RSSR	$LiAlH_4$	89
RSO_2Cl	RSH RSO_2H	$LiAlH_4$, $HI(P+I_2)$ $LiAlH_4/-20°$, Zn, Na_2S, Na_2SO_3	90 90
RSO_2OR'	RH RSH RSO_2H R'H R'OH	$H_2/Raney$ Ni, $C_{10}H_8 \cdot Na$ $LiAlH_4$ $H_2/Raney$ Ni, $C_{10}H_8 \cdot Na$ $H_2/Raney$ Ni, $LiAlH_4$, $LiBHEt_3$ $NaBH_4$ $H_2/Raney$ Ni, $LiAlH_4$, $C_{10}H_8 \cdot Na$, NaHg, NaI/Zn	90, 91 90, 91 90, 91 90, 91 90, 91 90, 91
RSO_2NHR'	RH RSH RSO_2H $R'NH_2$	$C_{10}H_8 \cdot Na$, Na/ROH, Zn, $SnCl_2$ HI/PH_4I $C_{10}H_8 \cdot Na$, Na/ROH $C_{10}H_8 \cdot Na$, Na/ROH, NaHg, Zn, $SnCl_2$, HI/PH_4I	91, 92 91, 92 91, 92 91, 92

[a] Benzyl sulfides are not desulfurized but reduced to mercaptans.
[b] From alkali salts.

Table 15   Reduction of amines and their derivatives

Starting compound	Product	Reducing agent	Page
RNH_2	RH	H_2/Ni^a, H_2NOSO_3H	92
$RCH=CHCH_2NH_2$	$RCH_2CH_2CH_2NH_2$	H_2/catalysts, Na/NH_3, MeOH	92
$RCH=CHNR'_2$	$RCH_2CH_2NH_2$	$AlHCl_2$, $NaBH_4$, $NaBH_3CN$, HCO_2H	92 92
	$RCH=CH_2 + NHR'_2$	AlH_3	92
$ArNH_2$, ArNHR, ArNRR'	Dihydro and tetrahydro aromatic amines	Na/NH_3, ROH, Na/ROH	93
	Hexahydro amines	H_2/PtO_2, H_2/Ni, $Li/EtNH_2$	92, 93
$ArCH_2NR_2$	$ArCH_3$	H_2/PdO, H_2/Pd	93
$ArCH_2\overset{\oplus}{N}R_3\overset{\ominus}{X}$	$ArCH_3$	NaHg	93
$Ar\overset{\oplus}{N}R_3\overset{\ominus}{X}$	ArH	Electroreduction	93
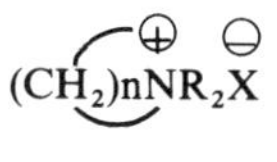 $(CH_2)n\overset{\oplus}{N}R_2\overset{\ominus}{X}$	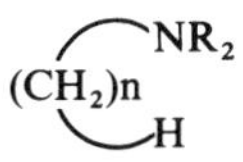 $(CH_2)n\overset{NR_2}{\underset{H}{}}$	H_2/Raney Ni, $NaAlH_4$	93
R_2NNO	R_2NH	$LiAlH_4$	94
	R_2NNH_2	$LiAlH_4$, Zn/AcOH	94
R_2NNO_2	R_2NNH_2	Electroreduction	94
R_3NO	R_3N	H_2/Pd^b, BH_3, Fe/AcOH, $TiCl_3$, $CrCl_2$	94 95
R_2NOR'	R_2NH	Zn/AcOH	95
$RNHNH_2$	RNH_2	H_2/Raney Ni	95
RNHNHR	RNH_2	H_2/Pd, $Na_2S_2O_4$ $(EtO)_2P(S)SH$	95 96
RN=NR	RNHNHR	H_2/Pd, N_2H_2, NaHg, AlHg, Zn/NH_3, H_2S	95, 96 95
	RNH_2	H_2/Pd, $Na_2S_2O_4$, $(EtO)_2P(S)SH$	96

^{a} Very exceptional.
^{b} Nitroamine oxide reduced selectively to nitro amine.

Table 16 Reduction of aldehydes

Starting compound	Product	Reducing agent	Page
RCHO	RCH$_2$OH	H$_2$/PtO$_2$, H$_2$/Raney Ni,	96, 97
		LiAlH$_4$, LiAlH(OR)$_3$,	96, 97
		Na$_2$AlH$_2$(OCH$_2$CH$_2$OMe)$_2$,LiBH$_4$,	96, 97
		NaBH$_4$, Bu$_4$NBH$_3$CN, Bu$_3$SnH,	96, 97
		NaH/FeCl$_3$, Fe/AcOH, Na$_2$S$_2$O4,	96, 97
		iso-PrOH/Al(O-iso-Pr)$_3$	96, 97
	RCH$_3$	ZnHg/HCl, N$_2$H$_4$/KOH	97
		1. TosNHNH$_2$, 2. NaBH$_4$	106
	RCH=CHR	TiCl$_3$/Mg	97
RCH=CHCHO	RCH$_2$CH$_2$CHO	H$_2$/Pd, H$_2$/Ni, H$_2$/HCo(CO)$_4$,	97, 98
		Et$_3$N·HCO$_2$H/Pd	98
	RCH=CHCH$_2$OH	H$_2$/PtO$_2$, H$_2$/Os, LiAlH$_4$,	98
		Mg(AlH$_4$)$_2$, LiBH$_4$, NaBH$_4$,	98
		NaBH(OMe)$_3$,	98
		9-BBN, Ph$_2$SnH$_2$,	98
		iso-PrOH/Al(O-iso-Pr)$_3$	98
	RCH$_2$CH$_2$CH$_2$OH	H$_2$/Ni, Electroreduction	98
ArCHO	ArCH$_2$OH	H$_2$/Pt, H$_2$/Pd, H$_2$/Ni, H$_2$/Raney Ni,	99, 100
		H$_2$/CuCr$_2$O$_4$, HCO$_2$H/Cu,	99, 100
		Ni-Al/NaOH, LiAlH$_4$, AlH$_3$,	99, 100
		LiBH$_4$,	99, 100
		NaBH$_4$, NaBH(OMe)$_3$, Bu$_4$NBH$_4$,	99, 100
		Bu$_4$NBH$_3$CN	99, 100
		Bu$_3$SnH, Ph$_2$SnH$_2$, Na$_2$S$_2$O$_4$, iso-	99, 100
		PrOH/Al(O-iso-Pr)$_3$, CH$_2$O	
	ArCH$_3$	H$_2$/Pd, H$_2$/Ni, H$_2$/CuCr$_2$O$_4$, AlH$_3$,	101
		Et$_3$SiH/BF$_3$, Electroreduction,	100, 101
		Li/NH$_3$, ZnHg/HCl, N$_2$H$_4$/KOH	101
	ArCH=CHAr	TiCl$_3$/Li	101
	Hydroaromatic methyl derivative	H$_2$/Pt, AcOH	101
ArCH=CHCHO	ArCH$_2$CH$_2$CHO	H$_2$/Pd, H$_2$/Ni, H$_2$/(Ph$_3$P)$_3$RhCl	101, 102
	ArCH=CHCH$_2$OH	H$_2$/Os, H$_2$/PtO$_2$, LiAlH$_4$, AlH$_3$,	102
		NaBH$_4$, NaBH(OMe)$_3$, NaBH$_3$CN,	102
		9-BBN, Bu$_4$NBH$_4$, BH$_3$·THF,	102
		Ph$_2$SnH$_2$, iso-PrOH/Al(O-iso-Pr)$_3$	102
	ArCH$_2$CH$_2$CH$_2$OH	H$_2$/Raney Ni, Electroreduction,	101, 102
		LiAlH$_4$	102
	ArCH$_2$CH$_2$CH$_3$	ZnHg/HCl	103
	ArCH$_2$CH=CH$_2$	1. TosNHNH$_2$, 2. BH$_3$	102, 103

Table 17   Reduction of derivatives of aldehydes

Starting compound	Product	Reducing agent	Page
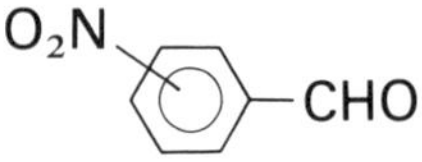 O$_2$N—C$_6$H$_4$—CHO	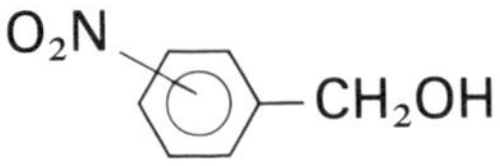 O$_2$N—C$_6$H$_4$—CH$_2$OH	AlH$_3$, iso-PrOH, Al(O-iso-Pr)$_3$	103
	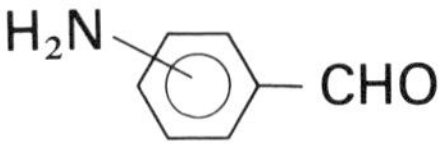 H$_2$N—C$_6$H$_4$—CHO	SnCl$_2$, TiCl$_3$, FeSO$_4$	103
	H$_2$N—C$_6$H$_4$—CH$_2$OH	H$_2$/catalysts	103
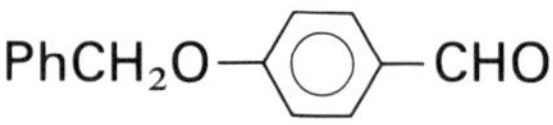 PhCH$_2$O—C$_6$H$_4$—CHO	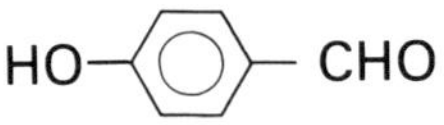 HO—C$_6$H$_4$—CHO	H$_2$/Pd	103
RCH(OR')$_2$	RCH$_2$OR'	AlH$_3$	103
RCH(OCOR')$_2$	RCH$_2$OCOR'	Fe/AcOH	105
RCH(O,S ring)	RCH$_2$OCH$_2$CH$_2$SH	Ca/NH$_3$	104, 105
RCH(S,S ring)	RCH$_2$SCH$_2$CH$_2$SH	Ca/NH$_3$	104, 105
	RCH$_3$	Raney Ni, Bu$_3$SnH, Na/NH$_3$[a]	104, 105
RCH=NR'	RCH$_2$NHR'	H$_2$/Pt, Ni, LiAlH$_4$, NaAlH$_4$, LiBH$_4$, NaBH$_4$, NaHg, K/C	105 105
RCH=NOH	RCH$_2$NH$_2$	H$_2$/Raney Ni, LiAlH$_4$, Na/ROH	106
RCH=NNHSO$_2$Ar	RCH$_3$	LiAlH$_4$, NaBH$_4$, BH$_3$/BzOH, NaBH$_3$CN	106 106

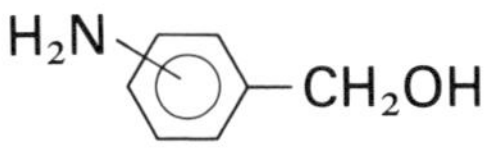

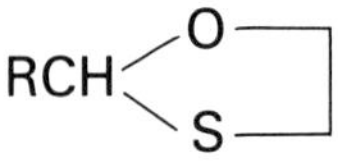

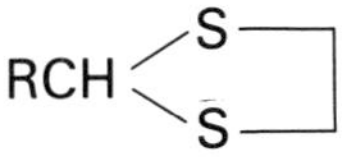

[a] Only if R = Ar.

Table 18   Reduction of ketones

Starting compound	Product	Reducing agent	Page
RCOR'	RCH(OH)R'	H_2/PtO_2, $H_2/RhO_2 \cdot PtO_2$, H_2/Ru,	107, 108
		H_2/Ni, $H_2/Raney\ Ni$, $H_2/CuCr_2O_4$,	107, 108
		Ni-Al/NaOH, $LiAlH_4$, $Mg(AlH_4)_2$,	107, 108
		$LiAlH(OMe_3)_3$, $AlHCl_2$, BH_3,	107, 108
		$LiBH_4$, $LiBHEt_3$, $NaBH_4$,	107, 108
		$NaBH(OMe)_3$, Bu_4NBH_4,	107
		Bu_4NBH_3CN, R_2SnH, Na/EtOH,	107, 108
		$Na_2S_2O_4$, iso-PrOH/Al(O-iso-Pr)$_3$,	107, 108
		Biochemical reduction	108
	RCH_2R'	Et_3SiH/BF_3, ZnHg/HCl, $N_2H_4/$	108
		KOH,	108
		1. $TosNHNH_2$, 2. $NaBH_4$ or	
		$NaBH_3CN$	
	RC(OH)R' | RC(OH)R'	Na	109
		MgHg, AlHg	109, 118
	RCR' ∥ RCR'	$TiCl_3/LiAlH_4$, $TiCl_3/Li$, $TiCl_3/K$,	109, 118
		$TiCl_3/Mg$, $TiCl_3/Zn$-Cu	109, 118
RCH=CHCOR' (R = aliphatic or aromatic residue)	RCH_2CH_2COR'	H_2/PtO_2, H_2/Pt, H_2/Pd, H_2/Rh, $H_2/$	119-121
		$(Ph_3P)_3RhCl$, Ni-Al/NaOH, Ni-Zn,	119, 121
		$AlH(OR)_2$, $AlH(NR_2)_2$, Ph_3SnH,	120, 121
		CuH, $NaHFe_2(CO)_8$,	120
		Electroreduction, $Li/PrNH_2$, C_8K,	120, 121
		ZnHg/HCl, Biochemical reduction	120
	RCH=CHCH(OH)R'	$LiAlH_4$, $LiAlH(OMe)_3$, AlH_3,	120, 121
		$AlH(iso$-Bu)$_2$, $LiBH_3Bu$, $NaBH_4$,	120, 121
		$NaBH_3CN$, 9-BBN,	120, 121
		iso-PrOH/Al(O-iso-Pr)$_3$	120, 121
	$RCH_2CH_2CH(OH)R'$	H_2/Ni, $H_2/CuCr_2O_4$, Ni-Al/NaOH,	121
		$LiAlH_4/THF$, $LiAlH(OMe_3)_3$,	121
		$NaBH_4$	121
	$RCH=CHCH_2R'$	1. $TosNHNH_2$, 2. $NaBH_3CN$ or	121
		$BH_3 \cdot BF_3$ or catechol borane	121
	$RCH_2CH_2CH_2R'$ $RCHCH_2COR'$ | $RCHCH_2COR'$	ZnHg/HCl, N_2H_4, $H_2/catalysts$	121
		NaHg	121, 122
RC≡CCOR'	RCH=CHCOR'	H_2/Pd^a	122
	RCH_2CH_2COR'	H_2/Pd^a	122
	$RC≡CCH(OH)R'$	$NaBH_3CN$, Bu_4NBH_3CN, Chiral	122
		boranes	
ArCOR'	ArCH(OH)R'	H_2/PtO_2, H_2/Pd, H_2/Ni, $H_2/$	109-111
		$CuCr_2O_4$, Ni-Al/NaOH, $LiAlH_4$,	109-111
		$LiBH_4$, $NaAlH_2(OCH_2CH_2OMe)_2$,	110, 111
		$LiBHEt_3$, $NaBH_4$, $NaBH(OMe)_3$,	110, 111
		Bu_4NBH_4, Bu_4NBH_3CN, Ph_2SnH_2,	110, 111
		$Na_2S_2O_4$, iso-PrOH/Al(O-iso-Pr)$_3$,	110, 111
		CH_2O^b, Biochemical reduction	111, 112
	$ArCH_2R'$	H_2/PtO_2, AcOH, H_2/Pd, $H_2/Raney$	112, 113
		Ni, $H_2/CuCr_2O_4$, H_2/MoS_2; $Et_3SiH/$	112, 113
		CF_3CO_2H, $AlHCl_2$, $NaBH_4/$	113
		CF_3CO_2H, Na/EtOH, Li/NH_3,	113
		ZnHg/HCl, N_2H_4/KOH, (HI + P)	113
	ArR'C(OH)C(OH)ArR'	Mg, AlHg, Zn, iso-PrOH/hv	112
	ArR'C=CArR'	$TiCl_3/Li$, K, Mg	112
	Completely saturated hydrocarbon	H_2/PtO_2, $H_2/RhO_2/AcOH$	113

^{a} Non-conjugated systems.
^{b} Only purely aromatic ketones.

Table 19   Reduction of substitution derivatives of ketones

Starting compound	Product	Reducing agent	Page
RCOCHR'X[a,b]	RCOCH$_2$R'[c]	H$_2$/Pd, Zn, ZnHg, CrCl$_2$, CeI, VCl$_2$, HBr, HI	123, 31,32, 123
	RCH(OH)CHR'X	LiAlH$_4$, NaBH$_4$, iso-PrOH/Al (O-iso-Pr)$_3$	122, 123
	RCH(OH)CH$_2$R'	iso-PrOH/Al(O-iso-Pr)$_3$[d]	123
	RCH=CHR'	N$_2$H$_4$/AcOK	123
RCOCHR'(NO$_2$)[b]	RCOCHR'(NH$_2$)	Fe[e], SnCl$_2$[e]	124
	RCH(OH)CHR'(NO$_2$)	LiAlH$_4$, NaBH$_4$, Ca(BH$_4$)$_2$, iso-PrOH/Al(O-iso-Pr)$_3$, Glucose[f]	123, 123, 36, 124
	RCH$_2$CHR'(NH$_2$)	H$_2$/Pd[g]	124
RCOCR'N$_2$	RCOCH$_2$R'	H$_2$/PdO + CuO or HCl, HI, EtOH[h], H$_3$PO$_2$[h]	124, 124, 125
	RCOCHR'NH$_2$	H$_2$/Pd	124
	RCH(OH)CHR'NH$_2$	LiAlH$_4$	124
	RCOCR=NNH$_2$	H$_2$/PdO	124
RCOCHR'N$_3$	RCH(OH)CHR'NH$_2$	LiAlH$_4$	125

[a] X = Cl, Br, I.
[b] The substituent is not necessarily in a position α to the carbonyl group.
[c] Only with very reactive halogen.
[d] Not general.
[e] Only with NO$_2$ in an aromatic ring.
[f] Nitro group reduced to azo group.
[g] Only with the keto group adjacent to and nitro group in the aromatic ring.
[h] For diazonium group in aromatic ring.

Table 20 Reduction of hydroxy ketones, diketones and quinones

Starting compound	Product	Reducing agent	Page
RCOCH(OH)R'[a]	RCH(OH)CH(OH)R'	H_2/CuCr$_2$O$_4$,	125
		Biochemical reduction	125
	RCOCH$_2$R'	Zn, ZnHg, Sn, HI	125
	RCH$_2$CH$_2$R'	ZnHg/HCl	126
RCOCOR'[a]	RCOCH(OH)R'	H_2/Ru[b], Zn, TiCl$_3$, VCl$_2$, H$_2$S,	126, 127
		P(OEt)$_3$, Ph$_2$C(OH)C(OH)Ph$_2$	126, 127
	RCOCH$_2$R'	ZnHg/HCl[c], HI, H$_2$S,	126, 127
		N$_2$H$_4$/KOH	127
	RCH$_2$CH$_2$R'	ZnHg/HCl,[c] 2 N$_2$H$_4$/KOH	127
	(CH$_2$)$_n$		
	RC=CR	TiCl$_3$/ZnCu[d]	128
RCOCHCHR' over \ / over O	RCOCH=CHR'	Zn/AcOH, CrCl$_2$	126
	RCH=CHCH(OH)R'	N$_2$H$_4$	126
RCOCH=CHCOR'	RCOCH$_2$CH$_2$COR'	Sn, CrCl$_2$	128
O=⟨benzene ring⟩=O	HO—⟨benzene ring⟩—OH	H_2/Pt,[e] H_2/Pd, LiAlH$_4$, SnCl$_2$,	128, 129
		TiCl$_3$, VCl$_2$, CrCl$_2$, H$_2$S, SO$_2$	129
	O=⟨cyclohexadiene with H,H⟩	Sn	129
	 HO,H / H,OH cyclohexadiene	LiAlH$_4$, LiBHEt$_3$, NaBH$_4$, 9-BBN	129 / 129
	 ⟨cyclohexane ring⟩	Al, Zn/ZnCl$_2$, SnCl$_2$, HI	129

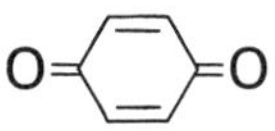

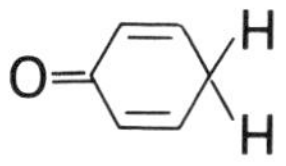

[a] Both functions are not necessarily vicinal.
[b] Only γ-diketone.
[c] Clemmensen reduction gives rearranged ketones and hydrocarbon.
[d] Only if $n \geq 2$.
[e] In mineral acid cyclohexanol was formed.

Table 21   Reduction of ketals, thioketals, ketimines and ketoximes

Starting compound	Product		Page
$RR'C(OR'')_2^a$	$RR'CHOR'$	$H_2/Pd, H_2/Rh, H_2/Ru, AlH_3$	130, 131
$RR'C(OR'')(SR'')^a$	$RCOR'$	Raney Ni	130
	$RR'CHOR''$	Ca/NH_3	130, 131
	$RR'CHSR''$	AlH_3	130, 131
$RR'C(SR'')_2$	$RR'CH_2$	Ni, Raney Ni,	130, 131
		Bu_3SnH, N_2H_4	131, 132
$RR'C=NH$	$RR'CHNH_2$	H_2/catalysts, $NaBH_4$,	132
		$KBH_4, NaAlH_2(OCH_2CH_2OMe)_2$	132
$RR'C=NOH$	$RR'CHNH_2$	$H_2/Pd, Pt, Rh, H_2/Raney Ni,$	132, 133
		$N_2H_4/Raney Ni, LiAlH_4, Na/ROH,$	132, 133
		$SnCl_2, TiCl_3$	132, 133

[a]Included are cyclic ketals and thioketals with five- or six-membered rings:

$$\begin{array}{c} R \\ R' \end{array}\!\!\!>\!\!C\!\!<\!\!\!\begin{array}{c} O-CH_2 \\ | \\ O-CH_2, \end{array} \qquad \begin{array}{c} R \\ R \end{array}\!\!\!>\!\!C\!\!<\!\!\!\begin{array}{c} S-CH_2 \\ | \\ S-CH_2 \end{array}$$

Table 22   **Reduction of carboxylic acids**

Starting compound	Product	Reducing agent	Page
RCO_2H	RCHO	$AlH(NR_2)_2$	137
		$Li/MeNH_2$	137
		Electroreduction,	137
		NaHg	137
	RCH_2OH	$H_2/Ru, Re_2O_7, Re_2S_7$	137
		$H_2/CuCr_2O_4$	137
		$LiAlH_4$	137
		$NaAlH_2(OCH_2CH_2OMe)_2$	137
		BH_3	137, 138
$RCH=CHCO_2H$	$RCH_2CH_2CO_2H$	H_2/catalysts	138
		NaHg	138
		$CrSO_4$	138
	$RCH=CHCH_2OH$	$LiAlH_4$	138
$RC\equiv CCO_2H$	$RCH=CHCO_2H$	H_2/Ni (*cis*)	138
		Na (*trans*)	138
		$CrSO_4$ (*trans*)	138
	$RCH=CHCH_2OH$	$LiAlH_4$ (*trans*)	138
$ArCO_2H$	RCO_2H^a	H_2/Rh^b	140
		$Li/NH_3, EtOH^c$	140
		$Na/NH_3, EtOH^c, NaHg^d$	140
	ArCHO	$AlH(NR_2)_2$	139
		$NaHg+RNH_2$	139
	$ArCH_2OH$	$LiAlH_4, NaAlH_4$	139
		$NaAlH_2(OCH_2CH_2OMe)_2$	139
		BH_3	139
		Electroreduction	139
	$ArCH_3$	$H_2/Ni, H_2/CuCr_2O_4, SiHCl_3/Pr_3N$	139
		$NaAlH_2(OCH_2CH_2OMe)_2^e$	139
$ArCH=CHCO_2H$	$ArCH_2CH_2CO_2H$	$H_2/Pd, H_2/Raney Ni, H_2/CuCr_2O_4$	140
		$H_2/(Ph_3P)_3RhCl$	140
		$NiAl/NaOH, NaHg, CrSO_4$	141
	$ArCH_2CH_2CH_2OH$	$LiAlH_4$	141
$ArC\equiv CCO_2H$	$ArCH=CHCO_2H$	H_2/Pt (*cis*), $CrSO_4$ (*trans*)	141
			141
	$ArCH_2CH_2CO_2H$	H_2/Pt	

[a]R = hydrogenated aromatic ring.
[b]Total hydrogenation of the ring.
[c]Partial or total hydrogenation of the ring.
[d]Only in anthracene series.
[e]Only with OH or NH_2 in *ortho* or *para* positions.

Table 23   Reduction of substitution derivatives of carboxylic acids

Starting compound	Product	Reducing agent	Page
$RCHXCO_2H^{a, b}$	$RCHXCH_2OH$	BH_3, AlH_3, $LiAlH_4$ (inverse technique)	141
$X-ArCO_2H^b$	$ArCO_2H$	H_2/Raney Ni, NaOH	141
	$X-ArCH_2OH$	AlH_3, BH_3	141
	$X-ArCH_3$	$SiHCl_3/Pr_3N$	141
$RCX=CHCO_2H^b$	$RCH=CHCO_2H$	$H_2/Pd(BaSO_4)^c$, NaHg, Zn	142
	$RCH_2CH_2CO_2H$	$H_2/Pd(C)^{c, d}$, Zn, Biochemical reduction	142
$O_2N-ArCO_2H$	$H_2N-ArCO_2H$	NH_4SH	142
	$O_2N-ArCH_2OH$	BH_3	142
$XN_2^{\ominus \oplus}-ArCO_2H$	$ArCO_2H$	$NaBH_4$, EtOH/hv	142
$RSS-ArCO_2H$	$HS-ArCO_2H$	Zn/AcOH	143
$\underset{O}{RCHCHCO_2H}$ (epoxide)	$\underset{OH}{RCHCH_2CO_2H}$ + $\underset{OH}{RCH_2CHCO_2H}$	$NaBH_4$	143
$RCOCO_2H^e$	$RCOCH_2OH$	BH_3	143
	$RCH(OH)CO_2H$	H_2/Raney Ni, H_2/Pd, $NaBH_4$, $SiHEt_3/CF_3CO_2H$, NaHg, Zn, Biochemical reduction	143 143 143
	RCH_2CO_2H	$SiHEt_3/CF_3CO_2H$; ZnHg, N_2H_4, HI/P	143, 144

[a]The substituent is not necessarily in the α-position.
[b]X = Cl, Br.
[c]Not general.
[d]Applies to X = F.
[e]The keto group is not necessarily in the α-position.

Table 24   Reduction of acyl chlorides and acid anhydrides

Starting compound	Product	Reducing agent	Page
RCOCl	RCHO	$H_2/Pd(BaSO_4)+S$, Et_3SiH/Pd, $LiAlH(OCMe_3)_3$, CuH, 1. $(EtO)_3P$, 2. $NaBH_4$	144, 145 145 145
	RCH_2OH	$LiAlH_4$, $NaAlH_4$, $NaAlH_2(OCH_2CH_2OMe)_2$, AlH_3, $NaBH_4$, $NaBH(OMe)_3$, Bu_4NBH_4, tert-BuMgCl	145, 146 146 146 146
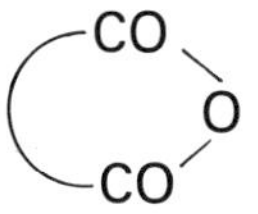 (cyclic anhydride: –CO–O–CO–)	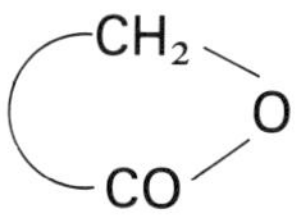 (–CH$_2$–O–CO–)	$H_2/CuCr_2O_4$, $H_2/(Ph_3)_3RhCl$, $LiAlH_4/-55°$, $NaBH_4$, $LiBH_4$, $LiBHEt_3$, $LiBH(CHMeEt)_3$, NaHg, Zn	146 147 147 147
	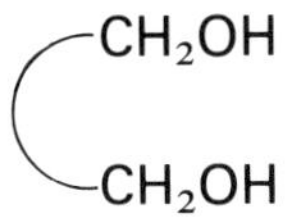 (–CH$_2$OH, –CH$_2$OH)	$LiAlH_4$, $NaAlH_2(OCH_2CH_2OMe)_2$	147

Table 25   Reduction of esters of carboxylic acids

Starting compound	Product	Reducing agent	Page
RCO_2R'	$RCHO$	$(LiAlH_4)$, $KAlH_4$,	148, 149
		$NaAlH_4$, $NaAlH_2(CHMeEt)_2$,	148, 149
		$AlH(CHMeEt)_2$,	148,149
		$NaAlH_2(OCH_2CH_2OMe)_2$,	148, 149
		$LiAlH(OMe)_3$	148
		$AlH(N\overset{\frown}{}NMe)_2$	148, 149
	RCH_2OR	BH_3, $SiHCl_3$	149, 150
	$RCO_2H + R'H$	H_2/Pd^a, Zn^a, $CrCl_2$	150, 151
	$RCH_2OH + ROH'$	H_2/Raney Ni, $H_2/CuCr_2O_4$,	153, 154
		$H_2/ZnCr_2O_4$, H_2/Re_2O_7,	153, 154
		$LiAlH_4$, $NaAlH_4$,	153, 154
		$LiAlH(OMe)_3$,	154
		$LiAlH(OCMe_3)_3$,	154
		$NaAlH_2(OCH_2CH_2OMe)_2$, AlH_3,	154–156
		$Mg(AlH_4)_2$, $LiBH_4$, $NaBH_4$,	154–156
	$\underset{\overset{\displaystyle RCO}{\displaystyle \vert}}{RCHOH}$	$Ca(BH_4)_2$, $NaBH(OMe)_3$, $LiBHEt_3$,	155, 156
		BH_3, $MgHBr$, Na, ROH	152

(lactone, CO–O ring)	(lactol, CHOH–O ring)	$LiAlH_4$, $NaBH_4$, $AlH(iso\text{-}Bu)_2$, NaHg	149
	(ether, CH₂–O ring)	1. $AlH(iso\text{-}Bu)_2$,	150
		2. $Et_3SiH/BF_3 \cdot Et_2O$,	150
		$SiHCl_3/(tert\text{-}BuO)_2$	
	(diol, CH₂OH / CH₂OH)	$LiAlH_4$	154

^{a} Benzyl-type only.

Table 26   Reduction of unsaturated esters

Starting compound	Product	Reducing agent	Page
$RCH=CHCO_2R'^a$	$RCH_2CH_2CO_2R'$	H_2, Pd, Raney Ni, $H_2/$	156, 157
		$CuCr_2O_4$, $H_2/(Ph_3P)_3RhCl$	156, 157
		$(NaBH_4)$, Bu_3SnH, $CrSO_4$	157
	$RCH=CHCH_2OH$	$H_2/ZnCrO_4$, $LiAlH_4$,	157
		$NaAlH_2(OCH_2CH_2OMe)_2$	157
	$RCH_2CH_2CH_2OH$	H_2/Raney Ni, $H_2/CuCr_2O_4$,	157, 158
		$LiAlH_4$,	158
		$NaAlH_2(OCH_2CH_2OMe)_2$,	158
		$NaBH_4$,^{b} $KBH_4/LiCl$, Na	158
$RC\equiv CCO_2R'$	$RCH=CHCO_2R'$	H_2/Pd (deact.), H_2O/Ph_3P	159
	$RCH_2CH_2CO_2R'$	H_2/PtO_2	159
	$RCH\equiv CHCH_2OH$	$LiAlH_4$	159
	$RCH=CHCH_2OH$	$LiAlH_4$ (excess)	159

^{a}R = alkyl or aryl.
^{b}Only if R = pyridyl or pyrimidyl.

Table 27   Reduction of derivatives of esters

Starting compound	Product	Reducing agent	Page
X-ArCO$_2$R$'^b$	X-ArCO$_2$R', ArCO$_2$R'	H$_2$/Pd, H$_2$/Raney Ni	159
RCHXCO$_2$R$'^{a,\ b}$	RCH$_2$CO$_2$R	H$_2$/Pd	159
	RCHXCH$_2$OH	(LiAlH$_4$); AlH$_3$, LiBH$_4$,	159
		Ca(BH$_4$)$_2$	159
	RCH$_2$CH$_2$OH	(LiAlH$_4$)	159
RCH(NO$_2$)CO$_2$R$'^b$	RCH(NH$_2$)CO$_2$R'	H$_2$/Pd	159
O$_2$N-ArCO$_2$R' ^{or}	H$_2$N-ArCO$_2$R'	Fe	159
	RCH(NO$_2$)CH$_2$OH or	LiAlH$_4$, AlH$_3$, BH$_3$	159, 160
	O$_2$N-ArCH$_2$OH		
RCN$_2$CO$_2$R'	RC(:NNH$_2$)CO$_2$R'	K$_2$SO$_3$	160
	RCH(NHNH$_2$)CO$_2$R'	NaHg	160
	RCH(NH$_2$)CO$_2$R'	H$_2$/PtO$_2$	160
RCH(N$_3$)CO$_2$R$'^b$	RCH(NH$_2$)CO$_2$R	H$_2$/Pd, AlHg	160

^{a} X = F, Cl, Br.
^{b} Not necessarily in the α-position.

Table 28   Reduction of hydroxy esters, amino esters, keto esters, oximino esters and acid-esters

Starting compound	Product	Reducing agent	Page
ArCH(OH)CO$_2$R	ArCH(OH)CH$_2$OH	LiAlH$_4$, LiBH$_4$, Ca(BH$_4$)$_2$	161
	ArCH$_2$CH$_2$OH	H$_2$/Ni, H$_2$/CuCr$_2$O$_4$	160
RCH(NH$_2$)CO$_2$R'	RCH(NH$_2$)CH$_2$OH	H$_2$/Raney Ni, LiAlH$_4$	161
RCOCO$_2$R'	RCH(OH)CO$_2$R'	H$_2$/PtO$_2$, Raney Ni, NaBH$_4$,	161
		NaHg, AlHg, *iso*-PrOH/	161
		Al(*O-iso*-Pr)$_3$, Yeast	161, 162
	RCH$_2$CO$_2$R'	2R''SH; Raney Ni desulf-	162
		urization, (NaBH$_3$CN)	162
	RCOCH$_2$OH	1. LiN (iso-Pr)$_2$, 2. LiAlH$_4$	162
		1. HC(OR)$_3$, 2. LiAlH$_4$	162, 163
RC(:NOH)CO$_2$R'	RCH(NH$_2$)CO$_2$R'	H$_2$/Pd, AlHg, Zn	163
(CH$_2$)$_n$⟨CO$_2$H / CO$_2$R	(CH$_2$)$_n$⟨CH$_2$OH / CO$_2$R	AlH$_3$, BH$_3$	163
	(CH$_2$)$_n$⟨CO$_2$H / CH$_2$OH	Na/ROH	163
RC(OR')$_3$	RCH(OR')$_2$	LiAlH$_4$	163
RCOSR'	RCH$_2$SR',	AlH$_3$,	163, 164
	RCHO	Raney Ni	
RC(S)SR'	RCH$_3$	Raney Ni	164

Table 29 Reduction of amides, lactams and imides

Starting compound	Product	Reducing agent	Page
RCONR'R"	RCHO^a	LiAlH$_4$, LiAlH(OEt)$_3$,	164, 165
		NaAlH$_2$(OCH$_2$CH$_2$OMe)$_2$,	164, 165
		Electroreduction, Na/NH$_3$	165
	RCH$_2$OH	(H$_2$/PtO$_2$), LiAlH$_4$,	166
		LiBHEt$_3$,	166
		NaAlH$_2$(OCH$_2$CH$_2$OMe)$_2$,	166
		NaBH$_4$	166
	RCH$_2$NR'R"	H$_2$/CuCr$_2$O$_4$, LiAlH$_4$,	167
		Mg(AlH$_4$)$_2$,	167
		NaAlH$_2$(OCH$_2$CH$_2$OMe)$_2$,	167
		LiAlH(OMe)$_3$, AlH$_3$, BH$_3$,	167, 168
		Electroreduction, (Na, NaHg)	167, 168
(ring) –CO / –NH	(ring) –CH$_2$ / –NH	H$_2$/PtO$_2$, Electroreduction,	168
		LiAlH$_4$,	168
		NaAlH$_2$(OCH$_2$CH$_2$OMe)$_2$,	168
		NaBH$_4$, BH$_3$	168
(ring) –CO, NH, –CO	(ring) –CH$_2$, NH, –CO	H$_2$/Pd, Ni, CuCr$_2$O$_4$, LiAlH$_4$,	168
		NaAlH$_2$(OCH$_2$CH$_2$OMe)$_2$,	168
		BH$_3$, Electroreduction	168, 169
	(ring) –CH$_2$, NH, –CH$_2$	LiAlH$_4$, Electroreduction	169
RCH=CHNHCOR'	RCH$_2$CH$_2$NHCOR'	H$_2$/PtO$_2$	169
RCH=CHCONHR'	RCH$_2$CH$_2$CONHR'	H$_2$/PtO$_2$	169
	RCH$_2$CH$_2$CH$_2$NHR'	(LiAlH$_4$, LiAlH(OR)$_3$, AlH$_3$)	170
O$_2$N——CONR$_2$	H$_2$N——CH$_2$NR$_2$	AlH$_3$	170
RCOCONHR'	RCH(OH)CONHR'	Biochemical reduction	170
	RCH(OH)CH$_2$NHR'	LiAlH$_4$	170
RO$_2$C——CONH$_2$	RO$_2$C——CH$_2$NH$_2$	BH$_3$	170

^{a} Only with R'=alkyl or aryl, R"=aryl.

Table 30   Reduction of amidines, thioamides, imidoyl chlorides and hydrazides

Starting compound	Product	Reducing agent	Page
RC(:NH)NH$_2$	{ RCH$_3$ RCHO	Na/NH$_3$	171
RCSNH$_2$	{ RCHO RCH$_2$NH$_2$	Raney Ni, Fe Raney Ni (excess), LiAlH$_4$, NaBH$_4$, NaBH$_3$CN, Electroreduction	171, 172 171 171 171
RCCl=NR′	{ RCH=NR′ RCHO	LiAlH(OCMe$_3$)$_3$ Mg, SnCl$_2$, CrCl$_2$	172 172
RCONHNHSO$_2$Ar	{ RCHO RCH$_3$	Na$_2$CO$_3$ LiAlH$_4$, NaBH$_4$, NaBH$_3$CN	172 172

Table 31   Reduction of nitriles

Starting compound	Product	Reducing agent	Page
RC≡N	RCHO	Raney Ni/Na$_3$PO$_2$, Raney Ni/ HCO$_2$H, NaAlH(OEt)$_3$, LiAlH(OEt)$_3$, LiAlH(OBu)$_3$, (AlH(*iso*-Bu)$_2$), SnCl$_2$	173 173 173 173
	RCH$_2$NH$_2$	H$_2$/Rh, H$_2$/Pt, H$_2$/Raney Ni, H$_2$/NiB$_2$, LiAlH$_4$, LiAlH(OMe)$_3$, AlH$_3$, BH$_3$, Na/ROH	173, 174 174 174
RCH=CHCN	RCH$_2$CH$_2$CN	H$_2$/PtO$_2$, H$_2$/Pd, H$_2$/Co, NaBH$_4$, CuH, Mg/ROH, Zn, Et$_3$N·HCO$_2$H	175 175 175
	RCH=CHCH$_2$NH$_2$ RCH$_2$CH$_2$CH$_2$NH$_2$	H$_2$/Raney Co[a] LiAlH$_4$	175 175
O$_2$N⌇⌇CN	H$_2$N⌇⌇CN	Fe, SnCl$_2$	175
RCOCN	RCOCH$_2$NH$_2$ RCH(OH)CH$_2$NH$_2$	SnCl$_2$ LiAlH$_4$	175 175

[a] The nitrile contained a β, γ-double bond.

PROCEDURES

1. CATALYTIC HYDROGENATION UNDER ATMOSPHERIC PRESSURE (FIG. 1)

A 25–250 ml ground glass flask A fitted with a magnetic stirring bar is charged with the catalyst, with the compound to be hydrogenated and with the solvent (if necessary.) The catalyst, its amount relative to the amount of the hydrogenated compound, and reaction conditions are summarized in Table 1 (p. 6), which gives general guidelines for hydrogenations of most common types of compounds. The solvent is chosen depending on the solubility of the compound to be hydrogenated. The ground glass joint of the hydrogenating flask is lightly greased or sealed with a Teflon* sleeve and the flask is attached to the hydrogenation apparatus by means of steel springs or rubber bands. The three-way stopcock B which connects the hydrogenating flask to a 50–500 ml graduated tube C, filled completely by a liquid from a reservoir D, is set so as to cut off the graduated tube and to connect the hydrogenating flask to a glass tube fitted with another three-way stopcock E and a manometer F. The stopcock E is opened to outlet G connected to an aspirator or a vacuum pump and the apparatus is evacuated. If a low-boiling solvent is used, the flask A must be cooled to prevent effervescence under reduced pressure. When the pressure in the apparatus has dropped to 5–20 mm the stopcock E is opened to outlet H through which hydrogen is cautiously introduced into the apparatus using a hydrogen regulator on a hydrogen tank.

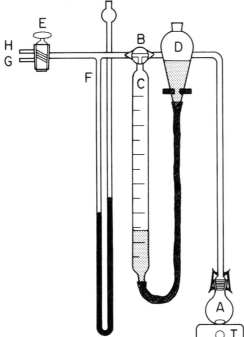

Fig. 1 Apparatus for hydrogenation at atmospheric pressure.

* "Teflon" is a trade name of E.I. Dupont De Nemours

When the pressure in the manometer has risen to about 1 atm the stopcock E is switched to the vacuum pump and the apparatus is evacuated again. Such flushing of the apparatus with hydrogen is necessary to displace air and may be repeated once more.

Then the stopcock E is set to connect the apparatus to the source of hydrogen again, hydrogen is introduced until its pressure reaches 1 atm or slightly more, and the stopcock B is opened so as to allow the hydrogen to enter the graduated tube C. Hydrogen displaces the liquid into the reservoir D. By lowering the reservoir D the height of the liquid is adjusted so that it is level in both the graduated tube C and the reservoir D. At this moment the stopcock E is turned to cut off the hydrogenating assembly A–D from the source of hydrogen, the volume of hydrogen at exactly atmospheric pressure is read, and the reservoir D is placed in its original position as shown in the illustration.

The magnetic stirrer I is started and the hydrogenation begins. Its progress is followed by the decrease in the volume of hydrogen enclosed in the tube C. When the consumption of hydrogen stops the final reading is taken after equalizing the liquid levels in both graduated tube C and the reservoir D. If the hydrogenation has not been finished after the volume of hydrogen has been used up, additional hydrogen is supplied in the way described for the initial filling.

After the hydrogenation has come to an end the exact final volume of hydrogen used for the hydrogenation is calculated from the volume read at atmospheric pressure, adjusted to normal conditions, i.e. 760 mm and 25° (298.15 K).

The excess hydrogen (if any) is bled through the valve E, the hydrogenating flask is disconnected, the catalyst is removed by centrifugation or filtration under the necessary precautions (p. 13), and the filtrate is worked up depending on the nature of the products.

Notes: All the ground glass joints and stopcocks must be gently greased and held in position by springs or rubber bands. The manometer need not be sealed on but may be connected to the apparatus by rubber or plastic hose. Also the hydrogenating flask may be attached to the apparatus by means of rubber or plastic tubing. This arrangement is preferable in hydrogenations involving Raney nickel, for which shaking of the flask is better than magnetic stirring as Raney nickel sticks to the magnetic stirring bar. In this case solvents which cause swelling of the plastic tubing must be avoided since they may extract organic material and contaminate the product of hydrogenation [68]. The manometric liquid used for an apparatus of any size is usually water. For precise work in a small apparatus (50–100 ml volume) mercury is preferred. For hydrogenations at slightly elevated pressures (1–4 atm) more complicated apparatus have been designed and described in the literature.

2. CATALYTIC HYDROGENATION WITH HYDROGEN GENERATED FROM SODIUM BOROHYDRIDE [4]

Hydrogenation *ex situ*

The special apparatus (Fig. 2) consists of two 500 ml Erlenmeyer flasks A and B having slightly rounded bottoms and fitted with Teflon®-coated magnetic stirring bars and injection ports, a mercury-filled bubbler back-up prevention valve C, and a mercury-filled automatic dispenser D with airtight-connected inserted delivery tube of a buret E.

The flask B (reactor) connected to the bubbler C is charged with 100 ml of ethanol, 5 ml of a 0.2 M solution of chloroplatinic acid in ethanol and 5 g of activated charcoal, is attached to flask A (generator), and the whole apparatus is flushed with nitrogen. The buret E is filled with 150 ml of 1 M sodium borohydride solution in 1 M sodium hydroxide. A 1 M solution of sodium borohydride in ethanol (20 ml) is injected into the reaction flask B with vigorous stirring to prepare the catalyst. Concentrated hydrochloric acid (5 ml) or glacial acetic acid (4 ml) is injected into the generator A followed by 30 ml of 1 M sodium borohydride solution (to flush the apparatus with hydrogen). A solution of the compound to be hydrogenated (0.5 mol of double bond or its equivalent) is injected into the reactor B. As soon as hydrogen starts reacting the pressure in the apparatus decreases and the solution of sodium borohydride is sucked in continuously from the buret E through the automatic dispenser D until the hydrogenation ends.

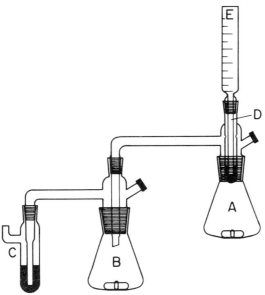

Fig. 2 Apparatus for hydrogenation with hydrogen generated from sodium borohydride.

Hydrogenation *in situ*

Hydrogenation of a compound may be carried out in the same vessel in which hydrogen is generated by decomposition of sodium borohydride. For this purpose the apparatus shown in Fig. 2 is simplified by leaving out the flask B.

The flask A connected to a bubbler C is charged with 100 ml of ethanol, 5 ml of 0.2 M solution of chloroplatinic acid in ethanol and with 5 g of activated charcoal and is flushed with nitrogen. The buret E is filled with 150 ml of 1 M ethanolic solution of sodium borohydride. A 1 M ethanolic solution (20 ml) of sodium borohydride is injected with vigorous stirring to prepare the catalyst followed by 20 ml of concentrated hydrochloric acid (or 10 ml of glacial acetic acid). Finally a solution of 0.5 mol of a compound to be hydrogenated is added through the injection port. The stock solution of sodium borohydride is drained continuously from the automatic dispenser D as soon as the absorption of hydrogen has started and ends automatically when the reaction has finished.

The advantage of hydrogenation with hydrogen generated from sodium borohydride is easy handling of relatively large amounts of compounds to be reduced in relatively small apparatus (much larger equipment would be needed for regular catalytic hydrogenation at atmospheric pressure). The disadvantages include the more expensive source of hydrogen and the more complicated isolation of products which cannot, as in regular catalytic hydrogenation, be obtained just by filtration and evaporation of the solvent but require extraction from the reaction mixture.

3. CATALYTIC HYDROGENATION UNDER ELEVATED PRESSURE

A medium-pressure apparatus is assembled from a stainless steel bomb, standard parts, valves, unions and copper or stainless steel tubing according to Fig. 4.

The stainless steel bomb A (30–500 ml, tested for 125 atm) is fitted with a magnetic stirring bar and charged with the catalyst, the compound to be hydrogenated and the solvent (if necessary). It is connected tightly to a tree carrying two valves and a pressure gauge by means of a union using Teflon® tape around the threads. After closing valve B the apparatus is connected at valve C by means of copper or stainless steel tubing to a regulator of a hydrogen tank. To test

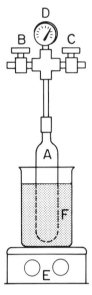

Fig. 4   Apparatus for hydrogenation at medium pressure.

the apparatus for leakage it is pressurized with hydrogen until the pressure gauge for 100 atm D reads 50–100 atm. At this moment, valve C is closed. If the pressure of hydrogen drops immediately and continues dropping within a few minutes, some of the connections are leaking and the leak must be stopped. If tightening of the connections does not prevent the leakage the apparatus must be disassembled and the connections cleaned or replaced. If the pressure does not show any decrease within 5 minutes, or if a small initial drop of pressure due to absorption of hydrogen by the catalyst does not continue, the valve B is connected to an aspirator or a vacuum pump, is slowly opened and the apparatus is evacuated to remove the air originally present. This flushing with hydrogen may be repeated once more.

Thereafter the valve B is closed, and through the valve C hydrogen is introduced until the pressure gauge D indicates the desired pressure. The pressure should be well below the maximum pressure for which the gauge and the bomb are designed. This precaution is necessary since, during exothermic hydrogenations, the temperature of the reaction may raise the pressure considerably, especially if the reaction mixture occupies a large portion of the container.

After both valves B and C have been closed and the pressure reading recorded, magnetic stirring is started. Progress of the reaction is followed by the pressure drop. When the pressure no longer falls the hydrogenation is finished, or is stopped at an intermediate stage (in the case of a multifunctional substrate). In the former case the hydrogen pressure is released by opening of the valve B, the bomb is disconnected, and the contents are worked up.

In the latter case of unfinished hydrogenation the hydrogenation may be tried at a higher temperature by applying an oil bath and by heating of the reaction mixture with continued stirring. The hydrogen pressure rises due to the rise in temperature. After the temperature has stabilized any pressure drop indicates that further hydrogenation is taking place. Stirring and heating may be continued until no more absorption of hydrogen takes place at the same temperature. After cooling the remaining hydrogen is released through the valve B and the apparatus is disassembled.

If during the hydrogenation at room temperature or at an elevated temperature the pressure of hydrogen has dropped to zero more hydrogen may be needed for completing the reduction. In this case the apparatus is refilled with hydrogen by opening the valve C connected to the source of hydrogen. After repressurizing the apparatus to the desired pressure the valve C is closed and the hydrogenation is continued.

After the hydrogenation has definitely finished the final reading of the hydrogen pressure is recorded and the excess of hydrogen is carefully bled off by opening the valve B. Only then is the bomb disconnected. The contents are filtered, the bomb is rinsed once or twice with the solvent, the washings are used to wash the catalyst on the filter, and the filtrate is worked up depending on the nature of the products.

Notes: As soon as the apparatus is pressurized for the hydrogenation and the valve C is closed the master valve on the hydrogen cylinder should be closed for safety reasons. While removing the catalyst precautions must be taken during the filtration of pyrophoric catalysts (p. 13).

4. PREPARATION OF PALLADIUM CATALYST [31]

Powdered sodium borohydride (0.19 g, 5 mmol) is added portionwise at room temperature over a period of 5–10 minutes to a stirred suspension of 0.443 g (2.5 mmol) of powdered palladium chloride in 40 ml of absolute methanol. Stirring is continued until the evolution of hydrogen has ceased (about 20 minutes). The solvent is decanted from the black settled catalyst which is washed twice or three times with a solvent to be used in hydrogenations.

5. REDUCTION WITH RANEY NICKEL

Hydrogenolysis of Halogens [536]

A halogen derivative (0.1 mol) dissolved in 30 ml of methanol is dissolved or dispersed in 100 ml of 1 N methanolic potassium hydroxide. After addition of a suspension of 19 g of methanol-wet and hydrogen-saturated Raney nickel in 70 ml of methanol, hydrogenation is carried out at room temperature and atmospheric pressure until the volume of hydrogen corresponding to the hydrogenolyzed halogen has been absorbed (25–500 minutes). The reaction mixture is filtered with suction, taking precautions necessary for handling pyrophoric catalysts (p. 13); the residue on the filter is washed several times with methanol, and the filtrate is worked up depending on the properties of the product.

With polyhalogen derivatives the amount of the potassium hydroxide must be equivalent to the number of halogens to be replaced by hydrogen.

6. DESULFURIZATION WITH RANEY NICKEL

Preparation of Aldehydes from Thiol Esters [1101]

A solution of 10 g (0.06 mol) of ethyl thiolbenzoate in 200 ml of 70% ethanol is refluxed for 6 hours with 50 g of Raney nickel. After the removal of the catalyst the filtrate is distilled and the aldehyde is isolated from the distillate by treatment with a saturated (40%) solution of sodium bisulfite. The yield of benzaldehyde–sodium bisulfite addition complex is 8.0 g (62%).

Deactivation of Raney Nickel [46]

For desulfurization of compounds containing reducible functions the Raney nickel (20 g) is deactivated prior to the desulfurization by stirring and refluxing with 60 ml of acetone for 2 hours. This removes the hydrogen which is adsorbed in Raney nickel.

7. PREPARATION OF NICKEL CATALYST [31]

To a stirred suspension of 1.24 g (5 mmol) of powdered nickel acetate in 50 ml of 95% ethanol in a 250 ml flask is added 5 ml of 1 M solution of sodium borohydride in 95% ethanol at room temperature. Stirring is continued until the evolution of gas has ceased, usually within 30 minutes. The flask with the colloidal material is used directly in the hydrogenation.

8. HOMOGENEOUS HYDROGENATION

Preparation of the Catalyst Tris(triphenylphosphine)rhodium Chloride [57]

Rhodium chloride trihydrate (1 g, 5.2 mmol) and triphenylphosphine (6 g, 23 mmol) are refluxed in 120 ml of ethanol. The solid tris(triphenylphosphine)rhodium chloride is filtered with suction and washed with ethanol and ether. Yield is 3.5 g (73%).

Reduction of Unsaturated Aldehydes to Saturated Aldehydes [58]

In a flask of a hydrogenation apparatus, 0.2 g (0.00216 mol) of tris(triphenylphosphine)rhodium chloride is dissolved in 75 ml of de-aerated benzene under hydrogen at 25° and 700 mm pressure. After the addition of 1.5 ml (0.0183 mol) of crotonaldehyde (2-butenal) hydrogenation is carried out with vigorous stirring for 16 hours. The solution is washed with three 10 ml portions of a saturated solution of sodium hydrogen sulfite containing some solid metabisulfite, each portion being left in contact with the benzene solution for about 1 hour. The combined aqueous layers and the solid are neutralized with sodium hydrogen carbonate; the solution is extracted with three 10 ml portions of ethyl ether; the combined ether solutions are dried over sodium sulfate and evaporated to give 1.0 ml (61.2%) of butanal.

9. PREPARATION OF LINDLAR CATALYST [36]

To a suspension of 50 g of purest precipitated calcium carbonate in 400 ml of distilled water is added 50 ml of a 5% solution of palladous chloride (PdCl$_2$) and the mixture is stirred for 5 minutes at room temperature and for 10 minutes at 80°. The hot suspension is shaken with hydrogen until no more absorption takes place, filtered with suction and the cake on the filter is washed thoroughly with distilled water. The cake is then stirred vigorously in 500 ml of distilled water; the suspension is treated with a solution of 5 g of lead acetate in 100 ml of water, stirred for 10 minutes at 20° and finally for 40 minutes in a boiling water bath. The catalyst is filtered with suction, washed thoroughly with distilled water and dried at 40–50° *in vacuo*.

10. REDUCTION BY CATALYTIC TRANSFER OF HYDROGEN [77]

In a flask fitted with an efficient stirrer and a nitrogen inlet 0.001 mol of a compound to be hydrogenated dissolved in 4 ml of ethanol or, preferably, glacial acetic acid is treated with an equal weight* of 10% palladium on charcoal and 0.94 ml (0.01 mol) of cyclohexadiene. The flask is kept at 25° and a slow stream of nitrogen is passed through the mixture. After at least 2 hours the mixture is filtered through Celite, the solid is washed with solvent and the filtrate is evaporated *in vacuo* to give 90–100% yield of the hydrogenated product.

11. PREPARATION OF ALANE (ALUMINUM HYDRIDE) [116]

A 300 ml flask fitted with a magnetic stirring bar, an inlet port closed by a septum, and a reflux condenser connected to a calcium chloride tube is charged with 100 ml of a 1.2 M solution of lithium aluminum hydride in tetrahydrofuran and 100 ml of anhydrous tetrahydrofuran. By means of a syringe, 5.88 g (0.060 mol) of 100% sulfuric acid is added slowly with vigorous stirring. The solution is stirred for 1 hour and then allowed to stand at room temperature to permit the precipitated lithium sulfate to settle. The clear supernatant solution is transferred by syringe to a bottle and stored under nitrogen. For precise work the titer of the solution is found by volumetric analysis for hydrogen.

*Per mole of transferred hydrogen.

12. PREPARATION OF LITHIUM TRI-*TERT*-BUTOXYALUMINO-HYDRIDE [*1011*]

A one-liter flask fitted with a stirrer, reflux condenser and separatory flask is charged with 7.6 g (0.2 mol) of lithium aluminum hydride and 500 ml of anhydrous ether. A solution of 44.4 g (0.6 mol) of anhydrous *tert*-butyl alcohol in 250 ml of ether is added slowly from the separatory funnel to the stirred contents of the flask. (The hydrogen evolved is vented to a hood.) During the addition of the last third of the alcohol a white precipitate is formed. The solvent is decanted and the flask is evacuated with heating on the steam bath to remove the residual ether and *tert*-butyl alcohol. The solid residue – lithium tri-*tert*-butoxyaluminohydride – is stored in bottles protected from atmospheric moisture. Solutions 0.2 M in reagent are prepared by dissolving the solid in diglyme.

13. ANALYSIS OF HYDRIDES AND COMPLEX HYDRIDES [*68, 82*]

Of many analytical methods described in the literature, the most simple and sufficiently accurate one is decomposition of lithium aluminum hydride suspended in dioxane by a dropwise addition of water and by measurement of the evolved hydrogen gas.

For this purpose the apparatus shown in Fig. 1 could be used, provided that the flask is replaced by a flask with a side arm closed by a septum. Then, with the stopcock closed so as to connect the flask only to the graduated tube, lithium aluminum hydride is weighed into the flask and covered with a large amount of dioxane. Alternatively, an exact aliquot of a solution of lithium aluminum hydride to be analyzed is pipetted into the flask. The loaded flask containing a magnetic stirring bar is connected to the apparatus and, with stirring, water is added dropwise slowly through the septum with a hypodermic syringe. The volume of the evolved hydrogen is measured after it does not change any more. Sufficient time must be allowed since the decomposition of the lithium aluminum hydride is exothermic and warms up the apparatus [*68*]. For accurate measurement, correction for the partial vapor pressure of dioxane and water must be considered.

14. REDUCTION WITH LITHIUM ALUMINUM HYDRIDE

Acidic Quenching. Reduction of Aldehydes and Ketones [*83*]

In a two-liter, three-necked flask equipped with a dropping funnel, a mechanical stirrer and an efficient reflux condenser protected from moisture by a calcium chloride tube, 19 g (0.5 mol) of lithium aluminum hydride is dissolved in 600 ml of anhydrous ether. From the dropping funnel, 200 g (1.75 mol) of heptanal is introduced at a rate producing gentle reflux. Ten minutes after the last addition and with continued stirring water is added dropwise and cautiously. The flask is cooled if, during the exothermic decomposition of excess hydride, refluxing becomes too vigorous. The mixture is poured into 200 ml of ice-water and treated with 1 liter of 10% sulfuric acid. The ether layer is separated, the aqueous layer is extracted with two 100 ml portions of ether, the combined ether solutions are dried and evaporated, and the residue is fractionated to give an 86% yield of heptanol, b.p. 175–175.5°/750 mm.

15. REDUCTION WITH LITHIUM ALUMINUM HYDRIDE

Alkaline Quenching. Reduction of Nitriles [*121*]

A solution of 12.5 g (0.10 mol) of caprylonitrile (octanenitrile) in 20 ml of anhydrous ethyl ether is slowly added to a stirred and ice-cooled solution of 3.8 g (0.10 mol, 100% excess) of lithium aluminum hydride in 200 ml of anhydrous ethyl ether. With continued cooling and vigorous stirring 4 ml of water, 3 ml of 20% solution of sodium hydroxide, and 14 ml of water are added in succession. The ether solution is decanted from the white granular residue, the residue is washed twice with ether, all the ether solutions are combined, the ether is distilled off, and the product is distilled at 53–54° at 6 mm to yield 11.5–11.9 g (89–92%) of octylamine.

16. REDUCTION WITH LITHIUM TRI-*TERT*-BUTOXYALUMINO-HYDRIDE

Reduction of Acyl Chlorides to Aldehydes [1011]

A 500 ml flask fitted with a stirrer, separatory funnel, low-temperature thermometer and nitrogen inlet and outlet is charged with a solution of 37.1 g (0.20 mol) of *p*-nitrobenzoyl chloride in 100 ml of diglyme. The flask is flushed with dry nitrogen and cooled to approximately −78° with a dry ice–acetone bath. A solution (200 ml) of 50.8 g (0.20 mol) of tri-*tert*-butoxyaluminohydride in diglyme is added over a period of 1 hour with stirring, avoiding any major rise in temperature. The cooling bath is removed and the flask is allowed to warm up to room temperature over a period of approximately 1 hour. The contents are then poured onto crushed ice. The precipitated *p*-nitrobenzaldehyde (with some unreacted acyl chloride) is filtered with suction until dry and then extracted several times with 95% ethanol. The combined extracts are evaporated to give 24.5 g (81%) of *p*-nitrobenzaldehyde, m.p. 103–104°, which after recrystallization from water or aqueous ethanol affords 20.3 g (67%) of pure product, m.p. 104–105°.

17. REDUCTION WITH ALANE (ALUMINUM HYDRIDE) *IN SITU*

Reduction of Nitriles to Amines [787]

A one-liter three-necked flask equipped with a gas-tight mechanical stirrer, a dropping funnel and an efficient reflux condenser connected to a dry-ice trap is charged with 3.8 g (0.1 mol) of lithium aluminum hydride and 100 ml of anhydrous ether. Through the dropping funnel a solution of 13.3 g (0.1 mol) of anhydrous aluminum chloride is added rapidly. Five minutes after the last addition of halide a solution of 19.3 g (0.1 mol) of diphenylacetonitrile in 200 ml of ether is added dropwise to the well-stirred solution. One hour after the last addition of the nitrile decomposition of the reaction mixture and excess halide is carried out by dropwise addition of water followed by 140 ml of 6 N sulfuric acid diluted with 100 ml of water. The clear mixture is transferred to a separatory funnel, the ether layer is separated and the aqueous layer is extracted with four 100 ml portions of ether. After cooling by ice-water the aqueous layer is alkalized by solid potassium hydroxide to pH 11, diluted with 600 ml of water, and extracted with four 100 ml portions of ether. The combined ether extracts are stirred with Drierite and evaporated under reduced pressure. The residue (21.5 g) is distilled to give 91% yield of 2,2-diphenylethylamine, b.p. 184°/17 mm, m.p. 44–45°.

18. REDUCTION WITH DIISOBUTYLALANE

Reduction of α,β-Unsaturated Esters to Unsaturated Alcohols [1151]

Diisobutylalane* (diisobutylaluminum hydride) (0.85 mol) is added under nitrogen over a period of 10 hours to a stirred solution of 34.4 g (0.2 mol) of diethyl fumarate in 160 ml of benzene. The temperature rises to 50°. After standing overnight at room temperature, the reaction mixture is decomposed by the addition of 76.8 g (2.4 mol) of methanol in 150 ml of benzene followed by 4.5 g (2.5 mol) of water. The aluminum salts are filtered with suction, washed several times with methanol, and the filtrate is distilled to give 12.3 g (70%) of *trans*-2-butene-1,4-diol, b.p. 86–88°/0.5 mm.

* Diisobutylalane is available as a 1 M solution in hexane, Dibal-H®. Dibal-H® is a trade name of the Aldrich Chemical Company.

19. REDUCTION WITH BORANE

Reduction of Carboxylic Acids to Alcohols [971]

An oven-dried 100 ml flask with a side arm closed with a septum is fitted with a magnetic stirring bar and a reflux condenser connected to a mercury bubbler. The flask is cooled to room temperature under nitrogen, charged with 4.36 g (0.025 mol) of adipic acid monoethyl ester followed by 12.5 ml of anhydrous tetrahydrofuran, and cooled to −18° by immersion in an ice-salt bath. Then 10.5 ml of 2.39 M (or 25 ml of 1 M) solution of borane in tetrahydrofuran (0.025 mol) is slowly added dropwise over a period of 19 minutes. The resulting clear reaction mixture is stirred well and the ice–salt bath is allowed to warm slowly to room temperature over a 16-hour period. The mixture is hydrolyzed with 15 ml of water at 0°. The aqueous phase is treated with 6 g of potassium carbonate (to decrease the solubility of the alcohol-ester in water), the tetrahydrofuran layer is separated and the aqueous layer is extracted three times with a total of 150 ml of ether. The combined ether extracts are washed with 30 ml of a saturated solution of sodium chloride, dried over anhydrous magnesium sulfate, and evaporated *in vacuo* to give 3.5 g (88%) of a colorless liquid which on distillation yields 2.98 g (75%) of ethyl 6-hydroxyhexanoate, b.p. 79°/0.7 mm.

20. REDUCTION WITH BORANE *IN SITU*

Reduction of Esters to Alcohols [738]

In a one-liter three-necked flask equipped with a mechanical stirrer, a dropping funnel, a thermometer and a reflux condenser, 8.5 g (0.225 mol) of sodium borohydride is dissolved in 250 ml of diglyme with stirring. After addition of 125 g (0.4 mol) of ethyl stearate a solution of 0.084 mol of aluminum chloride (42 ml of a 2 M solution in diglyme) is added through the dropping funnel while the mixture is stirred vigorously. The rate of addition is adjusted so that the temperature inside the flask does not rise above 50°. After all of the aluminum chloride has been added the reaction mixture is stirred for 1 hour at room temperature followed by heating on a steam cone for 0.5–1 hour.

After cooling to room temperature the mixture is poured onto a mixture of 500 g of crushed ice and 50 ml of concentrated hydrochloric acid. The precipitate is filtered with suction, washed with ice-water, pressed and dried *in vacuo*. Recrystallization from aqueous alcohol affords 98.2 g (91%) of 1-octadecanol, m.p. 58–59°.

21. REDUCTION WITH SODIUM BOROHYDRIDE

Reduction of Ketones to Alcohols [68]

A 125 ml flask fitted with a magnetic stirring bar, a reflux condenser, a thermometer and a separatory funnel is charged with 7.2 g (9 ml, 0.1 mol) of 2-butanone (methyl ethyl ketone). From the separatory funnel a solution of 1.5 g (0.04 mol, 60% excess) of sodium borohydride in 15 ml of water is added dropwise with stirring at such a rate as to raise the temperature of the reaction mixture to 40° and maintain it at 40–50°. Cooling with a water bath may be applied if the temperature rises above 50°. After the addition has been completed (approximately 30 minutes) the mixture is stirred until the temperature drops to 30°. It is then transferred to a separatory funnel and saturated with sodium chloride. The aqueous layer is drained and the organic layer is dried with anhydrous potassium carbonate. Distillation affords 5.5–6.0 g (73–81%) of 2-butanol, b.p. 90–95°.

22. REDUCTION WITH SODIUM CYANOBOROHYDRIDE

Reduction of Enamines to Amines [103]

To a solution of 0.4 g (0.002 mol) of ethyl 3-(N-morpholino)crotonate in 4 ml of methanol at 25° is added a trace of bromocresol green (pH 3.8–5.4) followed by the addition of a 2 N solution of hydrogen chloride in methanol until the color turns yellow. Then 0.13 g (0.002 mol) of sodium cyanoborohydride is added with stirring, and the methanolic hydrogen chloride is added dropwise to maintain the yellow color. After stirring at 25° for 1 hour the solution is poured into 5 ml of 0.1 N sodium hydroxide, saturated with sodium chloride and extracted with three 10 ml portions of ether. The combined extracts are evaporated *in vacuo*; the residue (0.315 g) is dissolved in 10 ml of 1 N hydrochloric acid; the solution is extracted with two 10 ml portions of ether; the aqueous layer is alkalized to pH > 9 with 6 N potassium hydroxide, saturated with sodium chloride and extracted with three 10 ml portions of ether. The combined extracts are dried over anhydrous magnesium sulfate and evaporated *in vacuo* to give 0.260 g (65%) of chromatographically pure ethyl 3-morpholinobutyrate.

23. REDUCTION WITH TRIETHYLSILANE

Ionic Reduction of Alkenes Capable of Forming Carbonium Ions [342]

To a mixture of 1.16 g (0.01 mol) of triethylsilane and 2.28 g (0.02 mol) of trifluoroacetic acid is slowly added 0.96 g (0.01 mol) of 1-methylcyclohexene. The mixture is maintained at 50° for 10 hours, then poured into water; the hydrocarbon layer is separated, the aqueous layer is extracted with ether, and the combined organic layers are neutralized, washed with water, dried and distilled to give 67% yield of methylcyclohexane.

24. REDUCTION WITH STANNANES

Hydrogenolysis of Alkyl Halides [508]

In a 50 ml flask fitted with a reflux condenser 5.5 g (0.0189 mol) of tributylstannane (tributyltin hydride) is added to 3.5 g (0.018 mol) of octyl bromide. An exothermic reaction ensues. Cooling is applied if the temperature rises above 50°. After 1 hour the mixture is distilled, giving 80% yield of octane and 90% yield of tributyltin bromide.

1-Bromo-2-phenylethane requires heating with tributylstannane at 100° for 4 hours to give an 85% yield of ethylbenzene and 87% of tributyltin bromide, and benzyl chloride requires heating with tributylstannane at 150° for 20 minutes to give a 78% yield of toluene and an 80% yield of tributyltin chloride.

25. ELECTROLYTIC REDUCTION

Partial Reduction of Aromatic Rings [132]

In an electrolytic cell (Fig. 5) consisting of platinum electrodes (2 cm × 5 cm in area) and cathode and anode compartments separated by an asbestos divider, each compartment is charged with 17 g (0.4 mol) of lithium chloride and 450 ml of anhydrous methylamine. Isopropylbenzene (12 g, 0.1 mol) is placed in the cathode compartment and a total of 50,000 coulombs (2.0 A, 90 V) is passed through the solution in 7 hours. After evaporation of the solvent the mixture is hydrolyzed by the slow addition of water and extracted with ether; the ether extracts are dried and evaporated to give 9.0 g (75%) of product boiling at 149–153° and consisting of 89% of a mixture of isomeric isopropylcyclohexenes and 11% of recovered isopropylbenzene.

Similar electrolysis without the divider affords 82% yield of a product containing 78% of 2,5-

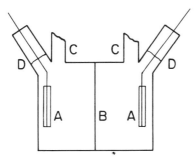

Fig. 5 Electrolytic cell. A, electrodes; B, diaphragm; C, reflux condensers; D, glass seals.

dihydroisopropylbenzene beside 6% of isopropylcyclohexene and 13% of recovered isopropylbenzene.

26. REDUCTION WITH SODIUM (BIRCH REDUCTION)

Partial Reduction of Aromatic Rings [399]

An apparatus is assembled from a 250 ml three-necked flask fitted with a magnetic stirring bar, an inlet tube reaching almost to the bottom and an efficient dry-ice reflux condenser connected to a calcium chloride tube. The remaining opening for addition of solids is stoppered, the reflux condenser is charged with dry ice and acetone, and the flask is immersed in a dry-ice acetone bath. From a tank of ammonia 100 ml of liquid ammonia is introduced. α-Naphthol (10 g, 0.07 mol) and 2.7 g (0.068 mol) of powdered sodamide are added followed by 12.5 g of *tert*-amyl alcohol and finally by 3.2 g (0.13 g-atom) of sodium in small pieces. After the blue color has disappeared the cooling bath and the reflux condenser are removed and ammonia is allowed to evaporate. Water (100 ml) is added and then the solution is extracted several times with ether and acidified with dilute hydrochloric acid. The precipitated oil solidifies and gives, after recrystallization from petroleum ether (b.p. 80–100°), 8.5 g (83.5%) of 5,8-dihydro-α-naphthol.

27. REDUCTION WITH SODIUM

Acyloin Condensation of Esters [1048]

A five-liter three-necked flask fitted with a high-speed stirrer (2000–2500 r.p.m.), a reflux condenser and a separatory funnel is charged with 115 g (5 g-atoms) of sodium and 3 liters of xylene and heated in an oil bath (or a heating mantle) to 105°. The air in the flask is displaced by nitrogen, and when the sodium has melted stirring is started and the sodium is dispersed in a finely divided state in xylene. From the separatory funnel, 535 g (2.5 mol) of methyl laurate is added at such a rate as to maintain the temperature below 110°. This requires approximately 1 hour. Stirring is continued for an additional 30 minutes. Small particles of unreacted sodium are then destroyed by the addition of 1–2 mol of methanol (40–80 ml). After cooling to 80°, 0.5–1.0 liter of water is added cautiously. The organic layer is separated, washed once or twice with water, neutralized with a slight excess of mineral acid, and washed free of acid with a solution of sodium bicarbonate. The xylene is removed by steam distillation and the oily residue is poured into a suitable vessel to solidify. The crude product contains 80–90% of acyloin and is purified by crystallization from 95% ethanol. (Precautions necessary for work with sodium and flammable liquids have to be observed.)

28. REDUCTION WITH SODIUM NAPHTHALENE

Cleavage of Sulfonamides to Amines [705]

In an Erlenmeyer flask capped with a rubber septum a mixture of naphthalene, 1,2-dimethoxy-ethane and enough sodium to yield an 0.5–1.0 M solution of anion radical is stirred with a glass-covered stirring bar for 1–1.5 hours, by which time the anion radical will have formed and scavenged the oxygen inside the flask. A solution of one-sixth to one-third of an equivalent of the sulfonamide in dimethoxyethane is injected by syringe and the mixture is stirred at room temperature for approximately 1 hour. Quenching with water produces amines in 68–94% isolated yields.

29. REDUCTION WITH SODIUM AMALGAM

Preparation of Sodium Amalgam [68]

Clean sodium (5 g, 0.217 g-atom) is cut into small cubes of 3–5 mm sides. Each individual cube is placed on the surface of 245 g of mercury in a mortar, pressed under the surface of the mercury and held there by means of a pestle. As soon as it dissolves, further cubes are treated in the same way without any delay. The dissolution is accompanied by a hissing sound and sometimes by slight effervescense. It is strongly exothermic and warms up the mercury well above room temperature. Warmer mercury dissolves sodium more readily. The addition of sodium must therefore be accelerated in order not to let the mercury cool. When all the sodium has dissolved the liquid is poured into another dish and allowed to cool, whereafter it is crushed to pieces and stored in a closed bottle. The whole operation must be carried out in an efficient hood. If a cube of sodium is not entirely submerged its dissolution may be accompanied by a flash of fire.

Reduction of α,β-Unsaturated Acids [68]

In a 250 ml thick-walled bottle with a ground glass stopper 7.4 g (0.05 mol) of cinnamic acid is dissolved in a solution of 2 g (0.05 mol) of sodium hydroxide in 65 ml of water until the reaction is alkaline to phenolphthalein. A total of 150 g of sodium amalgam containing 3 g (0.13 g-atom) of sodium is added in pieces. After each addition the bottle is shaken vigorously until the solid amalgam changes to liquid mercury. Only then is the next piece of the amalgam added. When all the amalgam has reacted the bottle is warmed for 10 minutes at 60–70°, the contents of the bottle are transferred into a separatory funnel, and the mercury layer is drained; the aqueous layer is transferred into an Erlenmeyer flask, and acidified with concentrated hydrochloric acid to Congo Red. Hydrocinnamic acid precipitates as an oil which soon crystallizes. Crystallization from 45° warm water, slightly acidified with hydrochloric acid, gives 5.5–6.0 g (73–80%) of hydrocinnamic acid, m.p. 48–49°.

30. REDUCTION WITH ALUMINUM AMALGAM

Reduction of Aliphatic–Aromatic Ketones to Pinacols [144]

In a one-liter flask fitted with a reflux condenser 31.8 g (0.265 mol) of acetophenone is dissolved in 130 ml of absolute ethanol and 130 ml of dry sulfur-free benzene, and 0.5 g of mercuric chloride and 8 g (0.296 g-atom) of aluminum foil are added. Heating of the mixture initiates a vigorous reaction which is allowed to proceed without external heating until it moderates. Then the flask is heated to maintain reflux until all of the aluminum has dissolved (2 hours). After cooling the reaction mixture is treated with dilute hydrochloric acid and the product is extracted with benzene. The combined organic layers are washed with dilute acid, with a solution of sodium carbonate, and with a saturated solution of sodium chloride; they are then dried with sodium sulfate and the solvent is evaporated under reduced pressure. A rapid vacuum distillation affords a fraction at 160–170° at 0.5 mm which, after dissolving in petroleum ether (b.p. 65–110°), gives 18.0 g (56%) of 2,3-diphenyl-2,3-butanediol, m.p. 100–123°.

31. REDUCTION WITH ZINC (CLEMMENSEN REDUCTION)

Reduction of Ketones to Hydrocarbons [*68, 758*]

In a one-liter flask fitted with a reflux condenser, 400 g (5.3 mol, 3.2 equivalent) of mossy zinc is treated with 800 ml of a 5% aqueous solution of mercuric chloride for 1 hour. The solution is decanted; then 100 g (0.835 mol) of acetophenone is added followed by as much hydrochloric acid diluted with the same volume of water as is needed to cover all the zinc. The mixture is refluxed for 6 hours during which time additional dilute hydrochloric acid is added in small portions. After cooling the upper layer is separated, washed free of acid, dried and distilled to give 70 g (79%) of ethylbenzene, b.p. 135–136°.

32. REDUCTION WITH ZINC IN ALKALINE SOLUTION

Reduction of Nitro Compounds to Hydrazo Compounds*

To a mixture of 50 g (0.4 mol) of nitrobenzene, 180 ml of 30% sodium hydroxide, 20 ml of water and 50 ml of ethanol, 100–125 g (1.5–1.9 g-atom) of zinc dust is added portionwise with efficient mechanical stirring until the red liquid turns light yellow. After stirring for an additional 15 minutes, 1 liter of cold water is added, the mixture is filtered with suction, the solids on the filter are washed with water, and the hydrazobenzene is extracted from the solids by boiling with 750 ml of ethanol. The mixture is filtered while hot, the filtrate is cooled, the precipitated crystalline hydrazobenzene is filtered with suction, and the mother liquor is used for repeated extraction of the zinc residue. Recrystallization of the crude product from an alcohol–ether mixture gives hydrazobenzene of m.p. 126–127°.

33. REDUCTION WITH ZINC AND SODIUM IODIDE

Reductive Cleavage of Sulfonates to Hydrocarbons [*700*]

A mixture of a solution of 0.300 g of an alkyl methanesulfonate or *p*-toluenesulfonate in 3–6 ml of 1,2-dimethoxyethane, 0.300 g of sodium iodide, 0.300 g of zinc dust, and 0.3 ml of water is stirred and refluxed for 4–5 hours. After dilution with ether the mixture is filtered; the solution is washed with water, with 5% aqueous hydrochloric acid, with 5% aqueous solution of potassium hydrogen carbonate, with 5% aqueous solution of sodium thiosulfate and with water. After drying with anhydrous sodium sulfate the solution is evaporated and the residue worked up, giving 26–84% yield of alkane.

34. REDUCTION WITH IRON

Partial Reduction of Dinitro Compounds [*165*]

A 500 ml three-necked round-bottomed flask fitted with a robust mechanical stirrer, a reflux condenser and an inlet is charged with 50 g (0.3 mol) of *m*-dinitrobenzene, 14 g of 80 mesh iron filings, 45 ml of water and 7.5 ml of a 33% solution of ferric chloride (or 50 ml of water and 3.7 g of sodium chloride). The flask is immersed in a preheated water bath, stirring is started at 35 r.p.m., and additional iron filings are added to the mixture in two 13 g portions after 30 minutes and 1 hour. After 2.5 hours of stirring and heating the reaction mixture is cooled and extracted with ether, giving an essentially quantitative yield of *m*-nitroaniline.

---

* E. Fischer's procedure, Beilstein's *Handbuch der organischen Chemie*, 4th edn, Hauptwerk, Vol. XV, p. 68 (1932).

35. REDUCTION WITH TIN

Preparation of Tin Amalgam [*174*]

To a solution of 15 g of mercuric chloride in 100 ml of water is added 100 g of 30 mesh tin metal with vigorous shaking. After a few minutes, when all the tin has acquired a shiny coating of mercury, the liquid is decanted and the tin amalgam is washed repeatedly with water until the washings are clear. It is stored under distilled water.

Reduction of Quinones to Hydroquinones [*174*]

To 5.7 g (0.0527 mol) of *p*-benzoquinone is added 10 g (approximately 0.072 mol) of tin amalgam and 50 ml of glacial acetic acid. The mixture is heated on a steam bath. After 3 minutes green crystals of quinhydrone precipitate but soon dissolve to give a light yellow solution. After 0.5 hour the solution is filtered, the solvent is removed *in vacuo*, and the residue is recrystallized from benzene–acetone to give 5.0 g (88%) of hydroquinone, m.p. 169–190°.

36. REDUCTION WITH STANNOUS CHLORIDE

Deoxygenation of Sulfoxides [*186*]

A solution of 0.005 mol of a sulfoxide, 2.26 g (0.01 mol) of stannous chloride dihydrate, and 2.0 ml of concentrated hydrochloric acid in 10 ml of methanol is refluxed for 2–22 hours. The mixture is cooled, diluted with 20 ml of water and extracted with two 25 ml portions of benzene. The dried extracts are evaporated *in vacuo* leaving virtually pure sulfide which can be purified by filtration of its solution through a short column of alumina and vacuum distillation or sublimation. Yields are 62–93%.

37. REDUCTION WITH CHROMOUS CHLORIDE

Preparation of Chromous Chloride [*188*]

Zinc dust (10 g) is shaken vigorously with a solution of 0.8 g of mercuric chloride and 0.5 ml of concentrated hydrochloric acid in 10 ml of water for 5 minutes. The supernatant liquid is decanted, 20 ml of water and 2 ml of concentrated hydrochloric acid is added to the residue, and 5 g of chromic chloride is added portionwise with swirling in a current of carbon dioxide. The dark blue solution is kept under carbon dioxide until used.

Reduction of Iodo Ketones to Ketones [*188*]

A solution of 1 g of 2-iodo-$\triangle^4$-3-ketosteroid in 50–100 ml of acetone is treated under carbon dioxide portionwise with 20 ml of the solution of chromous chloride. After 10–30 minutes water is added and the product is filtered or extracted with ether and recrystallized. Yield of pure $\triangle^4$-3-ketosteroid is 60–63%.

38. REDUCTION WITH TITANIUM TRICHLORIDE

Reduction of 2,4-Dinitrobenzaldehyde to 2-Amino-4-nitrobenzaldehyde [*590*]

To a solution of 9.25 g (0.06 mol) of titanium trichloride in 57 ml of boiled water is added 66 g of concentrated hydrochloric acid and the solution is diluted to 1 liter with boiled water. The mixture is heated to boiling, a current of carbon dioxide is passed through it, and a hot solution of 1.96 g (0.01 mol) of 2,4-dinitrobenzaldehyde in ethanol is added. The dark blue color fades almost instantaneously. Addition of an excess of sodium acetate precipitates 2-amino-4-nitro-benzaldehyde which crystallizes on cooling. Repeated extraction with benzene and evaporation of the extract gives 0.8 g (50%) of 2-amino-4-nitrobenzaldehyde.

39. REDUCTION WITH LOW-VALENCE TITANIUM

Intermolecular Reductive Coupling of Carbonyl Compounds [206]

A mixture of 0.328 g (0.0473 mol) of lithium and 2.405 g (0.0156 mol) of titanium trichloride in 40 ml of dry 1,2-dimethoxyethane is stirred and refluxed under argon for 1 hour. After cooling a solution of 0.004 mol of ketone in 10 ml of 1,2-dimethoxyethane is added and the refluxing is continued for 16 hours. The mixture is cooled to room temperature, diluted with petroleum ether, and filtered through Florisil on a sintered glass filter; the residue is cautiously quenched by a slow addition of methanol and the filtrate is concentrated under reduced pressure to give up to 97% yield of crude alkylidenealkane.

A modified procedure suitable for intramolecular reductive coupling is achieved using low-valence titanium prepared by reduction of titanium trichloride with a zinc–copper couple followed by the extremely slow addition of ketone to the refluxing reaction mixture (0.0003 mol over a 9-hour period by use of a motor-driven syringe pump) [560].

40. REDUCTION WITH VANADOUS CHLORIDE

Preparation of Vanadous Chloride [213]

Vanadium pentoxide (60 g, 0.33 mol) is covered with 250 ml of concentrated hydrochloric acid and 250 ml of water in a one-liter flask equipped with a gas outlet tube dipping under water. Amalgamated zinc is prepared by treatment of 90 g of 20 mesh zinc for 10 minutes with 1 liter of 1 N hydrochloric (or sulfuric) acid containing 0.1 mol of mercuric chloride, by decanting of the solution and washing the zinc thoroughly with water, with 1 N sulfuric acid and finally with water. The zinc is added to the mixture of vanadium pentoxide and hydrochloric acid and the mixture is allowed to stand for 12 hours to form a brown solution. An additional 200 ml of concentrated hydrochloric acid is added giving, after 24 hours, a clear purple solution of vanadous chloride. The solution is approximately 1 M and is stored over amalgamated zinc or, after filtration, in an atmosphere of nitrogen. In time the purple solution turns brown but the purple color is restored by addition of a small amount of concentrated hydrochloric acid.

Reduction of Azides to Amines [217]

A 1 M aqueous solution of vanadous chloride (10 ml) is added to a stirred solution of 1 g of an aryl azide in 10 ml of tetrahydrofuran. After the evolution of nitrogen has subsided the mixture is poured into 30 ml of aqueous ammonia; the resultant slurry is mixed with benzene, filtered with suction, and the filter cake is washed with benzene. The benzene layer of the filtrate is separated, the aqueous layer is extracted with benzene, the combined benzene solutions are dried, the solvent is evaporated and the residue is distilled to give 70–95% of the amine.

41. REDUCTION WITH HYDRIODIC ACID

Reduction of Acyloins to Ketones [917]

To a mixture of 0.372 g (0.012 mol) of red phosphorus and 4.57 g (0.036 mol) of iodine in 30 ml of carbon disulfide stirred for 10 minutes is added 2.12 g (0.01 mol) of benzoin (alone or dissolved in 10 ml of benzene) followed by 0.8 g (0.01 mol) of pyridine a few minutes later. The dark reaction mixture is allowed to stand for 3.5 hours at room temperature, then poured into aqueous thiosulfate. The organic layer is separated, the layer is extracted with 25 ml of benzene, and the combined organic solutions are dried with anhydrous magnesium sulfate and evaporated to give a crude product which, after passing through a short column of alumina followed by crystallization, affords 1.76 g (90%) of deoxybenzoin, m.p. 58–60°.

42. REDUCTION WITH HYDROGEN SULFIDE

Reduction of α-Diketones to Acyloins or Ketones [237]

Hydrogen sulfide is introduced into an ice-cooled solution of 0.2 mol of a 1,2-diketone and 0.02 mol of piperidine in 30 ml of dimethylformamide for 1–4 hours. Elemental sulfur is precipitated during the introduction of hydrogen sulfide. The mixture is acidified with dilute hydrochloric acid; the sediment is filtered with suction and dissolved in warm methanol; the undissolved sulfur is separated, and the product is isolated by evaporation of the methanol and purified by crystallization. If the product after the acidification is liquid it is isolated by ether extraction and distillation after drying of the ether extract with sodium sulfate. Yield of benzoin from benzil after 1 hour of treatment with hydrogen sulfide is quantitative.

If hydrogen sulfide is introduced for 4 hours into a solution of 0.2 mol of an α-diketone in methanol and 0.02 mol of pyridine, a monoketone is obtained. The yield of deoxybenzoin from benzoin is quantitative.

43. REDUCTION WITH SODIUM SULFITE

Partial Reduction of Geminal Polyhalides [254]

In a 250 ml flask fitted with a magnetic stirring bar, a separatory funnel and a column surmounted by a reflux condenser with an adapter for regulation of the reflux ratio is placed 29 g (0.105 mol) of 1,1-dibromo-1-chlorotrifluoroethane. The flask is heated at 65–72°, and 100 g of a solution containing 14.7% of sodium sulfite (0.117 mol) and 6.3% of sodium hydroxide (0.157 mol) is added. The rate of addition and the intensity of heating are regulated so that the product boiling at 50° condenses in the reflux condenser without flooding the column. The reflux ratio is set so as to keep the temperature in the column head at 50°. The distillation is continued until all the product has been removed. Yield of 1-bromo-1-chloro-2,2,2-trifluoroethane is 80–90%.

44. REDUCTION WITH SODIUM HYDROSULFITE (DITHIONATE)

Reduction of Nitro Compounds to Amino Compounds [257]

A suspension of 1 g (0.005 mol) of 2-chloro-6-nitronaphthalene in 60 ml of hot ethanol is treated gradually with a solution of 3.5 g (0.02 mol) of sodium hydrosulfite (dithionate) $(Na_2S_2O_4)$ in 16 ml of water. The mixture is refluxed for 1 hour, cooled and filtered from most of the inorganic matter. The filtrate is diluted with 2–3 volumes of water. The precipitate is redissolved by heating, the hot solution is filtered, and the almost pure halogeno-amine is separated after cooling and recrystallized from 40% aqueous pyridine to give an almost quantitative yield of 2-amino-6-chloronaphthalene, m.p. 123°.

45. REDUCTION WITH HYDRAZINE

Wolff–Kizhner Reduction of Ketones to Hydrocarbons [281]

In a flask fitted with a take-off adapter surmounted by a reflux condenser and a thermometer reaching to the bottom of the flask a mixture of 40.2 g (0.30 mol) of propiophenone, 40 g of potassium hydroxide, 300 ml of triethylene glycol and 30 ml of 85% hydrazine hydrate (0.51 mol) is refluxed for 1 hour. The aqueous liquor is removed by means of the take-off adapter until the temperature of the liquid rises to 175–178°, and refluxing is continued for 3 hours. The reaction mixture is cooled, combined with the aqueous distillate and extracted with ether. The ether extract is washed free of alkali, dried and distilled over sodium to give 29.6 g (82.2%) of propylbenzene, b.p. 160–163°.

46. REDUCTION WITH HYPOPHOSPHOROUS ACID

Replacement of Aromatic Primary Amino Groups by Hydrogen [594]

To a solution of 3.65 g (0.03 mol) of p-aminobenzylamine in 60 g (0.45 mol) of 50% aqueous hypophosphorous acid placed in a 200 ml flask 50 ml of water is added and the mixture is cooled to 5°. With stirring, a solution of 2.3 g (0.033 mol) of sodium nitrite in 10 ml of water is added dropwise over a period of 10–15 minutes. The mixture is stirred for a half-hour at 5° and allowed to come to room temperature and to stand for about 4 hours. After strong alkalization with sodium hydroxide the product is obtained by continuous extraction with ether, drying of the extract with potassium hydroxide, and distillation. The yield of benzylamine is 2.67 g (84%), b.p. 181–183°.

47. DESULFURIZATION WITH TRIALKYL PHOSPHITES

Desulfurization of Mercaptans to Hydrocarbons [672]

A mixture of 83 g (0.5 mol) of triethyl phosphite (previously distilled from sodium) and 73 g (0.5 mol) of octyl mercaptan (octanethiol) in a Pyrex flask fitted with an efficient column is irradiated with a 100 watt General Electric S-4 bulb at a distance of 12.5 cm from the flask. After 6.25 hours of irradiation the mixture is distilled to give 50.3 g (88%) of octane, b.p. 122–124.5°.

48. REDUCTION WITH ALCOHOLS

Meerwein–Ponndorf–Verley Reduction [68, 309]

A mixture of 20 g (0.1 mol) of aluminum isopropoxide, 0.1 mol of an aldehyde or a ketone and 100 ml of dry isopropyl alcohol is placed in a 250 ml flask surmounted by an efficient column fitted with a column head providing for variable reflux. The mixture is heated in an oil bath or by a heating mantle until the by-product of the reaction – acetone – starts distilling. The reflux ratio is adjusted so that the temperature in the column head is kept at about 55° (b.p. of acetone) and acetone only is collected while the rest of the condensate, mainly isopropyl alcohol (b.p. 82°), flows down to the reaction flask. When no more acetone is noticeable in the condensate based on the test for acetone by 2,4-dinitrophenylhydrazine the reflux regulating stopcock is opened and most of the isopropyl alcohol is distilled off through the column. The residue in the distilling flask is cooled, treated with 200 ml of 7% hydrochloric acid and extracted with benzene; the benzene extract is washed with water, dried and either distilled if the product of the reduction is volatile or evaporated *in vacuo* in the case of non-volatile or solid products. Yields of the alcohols are 80–90%.

Reduction with Alcohol on Alumina [755]

Neutral aluminum oxide (Woelm W-200) is heated in a quartz vessel at 400° at 0.06 mm for 24 hours. A glass-wool plug placed in the neck of the quartz flask prevents alumina from escaping from the vessel. About 5 g of the dehydrated alumina is transferred inside a nitrogen-filled glove bag to an oven-dried, tared 25 ml round-bottom flask containing a magnetic stirring bar. The flask is stoppered, removed from the glove bag, and approximately 5 ml of an inert solvent such as diethyl ether, chloroform, carbon tetrachloride or hexane containing 0.5 g of isopropyl alcohol is added. After stirring for 0.5 hour at 25°, 0.001 mol of an aldehyde or ketone in about 1 ml of solvent is added and stirring is continued for 2 hours at 25°. Methanol (5 ml) is added; the mixture is stirred for 15 minutes and then suction-filtered through Celite. The Celite is washed with 35 ml of methanol and the combined filtrates are evaporated at reduced pressure.

This method reduces aldehydo group preferentially to a keto group in yields up to 88% and with 95% selectivity.

49. REDUCTIVE AMINATION (LEUCKART REACTION)

Preparation of Secondary Amines from Ketones [321]

In a distilling flask fitted with a thermometer reaching nearly to the bottom are placed 4 mol of ethylammonium formate prepared by adding ethylamine to cold formic acid. The mixture is heated to 180–190° and maintained at this temperature as long as water distils. After cooling, 1 mol of acetophenone is added and the mixture is heated to boiling. The water and amine distil and are collected in concentrated hydrochloric acid. If any ketone codistils it is returned to the flask at intervals. The temperature is raised to 190–230° and maintained there for 4–8 hours. After cooling the mixture is diluted with 3–4 volumes of water which precipitates an oil. The aqueous layer is mixed with the hydrochloric acid used for collecting the distilled amine and the mixture is evaporated to give the hydrochloride of the excess amine used. The oil is refluxed for several hours with 150–200 ml of hydrochloric acid. After the hydrolysis is complete the solution is filtered through moist paper, the filtrate, if not clear, is extracted once or twice with ether, and the amine is liberated by basifying with ammonia and isolated by extraction with ether and distillation. Yield of ethyl-α-phenethylamine is 70%.

50. BIOCHEMICAL REDUCTION

Reduction of Diketones to Keto Alcohols [327]

In a two-liter rotating flask 0.154 g (0.94 mmol) of 1,2,3,4,5,6,7,8-octahydronaphthalene-1,5-dione is treated with 440 ml of phosphate buffer for pH 7, 4 g saccharose and 20 g of wet centrifuged mycelium of *Curvularia falcata* for 20 hours. Extraction with ether and evaporation of the extract affords 0.194 g of dark-brown resinous material which after chromatography on 11.6 g of aluminum oxide activity III gives on elution with benzene and benzene–ether (9:1) 0.120 g (77%) of (5S)-5-hydroxy-1-oxo-1,2,3,4,5,6,7,8-octahydronaphthalene, m.p. 93–94° (ether–heptane), $[\alpha]_D$ −55° (c 0.76, 1.05). The first benzene eluates contain 3% of (5S, 9R)-5-hydroxy-*trans*-decalone.

Reduction of 1,2-Diketones to 1,2-Diols [834]

A one-liter Erlenmeyer flask containing 250 ml of a sterile solution of 6% glucose, 4% peptone, 4% yeast extract and 4% malt extract is inoculated with a culture of *Cryptococcus macerans* and is shaken at 30° for 2 days. After addition of 0.100 g (0.545 mmol) of 2,2-pyridil the shaking is continued for 7 days. The suspension is then alkalized with 10% potassium hydroxide, and extracted three times with 250 ml portions of ethyl acetate. The combined extracts are dried over anhydrous sodium sulfate and evaporated *in vacuo* to give 80% yield of *threo-* and 5% of *erythro*-di(2-pyridyl)ethanediol. The mixture is separated by thick-layer chromatography over silica gel by elution with a 1:1 mixture of ethyl acetate and hexane to give 0.072 g (70%) of *threo*-di(2-pyridyl)ethanediol, m.p. 92–93° (after the crystallization from 50% aqueous ethanol), $[\alpha]_D^{25}$ −51.7° (c 2.44, EtOH).

LIST OF REFERENCES

The numbers in square brackets at the end of the references refer to the page of the book where the reference is quoted.

[1] Wilde, von, M.P., *Chem. Ber.* 1974, **7,** 352 [1].

[2] Sabatier P., Senderens J.B., *Compt. Rend.* 1897, **124,** 1358 [1].

[3] Ipatiew W., *Chem. Ber.* 1907, **40,** 1270, 1281 [1].

[4] Brown C.A., Brown H.C., *J. Am. Chem. Soc.* 1962, **84,** 2829; *J. Org. Chem.* 1966, **31,** 3989 [1,8,202].

[5] Gilman H., Jacoby A.L., Ludeman H., *J. Am. Chem. Soc.* 1938, **60,** 2336 [4,176].

[6] Mann R.S., Lien T.R., *J. Catal.* 1969, **15,** 1 [4].

[7] Horiuti J., Polanyi M., *Trans. Faraday Soc.* 1934, **30,** 1164 [4].

[8] Baker R.H., Schuetz R.D., *J. Am. Chem. Soc.* 1947, **69,** 1250 [4, 5, 7, 12, 47, 50, 80, 81, 92].

[9] Schuetz R.D., Caswell L.R., *J. Org. Chem.* 1962, **27,** 486 [4, 47, 48].

[10] Siegel S., Smith G.V., Dmuchovsky B., Dubbell D., Halpern W., *J. Am. Chem. Soc.* 1962, **84,** 3136 [4, 12, 48].

[11] Siegel S., Dunkel M., Smith G.V., Halpern W., Cozort J., *J Org. Chem.* 1966, **31,** 2802 [4].

[12] Siegel S., Dmuchovsky B., *J. Am. Chem. Soc.* 1962, **84,** 3132 [4].

[13] Brown C.A., *J. Org. Chem.* 1970, **35,** 1900 [4, 5, 8, 39, 40, 42].

[14] Lozovoi A.V., Dyakova M.K., *Zh. Obshchei Khim.* 1940, **10,** 1: *Chem. Abstr.* 1940, **34,** 4728 [4, 39, 40, 46, 49].

[15] Beamer R.L., Belding R.H., Fickling C.S., *J. Pharm. Sci.* 1969, **58,** 1142 [5].

[16] Baker G.L., Fritschel S.J., Stille J.R., Stille J.K., *J. Org. Chem.* 1981, **46,** 2954 [5].

[17] Kagan H.B., Dang T.-P., *J. Am. Chem. Soc.* 1972, **94,** 6429 [5].

[18] Morrison J.D., Masler W.F., *J. Org. Chem.* 1974, **39,** 270 [5].

[19] Valentine D., Jr, Sun R.C., Toth K., *J. Org. Chem.* 1980, **45,** 3703 [5, 138].

[20] Valentine D., Jr, Johnson K.K., Priester W., Sun R.C., Toth, K., Saucy G., *J. Org. Chem.* 1980, **45,** 3698 [5, 10].

[21] Augustine R.L., Migliorini D.C., Foscante R.E., Sodano C.S., Sisbarro M.J., *J. Org. Chem.* 1969, **34,** 1075 [5, 120].

[22] Skita A., Schneck A., *Chem. Ber.* 1922, **55,** 144 [5, 11].

[23] Adams R., Shriner R.L., *J. Am. Chem. Soc.* 1923, **45,** 2171 [5].

[24] Adams R., Voorhees V., Shriner R.L., *Org. Syn. Coll. Vol.* 1932, **1,** 463 [5].

[25] Maxted E.B., Akhtar S., *J. Chem. Soc.* **1959,** 3130 [5, 10].

[26] Baltzly R., *J. Org. Chem.* 1976, **41,** 920, 928, 933 [5, 10].

[27] Brown H.C., Brown C.A., *J. Am. Chem. Soc.* 1962, **84,** 1493, 1494, 1495 [6, 7].

[28] Baltzly R., *J. Am. Chem. Soc.* 1952, **74,** 4586 [6, 10, 79].

[29] Starr D., Hixon R.M., *Org. Syn. Coll. Vol.* 1943, **2,** 566 [7].

[30] Shriner R.L., Adams R., *J. Am. Chem. Soc.* 1924, **46,** 1683 [1, 99].

[31] Russell T.W., Duncan D.M., Hansen S.C., *J. Org. Chem.* 1977, **42,** 551 [7, 97, 98, 101, 102, 205].

[32] Mozingo R., *Org. Syn. Coll. Vol.* 1955, **3,** 685 [7].

[33] Woodward R.B., Sondheimer F., Taub D., Heusler K., McLamore W.M., *J. Am. Chem. Soc.*1952, **74,** 4223 [7].

[34] Weygand C., Meusel W., *Chem. Ber.* 1943, **76,** 498 [7, 119].

[35] Mosettig E., Mozingo R., *Org. Reactions* 1948, **4,** 362 [7, 10, 144].

[36] Lindlar H., *Helv. Chim. Acta* 1952, **35,** 446 [7, 10, 44, 45, 78, 206].

[37] Freifelder M., Yew Hay Ng, Helgren P.F., *J. Org. Chem.* 1965, **30,** 2485 [7].

[38] Nishimura S., *Bull. Chem. Soc. Japan* 1961, **34,** 32 [7, 47, 73, 80, 107, 109, 110, 112, 113].

[39] Nishimura S., Onoda T., Nakamura A., *Bull. Chem. Soc. Japan,* 1960, **33,** 1356 [7, 80].

[40] Ham G.E., Coker W.P., *J. Org. Chem.* 1964, **29,** 194 [7, 63, 66].

[41] Carnahan J.E., Ford T.A., Gresham W.F., Grigsby W.E., Hager G.F., *J. Am. Chem. Soc.* 1955, **77,** 3766 [7, 137].

[42] Broadbent H.S., Campbell G.C., Bartley W.J., Johnson J.H., *J. Org. Chem.* 1959, **24,** 1847 [7, 9, 137, 153, 154].

[43] Adkins H., Cramer H.I., *J. Am. Chem. Soc.* 1930, **52,** 4349 [8, 47, 53, 58, 79, 80, 93, 99, 101, 107, 108, 110, 111, 169, 174].

[44] Adkins H., Pavlic A.A., *J. Am. Chem. Soc.* 1946, **68,** 1471; 1947, **69,** 3039 [8, 157, 161].

[45] Adkins H., Billica H.R., *J. Am. Chem. Soc.* 1948, **70,** 695, *Org. Syn. Coll. Vol.* 1955, **3,** 176 [8, 47, 73, 80, 81, 96, 97, 99, 101, 102, 108, 110, 111, 132, 133, 140, 163, 173].

[46] Spero G.B., McIntosh A.V., Jr, Levin R.H., *J. Am. Chem. Soc.* 1948, **70,** 1907 [8, 104, 131, 205].

[47] Brown C.A., Ahuja V.K., *Chem. Commun.* **1973,** 553 [8, 9].

[48] Motoyama I., *Bull. Chem. Soc. Japan* 1960, **33,** 232 [9, 47, 80, 156].

[49] Brunet J.J., Gallois P., Caubere P., *J. Org. Chem.* 1980, **45,** 1937, 1946 [9, 39, 40, 43, 44, 45, 97, 99, 107, 110, 111, 117, 121].

[50] Adkins H., Connor R., *J. Am. Chem. Soc.* 1931, **53,** 1091 [9, 54, 55, 58, 73, 99, 100, 107, 108, 110, 111, 121, 140, 153].

[51] Connor R., Folkers, K., Adkins H., *J. Am. Chem. Soc.* 1932, **54,** 1138 [9, 153].

[52] Sauer J., Adkins H., *J. Am. Chem. Soc.* 1937, **59,** 1 [9, 153, 157].

[53] Broadbent H.S., Slaugh L.H., Jarvis N.L., *J. Am. Chem. Soc.* 1954, **76,** 1519 [9, 53, 73].

[54] Broadbent H.S., Whittle C.W., *J. Am. Chem. Soc.* 1959, **81,** 3587 [9, 49, 53, 137].

[55] Landa S., Macák J., *Collect. Czech. Chem. Commun.* 1958, **23,** 1322 [9, 112].

[56] Harmon R.E., Parsons J.L., Cooke D.W., Gupta S.K., Schoolenberg J., *J. Org. Chem.* 1969, **34,** 3684 [9, 70, 101, 102, 119, 121, 140, 156, 158].

[57] Birch A.J., Walker K.A.M., *J. Chem. Soc. C,* **1966,** 1894, [9, 206].

[58] Jardine F.H., Wilkinson G., *J. Chem. Soc. C,* **1967,** 270 [9, 10, 206].

[59] Adams R., Garvey B.S., *J. Am. Chem. Soc.* 1926, **48,** 477 [10, 98].

[60] Weygand C., Meusel W., *Chem. Ber.* 1943, **76,** 503 [10].

[61] Adkins H., Billica H.R., *J. Am. Chem. Soc.* 1948, **70,** 3118 [11, 12].

[62] Busch M., Stöve H., *Chem. Ber.* 1916, **49,** 1063 [11, 64, 67].

[63] Horner L., Schläfer L., Kämmerer H., *Chem. Ber.* 1959, **92,** 1700 [11, 63, 64, 65, 66, 67, 81].

[64] Ubbelohde L., Svanoe T., *Angew. Chem.* 1919, **32,** 257 [11, 12].

[65] Adkins H., Billica H.R., *J. Am. Chem. Soc.* 1948, **70,** 3121 [11, 154].

[66] Hudlický M., *J. Fluorine Chem.* 1979, **14,** 189 [11, 66, 142].

[67] Lukeš R., Kovář J., Bláha K., *Collect. Czech. Chem. Commun.* 1956, **21,** 1475 [11].

[68] Hudlický M., *personal observations; unpublished results* [12, 20, 72, 118, 140, 141, 142, 148, 159, 207, 209, 212, 213, 217].

[69] Leggether B.E., Brown R.K., *Can. J. Chem.* 1960, **38,** 2363 [13].

[70] Kuhn L.P., *J. Am. Chem. Soc.* 1951, **73,** 1510 [13].

[71] Davies R.R., Hodgson H.H., *J. Chem. Soc.* **1943,** 281 [13, 36, 73, 74, 99, 100].

[72] Weiz J.R., Patel B.A., Heck R.F., *J. Org. Chem.* 1980, **45,** 4926 [13, 36, 44, 45].

[73] Terpko M.O., Heck R.F., *J. Org. Chem.* 1980, **45,** 4992 [13, 36, 73, 74].

[74] Olah G.A., Prakash G.K.S., *Synthesis* **1978,** 397 [13, 49].

[75] Rebeller M., Clément G., *Bull. Soc. Chem. Fr.* **1964,** 1302 [13, 84, 85].

[76] Eberhardt M.K., *Tetrahedron* 1967, **23,** 3029 [13, 49].

[77] Felix A.M., Heimer E.P., Lambros T.J., Tzougraki C., Meyerhofer J., *J. Org. Chem.* 1978, **43,** 4194 [13, 151, 206].

[78] Orchin M., *J. Am. Chem. Soc.* 1944, **66,** 535 [13, 51].

[79] Nishiguchi T., Tachi K., Fukuzumi K., *J. Org. Chem.* 1975, **40,** 237 [13].

[80] Citron J.D., *J. Org. Chem.* 1969, **34,** 1977 [13, 144, 145].

[81] Brieger G., Nestrick T.J., *Chem. Rev.* 1974, **74,** 567 [13].

[*82*] Finholt A. E., Bond A.C., Schlesinger H. I., *J. Am. Chem. Soc.* 1947, **69,** 1199 [13, 14, 207].

[*83*] Nystrom R.F., Brown W.G., *J. Am. Chem. Soc.* 1947, **69,** 1197 [13, 21, 22, 96, 97, 98, 99, 100, 107, 108, 110, 145, 146, 147, 154, 156, 207].

[*84*] Schlesinger H.I., Brown H.C., Finholt A.E., *J. Am. Chem. Soc.* 1953, **75,** 205 [14].

[*85*] Vít J., Procházka V., Donnerová Z., Čásenský B., Hrozinka I., Czech. Pat. 120,100 (1966); *Chem. Abstr.* 1967, **67,** 23634 [14].

[*86*] Zweifel G., Brown H.C., *Org. Reactions* 1963, **13,** 1 [14].

[*87*] Bruce M.I., *Chem. Ber.* 1975, **11,** 237 [14].

[*88*] Piťha J., Heřmánek S., Vít J., *Collect. Czech. Chem. Commun.* 1960, **25,** 736 [14, 139, 146, 154].

[*89*] Plešek J., Heřmánek S., *Collect. Czech. Chem. Commun.* 1966, **31,** 3060 [14, 107, 154, 156].

[*90*] Schlesinger H.I., Brown H.C., *J. Am. Chem. Soc.* 1940, **62,** 3429 [14].

[*91*] Banus M.D., Bragdon R.W., Hinckley A.A., *J. Am. Chem. Soc.* 1954, **76,** 3848 [14].

[*92*] Tolman V., Vereš K., *Collect. Czech. Chem. Commun.* 1972, **37,** 2962 [14, 159, 161].

[*93*] Brändström A., Junggren U., Lamm B., *Tetrahedron Lett.* **1972,** 3173 [14, 15].

[*94*] Brown H.C., Weissman P.M., *J. Am. Chem. Soc.* 1965, **87,** 5614 [15, 19, 154, 167, 174].

[*95*] Brown H.C., Tsukamoto A., *J. Am. Chem. Soc.* 1964, **86,** 1089 [15, 165].

[*96*] Brown H.C., McFarlin R.F., *J. Am. Chem. Soc.* 1958, **80,** 5372 [15, 145, 154].

[*97*] Vít J., Čásenský B., Macháček J., Fr. Pat. 1,515,582 (1967); *Chem. Abstr.* 1969, **70,** 115009 [15].

[*98*] Čásenský B., Macháček J., Abrham K., *Collect. Czech. Chem. Commun.* 1972, **37,** 1178 [15].

[*99*] Brown H.C., Mead E.J., *J. Am. Chem. Soc.* 1953, **75,** 6263 [15, 98, 99, 100, 102, 107, 110, 111, 146, 155].

[*100*] Brown H.C., Kim S.C., Krishnamurthy S., *J. Org. Chem.* 1980, **45,** 1 [15, 87, 107, 110, 129, 147, 155, 166, 174].

[*101*] Brown H.C., Krishnamurthy S., *J. Am. Chem. Soc.* 1972, **94,** 7159 [15].

[*102*] Brown C.A., *J. Am. Chem. Soc.* 1973, **95,** 4100: *J. Org. Chem.* 1974, **39,** 3913 [15].

[*103*] Borch R.F., Bernstein M.D., Durst H.D., *J. Am. Chem. Soc.* 1971, **93,** 2897 [15, 92, 134, 135, 136, 210].

[*104*] Whitesides G.M., San Filipo J., *J. Am. Chem. Soc.* 1970, **92,** 6611 [15, 16].

[*105*] Brown H.C., Knights, E.F., Scouten C.G., *J. Am. Chem. Soc.* 1974, **96,** 7765 [15, 16].

[*106*] Brown H.C., Mandal A.K., Yoon N.M., Singaram B., Schwier J.R., Jadhav P.K., *J. Org. Chem.* 1982, **47,** 5069 [15].

[*107*] Brown H.C., Jadhav P.K., Mandal A.K., *J. Org. Chem.* 1982, **47,** 5074 [15].

[*108*] Brown H.C., Desai M.C., Jadhav P.K., *J. Org. Chem.* 1982, **47,** 5065 [15, 16, 111].

[*109*] Midland M.M., Greer S., Tramontano A., Zderic S.A., *J. Am. Chem. Soc.* 1979, **101,** 2352 [15, 16, 96, 99, 100, 111].

[*110*] Midland M. M., Kazubski A., *J. Org. Chem.* 1982, **47,** 2814 [15, 16, 122].

[*111*] Ashby E.C., Lin J.J., Goel A.B., *J. Org. Chem.* 1978, **43,** 757 [15, 44].

[*112*] Citron J.D., Lyons J.E., Sommer L.H., *J. Org. Chem.* 1969, **34,** 638 [16, 68].

[*113*] Neumann W.P., Niermann H., *Ann. Chem.* 1962, **653,** 164 [16].

[*114*] Kuivila H.G., Beumel O.F., Jr, *J. Am. Chem. Soc.* 1958, **80,** 3798; 1961, **83,** 1246 [16, 98, 99, 100, 102, 107, 110, 121].

[*115*] Sorrell T.N., Pearlman P.S., *J. Org. Chem.* 1980, **45,** 3449 [16, 145].

[*116*] Brown H.C., Yoon N.M., *J. Am. Chem. Soc.* 1966, **87,** 1464 [19, 206].

[*117*] Mousseron M., Jacquier R., Mousseron-Canet M., Zagdoun R., *Bull. Soc. Chim. Fr.* **1952,** 1042; *Compt. Rend.* 1952, **235,** 177 [19, 175].

[*118*] Brown W.G., *Org. Reactions* 1951, **6,** 469 [19, 21].

[*119*] Lane C.F., *Chem. Rev.* 1976, **76,** 773 [19].

[*120*] Uffer A., Schlitter E., *Helv. Chim. Acta* 1948, **31,** 1397 [21, 167].

[*121*] Amundsen L.H., Nelson L.S., *J. Am. Chem. Soc.* 1951, **73,** 242 [22, 174, 207].

[*122*] Hansley V.L., *Ind. Eng. Chem.* 1951, **43,** 1759 [23, 27].

[*123*] Caine D., *Org. Reactions* 1976, **23,** 1 [23].

[*124*] Hardegger E.P., Plattner P.A., Blank F., *Helv. Chim. Acta* 1944, **27,** 793 [24].

[*125*] Storck G., White, W.N., *J. Am. Chem. Soc.* 1956, **78,** 4604 [24, 27].

[*126*] Sicher J., Svoboda M., Závada J., *Collect. Czech. Chem. Commun.* 1965, **30,** 421 [24, 45].

[*127*] Campbell K.N., Young E.E., *J. Am. Chem. Soc.* 1943, **65,** 965 [24, 45].

[*128*] Horner L., Röder H., *Chem. Ber.* 1968, **101,** 4179 [24].

[*129*] Benkeser R. A., Kaiser E.M., Lambert R.F., *J. Am. Chem. Soc.* 1964, **86,** 5272 [24, 48].

[*130*] Benkeser R.A., Watanabe H., Mels S.J., Sabol M.A., *J. Org. Chem.* 1970, **35,** 1210 [24, 165].

[*131*] Tafel J., *Chem. Ber.* 1900, **33,** 2209 [25].

[*132*] Benkeser R.A., Kaiser E.M., *J. Am. Chem. Soc.* 1963, **85,** 2858 [25, 48, 210].

[*133*] Junghans K., *Chem. Ber.* 1973, **106,** 3465; 1974, **107,** 3191 [25].

[*134*] Ruff, O., Geisel E., *Chem. Ber.* 1906, **39,** 828 [26].

[*135*] Hansley V.L., *Ind. Eng. Chem.* 1947, **39,** 55 [27, 152, 175].

[136] Ferguson L.N., Reid J.C., Calvin M., *J. Am. Chem. Soc.* 1946, **68,** 2502 [27, 139].

[137] Cason J., Way R.L., *J. Org. Chem.* 1949, **14,** 31 [27, 64].

[138] Levina R.Y., Kulikov S.G., *J. Gen. Chem. USSR* 1946, **16,** 117; *Chem. Abstr.* 1947, **41,** 115 [27].

[139] Bryce-Smith D., Wakefield B.J., *Org. Syn. Coll. Vol.* 1973, **5,** 998 [27].

[140] Adams R., Adams E.W., *Org. Syn. Coll. Vol.* 1932, **1,** 459 [27, 53, 109].

[141] Corey E.J., Chaykovsky M., *J. Am. Chem. Soc.* 1965, **87,** 1345, 1351 [27, 88, 89].

[142] Schreibmann P., *Tetrahedron Lett.* **1970,** 4271 [27].

[143] Kuhn R., Fischer H., *Chem. Ber.* 1961, **94,** 3060 [27, 49].

[144] Newman M.S., *J. Org. Chem.* 1961, **26,** 582 [27, 112, 212].

[145] Inoi T., Gericke P., Horton W.J., *J. Org. Chem.* 1962, **27,** 4597 [27, 65].

[146] Calder A., Hepburn S.P., *Org. Syn.* 1972, **52,** 77 [27, 69].

[147] Wislicenus H., *Chem. Ber.* 1896, **29,** 494 [27, 72].

[148] Wislicenus H., *J. Prakt. Chem.* 1896, [2], **54,** 18, 54, 65 [27, 95, 96, 161].

[149] Shin C., Yonezawa Y., Yoshimura J., *Chem. Lett.* **1976,** 1095 [27, 76, 160].

[150] Drinkwater D.J., Smith P.W.G., *J. Chem. Soc. C,* **1971,** 1305 [27, 163].

[151] Braun W., Mecke R., *Chem. Ber.* 1966, **99,** 1991 [27, 129].

[152] Pascali V., Umani-Ronchi A., *Chem. Commun.* **1973,** 351 [27, 89].

[153] Schroeck C.W., Johnson C.R., *J. Am. Chem. Soc.* 1971, **93,** 5305 [27, 89].

[154] Fischer E., Groh R., *Ann. Chem.* 1911, **383,** 363 [27, 134].

[155] Yamamura S., Hirata Y., *J. Chem. Soc. C* **1968,** 2887 [28, 118].

[156] Dekker J., Martins F.J.C., Kruger J.A., *Tetrahedron Lett.* **1975,** 2489 [28, 128].

[157] Rieke R.D., Uhm S.J., Hudnall P.M. *Chem. Commun.* **1973,** 269, [28, 64, 68].

[158] Gardner J.H., Naylor C.A., *Org. Syn. Coll. Vol.* 1943, **2,** 526 [28].

[159] Leonard N.J., Barthel E., Jr, *J. Am. Chem. Soc.* 1950, **72,** 3632 [28, 118, 126].

[160] Martin E.L., *Org. Reactions* 1942, **1,** 155 [28, 97, 102, 103, 108, 113, 118, 121].

[161] Risinger G.E., Mach E.E., Barnett K.W., *Chem. Ind. (London)* **1965,** 679 [28, 161].

[162] Burdon J., Price R.C., *Private communication* [28].

[163] Vedejs E., *Org. Reactions* 1975, **22,** 401 [28].

[164] Kreiser W., *Ann. Chem.* 1971, **745,** 164 [28, 126, 127].

[165] Lyons R.E., Smith L.T., *Chem. Ber.* 1927, **60,** 173 [29, 73, 74, 213].

[166] Hodgson H.H., Whitehurst J.S., *J. Chem. Soc.* **1945,** 202 [29, 73, 74].

[167] Clarke H.T., Dreger E.E., *Org. Syn. Coll. Vol.* 1932, **1,** 304 [29].

[168] Matsumura K., *J. Am. Chem. Soc.* 1930, **52,** 4433 [29].

[*169*] Rakoff H., Miles B.H., *J. Org. Chem.* 1961, **26,** 2581 [29].

[*170*] den Hertog J., Overhoff J., *Rec. Trav. Chim. Pays-Bas* 1950, **69,** 468 [29, 95].

[*171*] Blomquist A.T., Dinguid L.I., *J. Org. Chem.* 1947, **12,** 718, 723 [29, 86].

[*172*] Kindler K., *Ann. Chem.* 1923, **431,** 187 [29, 171, 172].

[*173*] Carter P.H., Craig J.G., Lack R.E., Moyle M., *Org. Syn. Coll. Vol.* 1973, **5,** 339 [30, 125].

[*174*] Schaefer J.P., *J. Org. Chem.* 1960, **25,** 2027 [30, 128, 129, 214].

[*175*] Meyer K.H., *Org. Syn. Coll. Vol.* 1932, **1,** 60 [30, 129].

[*176*] Hartman W.W., Dickey J.B., Stampfli J.G., *Org. Syn. Coll. Vol.* 1943, **2,** 175 [30, 81].

[*177*] Braun, V. J., Sobecki W., *Chem. Ber.* 1911, **44,** 2526, 2533 [30, 70].

[*178*] Oelschläger H., Schreiber O., *Ann. Chem.* 1961, **641,** 81 [30, 124].

[*179*] Reissert A., Grube F., *Chem. Ber.* 1909, **42,** 3710 [30, 175].

[*180*] Russig F., *J. Prakt. Chem.* 1900, [2], **62,** 30 [30, 129].

[*181*] Badger G.M., Gibb A.R.M., *J. Chem. Soc.* **1949,** 799 [30, 129].

[*182*] Sonn A., Müller E., *Chem. Ber.* 1919, **52,** 1927 [30, 172].

[*183*] Stephen H., *J. Chem. Soc.* 1925, **127,** 1874 [30, 172, 173].

[*184*] Bischler A., *Chem. Ber.* 1889, **22,** 2801 [30, 76].

[*185*] Pschorr R., *Chem. Ber.* 1902, **35,** 2729, 2738 [30].

[*186*] Ho T.-L., Wong C.M., *Synthesis* **1973,** 206 [30, 88, 214].

[*187*] Hanson J.R., Mehta S., *J. Chem. Soc. C* **1969,** 2349 [30, 129].

[*188*] Rosenkranz G., Mancer O., Gatica J., Djerassi C., *J. Am. Chem. Soc.* 1950, **72,** 4077 [30, 123, 214].

[*189*] Castro C.E., *J. Am. Chem. Soc.* 1961, **83,** 3262 [30].

[*190*] Akita Y., Inaba M., Uchida H., Ohta A., *Synthesis* **1977,** 792 [30, 70, 88, 93].

[*191*] Akita Y., Misu K., Watanabe T., Ohta A., *Chem. Pharm. Bull. Japan,* 1976, **24,** 1839 [30, 95].

[*192*] Allen W.S., Bernstein S., Feldman L.I., Weiss M.J., *J. Am. Chem. Soc.* 1960, **82,** 3696 [30, 83, 126].

[*193*] Castro C.E., Kray W.C., Jr, *J. Am. Chem. Soc.* 1963, **85,** 2768 [30, 64, 66].

[*194*] Hanson J.R., Premuzic E., *Tetrahedron* 1967, **23,** 4105 [30, 70].

[*195*] Castro C.E., Stephens R.D., *J. Am. Chem. Soc.* 1964, **86,** 4358 [30, 50, 78, 138, 141].

[*196*] Hanson J.R., Premuzic E., *J. Chem. Soc. C,* **1969,** 1201 [30, 128].

[*197*] Nozaki H., Arantani T., Noyori T., *Tetrahedron* 1967, **23,** 3645 [30, 64].

[*198*] Williamson K.L., Hsu Y., Young E.I., *Tetrahedron* 1968, **24,** 6007 [30, 66].

[*199*] McCall J.M., Ten Brink R.E., *Synthesis* **1975,** 335 [30, 95].

[*200*] Timms G.H., Wildsmith E., *Tetrahedron Lett.* **1971,** 195 [30, 133].

[*201*] Ho T.-L., Wong C.M., *Synthesis* **1974,** 45 [30, 73].

[202] McMurry J.E., Silvestri M., *J. Org. Chem.* 1975, **40,** 1502 [30, 134].

[203] Ho T.-L., Wong C.M., *Syn. Commun.* 1973, **3,** 37 [30, 88].

[204] McMurry J.E., Silvestri M.G., Fleming M.P., Hoz T., Grayston M.W., *J. Org. Chem.* 1978, **43,** 3249 [30, 79, 81, 83, 84].

[205] Fleming M.P., McMurry J.E., *J. Am. Chem. Soc.* 1974, **96,** 4708 [30, 109].

[206] McMurry J.E., Fleming M.P., Kees K.L., Krepski L.R., *J. Org. Chem.* 1978, **43,** 3255 [30, 77, 97, 101, 109, 112, 118, 119, 128, 215].

[207] Tyrlik S., Wolochowicz I., *Bull. Soc. Chim. Fr.* **1973,** 2147 [30, 31, 97, 109, 112,118].

[208] Tyrlik S., Wolochowicz I., *Chem. Commun.* **1975,** 781 [30].

[209] Castedo L., Saá J.M., Suan R., Tojo G., *J. Org. Chem.* 1981, **46,** 4292 [30, 31, 101, 109].

[210] McMurry J.E., Kees K.L., *J. Org. Chem.* 1977, **42,** 2655 [30, 128].

[211] McMurry J.E., *Acc. Chem. Res.* 1974, **7,** 281 [31].

[212] Ho T.-L., *Syn. Commun.* 1979, **9,** 241 [31].

[213] Slaugh L.H., Raley J.H., *Tetrahedron* 1964, **20,** 1005 [31, 215].

[214] Ho T.-L., Olah G.A., *Synthesis* **1976,** 807 [31, 123].

[215] Ho T.-L., Olah G.A., *Synthesis* **1976,** 815 [31, 126, 127, 129].

[216] Olah G.A., Prakash S.G., Ho T.-L., *Synthesis* **1976,** 810 [31, 88].

[217] Ho T.-L., Henniger M., Olah G.A., *Synthesis* **1976,** 815 [31, 76, 215].

[218] Smith L.I., Opie J.W., *Org. Syn. Coll. Vol.* 1948, **3,** 56 [31, 73, 103].

[219] Fieser L. F., Heyman H., *J. Am. Chem. Soc.* 1942, **64,** 376, 380 [31, 75].

[220] Adams R., Marvel C.S., *Org. Syn. Coll. Vol.* 1932, **1,** 358 [31, 64, 65].

[221] Bigelow H.E., Palmer A., *Org. Syn. Coll. Vol.* 1943, **2,** 5 [31, 72].

[222] Renaud R.N., Stephens J.C., *Can. J. Chem.* 1974, **52,** 1229 [31, 32, 129].

[223] Blatt A.H., Tristram E.W., *J. Am. Chem. Soc.* 1952, **74,** 6273 [32, 75].

[224] Olah G.A., Arvanaghi M., Vankar Y.D., *J. Org. Chem.* 1980, **45,** 3531 [32].

[225] Gunther F.A., Blinn R.C., *J. Am. Chem. Soc.* 1950, **72,** 4282 [32, 79].

[226] Konieczny M., Harvey R.G., *J. Org. Chem.* 1979, **44,** 4813 [32, 79].

[227] Bradsher C.K., Vingiello F.A., *J. Org. Chem.* 1948, **13,** 786 [32, 113].

[228] Platt K.L., Oesch F., *J. Org. Chem.* 1981, **46,** 2601 [32, 144].

[229] Bickel C.L., Morris R., *J. Am. Chem. Soc.* 1951, **73,** 1786 [32, 123].

[230] Wagner A.W., *Chem. Ber.* 1966, **99,** 375 [32, 90].

[231] Pojer P.M., Ritchie E., Taylor W.C., *Austral. J. Chem.* 1968, **21,** 1375 [32, 124].

[232] Smith P.A.S., Brown B.B., *J. Am. Chem. Soc.* 1951, **73,** 2438 [32, 76].

[233] Bastian J.M., Ebnöter A., Jucker E., Rissi E., Stoll A.P., *Helv. Chim. Acta* 1971, **54,** 277 [32].

[234] Newman M.S., *J. Am. Chem. Soc.* 1951, **73,** 4994 [32].

[235] Kellogg R.M., Schaap A.P., Harper E.T., Wynberg H., *J. Org. Chem.* 1968, **33,** 2902 [32].

[236] Brady O.L., Day J.N.D., Reynolds C.V., *J. Chem. Soc.* **1929**, 2264 [32, 73, 74].

[237] Mayer R., Hiller G., Nitzschke M., Jentzsch J., *Angew. Chem.* 1963, **75**, 1011 [32, 33, 126, 127, 216].

[238] Hartman W.W., Silloway H.L., *Org. Syn. Coll. Vol.* 1955, **3**, 82 [32, 74, 81].

[239] Robertson R., *Org. Syn. Coll. Vol.* 1932, **1**, 52 [32, 73, 142].

[240] Boon W.R., *J. Chem. Soc.* **1949**, S 230 [32, 133].

[241] Ueno K., *J. Am. Chem. Soc.* 1952, **74**, 4508 [32, 33, 73, 96].

[242] Beard H.G., Hodgson H.H., *J. Chem. Soc.* **1944**, 4 [32].

[243] Kober E., *J. Org. Chem.* 1961, **26**, 2270 [32].

[244] Pappas J.J., Keaveney W.P., Gancher E., Berger M., *Tetrahedron Lett.* **1966**, 4273 [32, 86].

[245] Kremers E., Wakeman N., Hixon R.M., *Org. Syn. Coll. Vol.* 1932, **1**, 511 [33, 75].

[246] Weil H., Traun M., Marcel S., *Chem. Ber.* 1922, **55**, 2664, 2671 [33, 96].

[247] Lieber E., Sherman E., Henry R.A., Cohen J., *J. Am. Chem. Soc.* 1951, **73**, 2327 [33, 76].

[248] Schmidt J., Kampf A., *Chem. Ber.* 1902, **35**, 3124 [33, 129].

[249] Kozlov V.V., Smolin D.D., *Zh. Obshchei Khim.* 1949, **19**, 740; *Chem. Abstr.* 1950, **44**, 3479 [33, 90].

[250] Dodgson J.W., *J. Chem. Soc.* **1914**, 2435 [33, 129].

[251] Hock H., Lang S., *Chem. Ber.* 1943, **76**, 169; 1944, **77**, 257, 262 [33, 85].

[252] Fuller A.T., Tonkin I.M., Walker J., *J. Chem. Soc.* **1945**, 633 [33, 90].

[253] Bamberger E., Kraus E., *Chem. Ber.* 1896, **29**, 1829, 1834 [33, 76].

[254] Madai H., Müller R., *J. Prakt. Chem.* 1963, [4], **19**, 83 [33, 64, 65, 216].

[255] Conant J.B., Corson B.B., *Org. Syn. Coll. Vol.* 1943, **2**, 33 [33, 75, 81].

[256] Prelog V., Wiesner K., *Helv. Chim. Acta* 1948, **31**, 870 [33, 73, 81].

[257] Hodgson H.H., Ward E.R., *J. Chem. Soc.* **1947**, 327 [33, 73, 75, 216].

[258] Fieser L.F., *Org. Syn. Coll. Vol.* 1943, **2**, 35, 39 [33, 96].

[259] Adams R., Blomstrom D.C., *J. Am. Chem. Soc.* 1953, **75**, 3405 [33, 76].

[260] Ehrlich P., Bertheim A., *Chem. Ber.* 1912, **45**, 756 [33, 176].

[261] Schmidt H., *Ann. Chem.* 1920, **421**, 221 [33, 176].

[262] de Vries J.G., Kellogg R.M., *J. Org. Chem.* 1980, **45**, 4126 [33, 96, 97, 99, 100, 107, 108, 110, 111, 113].

[263] Mager H.I.X., Berends W., *Rec. Trav. Chim. Pays-Bas* 1960, **79**, 282 [33].

[264] Hamersma J.W., Snyder E.I., *J. Org. Chem.* 1965, **30**, 3985 [33, 34, 95, 96].

[265] Thiele J., *Ann. Chem.* 1892, **271**, 127 [33].

[266] Corey E.J., Mock W.L., *J. Am. Chem. Soc.* 1962, **84**, 685 [33, 34].

[*267*] Diels O., Schmidt S., Witte W., *Chem. Ber.* 1938, **71,** 1186 [33].

[*268*] Wolinsky J., Schultz T., *J. Org. Chem.* 1965, **30,** 3980 [33, 42].

[*269*] Ohno M., Okamoto M., *Org. Syn. Coll. Vol.* 1973, **5,** 281 [33, 34, 42].

[*270*] van Tamelen E.E., Dewey R.S., *J. Am Chem. Soc.* 1961, **83,** 3729 [33, 34].

[*271*] Hünig S., Müller H.R., Thier W., *Angew. Chem. Intern. Ed. Engl.* 1965, **4,** 271 [33].

[*272*] Corey E.J., Pasto D.J., Mock W.L., *J. Am. Chem. Soc.* 1961, **83,** 2957 [33].

[*273*] Corey E.J., Mock W.L., Pasto D.J., *Tetrahedron Lett.* **1961,** 347 [33, 50].

[*274*] Corey E.J., Mock W.L., *J. Am. Chem. Soc.* 1962, **84,** 685 [33].

[*275*] Hanuš J., Voříšek J., *Collect. Czech. Chem. Commun.* 1929, **1,** 223 [34].

[*276*] Müller E., Kraemer-Willenberg H., *Chem. Ber.* 1924, **57,** 575 [34, 49].

[*277*] Fletcher T.L., Namkung M.J., *J. Org. Chem.* 1958, **23,** 680 [34, 73, 75].

[*278*] Mosby W.L., *Chem. & Ind.* **1959,** 1348; *J. Org. Chem.* 1959, **24,** 421 [34, 73, 74].

[*279*] Furst A., Berlo R.C., Hooton S., *Chem. Rev.* 1965, **65,** 51 [34].

[*280*] Todd D., *Org. Reactions* 1948, **4,** 378 [34, 97, 102, 108, 113, 118].

[*281*] Huang-Minlon, *J. Am. Chem. Soc.* 1946, **68,** 2487 [34, 108, 112, 113, 118, 144, 216].

[*282*] Soffer M.D., Soffer M.B., Sherk K.W., *J. Am. Chem. Soc.* 1945, **67,** 1435 [34, 113, 118].

[*283*] Buu-Hoi N.P., Hoan Ng, Xuong N.D., *Rec. Trav. Chim. Pays-Bas* 1952, **71,** 285 [34].

[*284*] McFadyen J.S., Stevens T.S., *J. Chem. Soc.* **1936,** 584 [34, 172].

[*285*] Mosettig E., *Org. Reactions* 1954, **8,** 232 [34, 172, 173].

[*286*] Work J.B., *Inorg. Syn.* 1946, **2,** 141 [34].

[*287*] Robertson A.V., Witkop B., *J. Am. Chem. Soc.* 1962, **84,** 1697 [35].

[*288*] Lutz R.E., Allison R.K. *et al., J. Org. Chem.* 1947, **12,** 617, 681 [35, 75, 125].

[*289*] Kornblum N., *Org. Reactions* 1944, **2,** 262 [35, 36, 75].

[*290*] Horner L., Jurgeleit W., *Ann. Chem.* 1955, **591,** 138 [35, 85].

[*291*] Davis R.E., *J. Org. Chem.* 1958, **23,** 1767 [35, 87].

[*292*] Horner L., Hoffman H., *Angew. Chem.* 1956, **68,** 473 [35].

[*293*] Clive D.L.J., Menchen S.M., *J. Org. Chem.* 1980, **45,** 2347 [35, 83, 84].

[*294*] Schuetz R.D., Jacobs R.L., *J. Org. Chem.* 1958, **23,** 1799; 1961, **26,** 3467 [35, 87].

[*295*] Brink M., *Synthesis* **1975,** 807 [35, 75].

[*296*] Mori T., Nakahara T., Nozaki H., *Can. J. Chem.* 1969, **47,** 3266 [35, 127].

[*297*] Saegusa T., Kobayashi S., Kimura Y., Yokoyama T., *J. Org. Chem.* 1977, **42,** 2797 [35].

[*298*] Bunyan P.J., Cadogan J.I.G., *J. Chem. Soc.* **1963,** 42 [35, 75].

[299] Hiraa T., Masunaga T., Ohshiro Y., Agawa T., *J. Org. Chem.* 1981, **46,** 3745 [35].

[300] Clive D.L.J., Beaulieu P.L., *J. Org. Chem.* 1982, **47,** 1124 [35].

[301] Pinnick H.W., Reynolds M.A., McDonald R.T., Jr, Brewster W.D., *J. Org. Chem.* 1980, **45,** 930 [35, 89].

[302] Newman M.S., Blum S., *J. Am. Chem. Soc.* 1964, **86,** 5598 [35, 103].

[303] Harpp D.N., Gleason J.G., Snyder J.P., *J. Am. Chem. Soc.* 1968, **90,** 4181 [35, 87].

[304] Barclay L.B., Farrar M.W., Knowles W.S., Raffelson H., *J. Am. Chem. Soc.* 1954, **76,** 5017 [35].

[305] Hodgson H.H., Turner H.S., *J. Chem. Soc.* **1942,** 748; **1943,** 86 [35, 75].

[306] Kato M., Tamano T., Miwa T., *Bull. Chem. Soc. Japan* 1975, **48,** 291 [35, 75, 142].

[307] van Leusen A.M., Smid P.M., Strating J., *Tetrahedron Lett.* **1967,** 1165 [35, 125].

[308] Bachmann W.E., *Org. Syn. Coll. Vol.* 1943, **2,** 71 [35, 112].

[309] Wilds A.L., *Org. Reactions* 1944, **2,** 178 [35, 96, 103, 107, 110, 121, 217].

[310] Pinkey D.T., Rigby R.D.G., *Tetrahedron Lett.* **1969,** 1267 [36, 68].

[311] Winstein S., *J. Am. Chem. Soc.* 1939, **61,** 1610 [36, 123].

[312] Werner E.A., *J. Chem. Soc.* **1917,** 844 [36, 136].

[313] Davidson D., Weiss M., *Org. Syn. Coll. Vol.* 1943, **2,** 590 [36, 100].

[314] Mole T., *J. Chem. Soc.* **1960,** 2132 [36, 110].

[315] Johns R.B., Markham K.R., *J. Chem. Soc.* **1962,** 3712 [36, 124].

[316] Newbold B.T., LeBlanc R.P., *J. Chem. Soc.* **1965,** 1547 [36, 72].

[317] Nanjo K., Suzuki K., Sekiya M., *Chem. Lett.* **1976,** 1169 [36, 70, 98, 175].

[317a] Cortese N.A., Heck R.F., *J. Org. Chem.* 1977, **42,** 3491; 1978, **43,** 3985 [67, 98].

[318] Lukeš R., Pliml J., *Collect. Czech. Chem. Commun.* 1950, **15,** 464 [36, 56].

[319] Kocián O., Ferles M., *Collect. Czech. Chem. Commun.* 1979, **44,** 1167 [36, 58, 113].

[320] de Benneville P.L., Macartney J.H., *J. Am. Chem. Soc.* 1950, **72,** 3073 [36, 92].

[321] Novelli A., *J. Am. Chem. Soc.* 1939, **61,** 520 [36, 136, 218].

[322] Moore M.L., *Org. Reactions* 1949, **5,** 301 [36, 53, 134, 135, 136].

[323] Blicke F.F., Powers L.D., *J. Am. Chem. Soc.* 1929, **51,** 3378 [36, 110].

[324] Greenwood F.L., Whitmore F.C., Crooks H.M., *J. Am. Chem. Soc.* 1938, **60,** 2028 [36, 146].

[325] Respess W.L., Tamborski C., *J. Organometal. Chem.* 1969, **18,** 263 [36].

[326] Meerwein H., Hinz G., Majert H., Sönke H., *J. Prakt. Chem.* 1936, [2], **147,** 226 [37].

[327] Baumann P., Prelog V., *Helv. Chim. Acta* 1959, **42**, 736 [37, 127, 218].

[328] Herzog H. L., Jevnik M. A., Perlman P. L., Nobile A., Hershberg L. B., *J. Am. Chem. Soc.* 1953, **75**, 266 [37].

[329] Mamoli L., Schramm G., *Chem. Ber.* 1938, **71**, 2698 [37, 125].

[330] Hashimoto H., Simon H., *Angew. Chem. Internat. Ed. Engl.* 1975, **14**, 106 [37, 142].

[331] Meier R., Böhler F., *Chem. Ber.* 1956, **89**, 2301 [72].

[332] Willstätter R., Bruce J., *Chem. Ber.* 1907, **40**, 3979, 4456 [39].

[333] Zelinsky N.D., Kazanski B.A., Plate A.F., *Chem. Ber.* 1933, **66**, 1415 [39].

[334] Kieboom A.P.G., Van Benschop H.J., Van Bekkum H., *Rec. Trav. Chim. Pays-Bas* 1976, **95**, 231 [39].

[335] Kursanov D.N., Parnes Z.N., Loim N.M., *Synthesis* **1974**, 633 [39, 41, 113].

[336] Jardine I., McQuillin F.J., *J. Chem. Soc. C* **1966**, 458 [39].

[337] Berkowitz L.M., Rylander P.N., *J. Org. Chem.* 1959, **24**, 708 [39].

[338] Brown C.A., *Chem. Commun.* **1969**, 952 [40].

[339] Siegel S., Smith G.V., *J. Am. Chem. Soc.* 1960, **82**, 6082, 6087 [40].

[340] Tyman J.H.P., Wilkins S.W., *Tetrahedron Lett.* **1973**, 1773 [41].

[341] Brown H.C., Murray K.J., *J. Am. Chem. Soc.* 1959, **81**, 4108 [41].

[342] Kursanov D.N., Parnes Z.N., Bassova G.I., Loim N.M., Zdanovich V.I., *Tetrahedron* 1967, **23**, 2235 [41, 210].

[343] Carey F.A., Tremper H.S., *J. Org. Chem.* 1971, **36**, 758 [41, 77].

[344] Junghans K., *Chem. Ber.* 1973, **106**, 3465; 1974, **107**, 3191 [41, 49].

[345] Whitesides G.M., Ehmann W.J., *J. Org. Chem.* 1970, **35**, 3565 [41, 44].

[346] Corey E.J., Cantrall E.W., *J. Am. Chem. Soc.* 1959, **81**, 1745 [41].

[347] van Tamelen E.E., Timmons R.J., *J. Am. Chem. Soc.* 1962, **84**, 1067 [42].

[348] Newhall W.F., *J. Org. Chem.* 1958, **23**, 1274 [42].

[349] Brown H.C., Brown C.A., *J. Am. Chem. Soc.* 1963, **85**, 1005 [42, 43, 44].

[350] Hubert A.J., *J. Chem. Soc. C*, **1967**, 2149 [42].

[351] Midgley T., Henne A.L., *J. Am. Chem. Soc.* 1929, **51**, 1293 [42].

[352] Cope A.C., Hochstein F.A., *J. Am. Chem. Soc.* 1950, **72**, 2515 [43].

[353] Ziegler K., Jakob L., Wollthan H., Wenz A., *Ann. Chem.* 1934, **511**, 64 [43].

[354] Ziegler K., Häffner F., Grimm H., *Ann. Chem.* 1937, **528**, 101 [43].

[355] Ziegler K., Wilms H., *Ann. Chem.* 1950, **567**, 1, 36 [43].

[356] Bance S., Barber H.J., Woolmann A.M., *J. Chem. Soc.* **1943**, 1 [43, 49].

[357] Lumb P.B., Smith J.C., *J. Chem. Soc.* **1952**, 5032 [43].

[358] Howton D.R., Davis R.H., *J. Org. Chem.* 1951, **16**, 1405 [43, 138].

[359] Campbell K.N., O'Connor M.J., *J. Am. Chem. Soc.* 1939, **61**, 2897 [43, 46, 50].

[360] Campbell K.N., Eby L.T., *J. Am. Chem. Soc.* 1941, **63**, 216, 2683 [43, 44, 45].

[361] Brown C.A., Ahuja V.K., *Chem. Commun.* **1973,** 553 [43, 44].

[362] Bellas T.E., Brownlee R.G., Silverstein R.M., *Tetrahedron* 1969, **25,** 5149 [43].

[363] Savoia D., Tagliviani E., Trombini C., Umani-Ronchi A., *J. Org. Chem.* 1981, **46,** 5340, 5344 [43].

[364] Newman M.S., Waltcher I., Ginsberg H.F., *J. Org. Chem.* 1952, **17,** 962 [43, 78].

[365] Hennion G.F., Schroeder W.A., Lu R.P., Scanlon W.B., *J. Org. Chem.* 1956, **21,** 1142 [43, 78].

[366] Harper S.H., Smith R.J.D., *J. Chem. Soc.* **1955,** 1512 [43, 44, 78].

[367] Blomquist A.T., Burge R.E., Jr, Sucsy A.C., *J. Am. Chem. Soc.* 1952, **74,** 3636 [44].

[368] Isler O., Huber W., Ronco A., Kofler M., *Helv. Chim. Acta* 1947, **30,** 1911 [44, 45, 78].

[369] Cram D.J., Allinger N.L., *J. Am. Chem. Soc.* 1956, **78,** 2518 [44].

[370] Tedeschi R.J., Clark G., Jr, *J. Org. Chem.* 1962, **27,** 4323 [44, 78].

[371] Magoon E.F., Slaugh L.H., *Tetrahedron* 1967, **23,** 4509 [44].

[372] Slaugh L.H., *Tetrahedron* 1966, **22,** 1741 [44].

[373] Zweifel G., Steele R.B., *J. Am. Chem. Soc.* 1967, **89,** 5085 [44].

[374] Wilke G., Müller H., *Chem. Ber.* 1956, **89,** 444 [44].

[375] Cope A.C., Berchtold G.A., Peterson P.E., Sharman S.H., *J. Am. Chem. Soc.* 1960, **82,** 6370 [44].

[376] Brown H.C., Zweifel G., *J. Am. Chem. Soc.* 1959, **81,** 1512; 1961, **83,** 3834 [44].

[377] Ashby E.C., Lin J.J., *Tetrahedron Lett.* **1977,** 4481 [44].

[378] Chum P.W., Wilson S.E., *Tetrahedron Lett.* **1976,** 15 [44].

[379] Benkeser R.A., Tinchner C.A, *J. Org. Chem.* 1968, **33,** 2727 [45, 49, 50].

[380] Dear R.E.A., Pattison F.L.M., *J. Org. Chem.* 1963, **85,** 622 [45].

[381] Svoboda M., Závada J., Sicher J., *Collect. Czech. Chem. Commun.* 1965, **30,** 413 [45].

[382] Blomquist A.T., Liu L.H., Bohrer J.C., *J. Am. Chem. Soc.* 1952, **74,** 3643 [45].

[383] Chanley J.D., Sobotka H., *J. Am. Chem. Soc.* 1949, **71,** 4140 [45, 78].

[384] Oppolzer, W., Fehr C., Warneke J., *Helv. Chim. Acta* 1977, **60,** 48 [45, 78].

[385] Dobson N.A., Raphael R.A., *J. Chem. Soc.* **1955,** 3558 [46].

[386] Hershberg E.B., Oliveto E.P., Gerold C., Johnson L., *J. Am. Chem. Soc.* 1951, **73,** 5073 [46, 122].

[387] Weissberger A., *J. Chem. Soc.* **1935,** 855 [46, 50].

[388] Adams R., Marshall I.R., *J. Am. Chem. Soc.* 1928, **50,** 1970 [46, 47].

[389] Nishimura S., *Bull. Chem. Soc. Japan* 1959, **32,** 1155 [46, 79].

[390] Stocker J.H., *J. Org. Chem.* 1962, **27,** 2288 [47].

[*391*] Zelinsky N.D., Margolis E.I., *Chem. Ber.* 1932, **65**, 1613 [47].

[*392*] Hückel W., Wörffel U., *Chem. Ber.* 1955, **88**, 338 [48].

[*393*] Slaugh L.H., Raley J.H., *J. Org. Chem.* 1967, **32**, 369 [48].

[*394*] Campbell K.N., McDermott J.P., *J. Am. Chem. Soc.* 1945, **67**, 282 [48].

[*395*] Paquette L.A., Barrett J.N., *Org. Syn. Coll. Vol.* 1973, **5**, 467 [48].

[*396*] Slaugh L.H., Raley J.H., *J. Org. Chem.* 1967, **32**, 2861 [48, 50].

[*397*] Benkeser R.A., Robinson R.E., Sauve D.M., Thomas O.H., *J. Am. Chem. Soc.* 1955, **77**, 3230 [48].

[*398*] Reggel L., Friedel R.A., Wender I., *J. Org. Chem.* 1957, **22**, 891 [48, 51].

[*399*] Birch A.J., *J. Chem. Soc.* **1944**, 430 [48, 80, 81, 82, 140, 211].

[*400*] Hückel W., Bretschneider H., *Ann. Chem.* 1939, **540**, 157 [48, 50, 51].

[*401*] Erman W.T., Flautt T.J., *J. Org. Chem.* 1962, **27**, 1526 [49].

[*402*] Hückel W., Schwen R., *Chem. Ber.* 1956, **89**, 150 [49].

[*403*] Goodman I., *J. Chem. Soc.* **1951**, 2209 [49, 51].

[*404*] Benkeser R.A., Arnold C., Jr, Lambert R.F., Thomas O.H., *J. Am. Chem. Soc.* 1955, **77**, 6042 [49].

[*405*] Adkins H., Reid W.A., *J. Am. Chem. Soc.* 1941, **63**, 741 [50].

[*406*] Willstätter R., Seitz F., *Chem. Ber.* 1923, **56**, 1388; 1924, **57**, 683 [50].

[*407*] Stuhl L.S., Rakowski DuBois M., Hirsekorn F.J., Bleeke J.R., Stevens A.E., Muetterties E.L., *J. Am. Chem. Soc.* 1978, **100**, 2405 [50, 51].

[*408*] Cook E.S., Hill A.J., *J. Am. Chem. Soc.* 1940, **62**, 1995 [50, 51].

[*409*] Birch A.J., Murray A.R., Smith H., *J. Chem. Soc.* **1951**, 1945 [50, 51].

[*410*] Hückel W., Schlee H., *Chem. Ber.* 1955, **88**, 346 [50, 51].

[*411*] Benkeser R.A., Kaiser E.M., *J. Org. Chem.* 1964, **29**, 955 [50].

[*412*] ·Hückel W., Wörffel U., *Chem. Ber.* 1956, **89**, 2098 [50, 51].

[*413*] Bass K.C., *Org. Syn. Coll. Vol.* 1973, **5**, 398 [51].

[*414*] Crossley N.S., Henbest H.B., *J. Chem. Soc.* **1960**, 4413 [51].

[*415*] Durland J.R., Adkins H., *J. Am. Chem. Soc.* 1937, **59**, 135; 1938, **60**, 1501 [52, 53].

[*416*] Burger A., Mosettig E., *J. Am. Chem. Soc.* 1935, **57**, 2731; 1936, **58**, 1857 [53].

[*417*] Phillips D.D., *Org. Syn. Coll. Vol.* 1963, **4**, 313 [52, 53].

[*418*] Bamberger E., Lodter W., *Chem. Ber.* 1887, **20**, 3073 [52, 53].

[*419*] Starr D., Hixon R.M., *Org. Syn. Coll. Vol.* 1943, **2**, 566 [53].

[*420*] Connor R., Adkins H., *J. Am. Chem. Soc.* 1932, **54**, 4678 [53, 77, 80, 81, 156, 158, 160, 161].

[*421*] ·Tarbell D.S., Weaver C., *J. Am. Chem. Soc.* 1941, **63**, 2939 [53].

[*422*] Entel J., Ruof C.H., Howard H.C., *J. Am. Chem. Soc.* 1951, **73**, 4152 [53].

[*423*] Papa D., Schwenk T., Ginsberg H.F., *J. Org. Chem.* 1951, **16**, 253 [53].

[424] Borowitz I.J., Gonis G., Kelsey R., Rapp R., Williams G.J., *J. Org. Chem.* 1966, **31**, 3032 [53].

[425] Campaigne E., Diedrich J.L., *J. Am. Chem. Soc.* 1951, **73**, 5240 [53].

[426] Greenfield H., Metlin S., Orchin M., Wender I., *J. Org. Chem.* 1958, **23**, 1054 [53].

[427] Craig L.C., Hixon R.M., *J. Am. Chem. Soc.* 1930, **52**, 804; 1931, **53**, 188 [53].

[428] Signaigo F.K., Adkins H., *J. Am. Chem. Soc.* 1936, **58**, 709 [53, 54, 112, 113, 153, 154].

[429] Cantor P.A., VanderWerf C.A., *J. Am. Chem. Soc.* 1958, **80**, 970 [53, 54].

[430] Adkins H., Coonradt H.L., *J. Am. Chem. Soc.* 1941, **63**, 1563 [53, 54, 57, 59].

[431] Overberger C.G., Palmer L.C., Marks B.S., Byrd N.R., *J. Am. Chem. Soc.* 1955, **77**, 4100 [54].

[432] Šorm F., *Collect. Czech. Chem. Commun.* 1947, **12**, 248 [54, 113].

[433] Lukeš R., Trojánek J., *Collect. Czech. Chem. Commun.* 1953, **18**, 648 [54].

[434] Freifelder M., *J. Org. Chem.* 1963, **28**, 602 [54].

[435] Leonard N.J., Barthel E., Jr, *J. Am. Chem. Soc.* 1949, **71**, 3098 [54, 169].

[436] Freifelder M., Wright H.B., *J. Med. Chem.* 1964, **7**, 664 [54].

[437] Adkins H., Kuick L.F., Farlow M., Wojcik B., *J. Am. Chem. Soc.* 1934, **56**, 2425 [54, 55].

[438] Freifelder M., Stone G.R., *J. Org. Chem.* 1961, **26**, 3805 [54, 55].

[439] Freifelder M., *Advan. Catalysis* 1963, **14**, 203 [55, 58].

[440] Lansbury P.T., *J. Am. Chem. Soc.* 1961, **83**, 429 [55].

[441] Lansbury P.T., Peterson J.O., *J. Am. Chem. Soc.* 1963, **85**, 2236 [55, 108, 110].

[442] Marvel C.S., Lazier W.A., *Org. Syn. Coll. Vol.* 1932, **1**, 99 [56].

[443] Ferles M., Tesařová A., *Collect. Czech. Chem. Commun.* 1967, **32**, 1631 [56, 101].

[444] Ladenburg A., *Chem. Ber.* 1889, **25**, 2768 [56].

[445] Papa D., Schwenk E., Klingsberg E., *J. Am. Chem. Soc.* 1951, **73**, 253 [56].

[446] Daeniker H.U., Grob C.A., *Org. Syn. Coll. Vol.* 1973, **5**, 989 [56].

[447] Holík M., Ferles M., *Collect. Czech. Chem. Commun.* 1967, **32**, 3067 [56].

[448] Ferles M., Attia A., Šilhánková A., *Collect. Czech. Chem. Commun.* 1973, **38**, 615 [56, 93].

[449] Ferles M., Holík M., *Collect. Czech. Chem. Commun.* 1966, **31**, 2416 [56, 170].

[450] Profft E., Linke H.W., *Chem. Ber.* 1960, **93**, 2591 [56].

[451] Ferles M., *Collect. Czech. Chem. Commun.* 1958, **23**, 479 [56].

[452] Ferles M., *Collect. Czech. Chem. Commun.* 1959, **24**, 2221 [56].

[453] Ochiai E., Tsuda K., Ikuma S., *Chem. Ber.* 1936, **69**, 2238 [56].

[454] Späth E., Kuffner F., *Chem. Ber.* 1935, **68**, 494 [56].

[455] Wibaut J.P., Oosterhuis A.G., *Rec. Trav. Chim. Pays-Bas* 1933, **52**, 941 [56].

[456] Smith A., Utley J.H.P., *Chem. Commun.* **1965**, 427 [56, 92].

[457] Gribble G.W., Lord P.D., Skotnicki J., Dietz S.E., Eaton J.T., Johnson J.L., *J. Am. Chem. Soc.* 1974, **96**, 7812 [56, 57, 171].

[458] Berger J.G., *Synthesis* **1974**, 508 [56, 57].

[459] Kikugawa Y., Saito K., Yamada S., *Synthesis* **1978**, 447 [56, 57, 58].

[460] O'Brien S., Smith D.C.C., *J. Chem. Soc.* **1960**, 4609 [56, 57].

[461] Remers W.A., Gibs G.J., Pidacks C., Weiss M.J., *J. Org. Chem.* 1971, **36**, 279 [56].

[462] Dolby L.J., Gribble G.W., *J. Heterocycl. Chem.* 1966, **3**, 124 [56, 57].

[463] Young D.V., Snyder H.R., *J. Am. Chem. Soc.* 1961, **83**, 3160 [57].

[464] King F.E., Barltrop J.A., Walley R.J., *J. Chem. Soc.* **1945**, 277 [57, 132, 133].

[465] Boekelheide V., Chu-Tsin Liu, *J. Am. Chem. Soc.* 1952, **74**, 4920 [57].

[466] Lowe O.G., King L.C., *J. Org. Chem.* 1959, **24**, 1200 [57].

[467] Galbraith A., Small T., Barnes R.A., Boekelheide V., *J. Am. Chem. Soc.* 1961, **83**, 453 [57].

[468] Hückel W., Hagedorn L., *Chem. Ber.* 1957, **90**, 752 [58].

[469] Hückel W., Stepf F., *Ann. Chem.* 1927, **453**, 163 [58].

[470] Gribble G.W., Heald P.W., *Synthesis* **1975**, 650 [58, 171].

[471] Vierhapper F.W., Eliel E.L., *J. Org. Chem.* 1975, **40**, 2729 [58, 59].

[472] Kocián O., Ferles M., *Collect. Czech. Chem. Commun.* 1978, **43**, 1413 [58].

[473] Birch A.J., Nasipuri D., *Tetrahedron* 1959, **6**, 148 [58].

[474] Witkop B., *J. Am. Chem. Soc.* 1948, **70**, 2617 [58].

[475] Boekelheide V., Gall W.G., *J. Am. Chem. Soc.* 1954, **76**, 1832 [58].

[476] Bohlmann F., *Chem. Ber.* 1952, **85**, 390 [59, 60, 61].

[477] Masamune T., Ohno M., Koshi M., Ohushi S., Iwadare T., *J. Org. Chem.* 1964, **29**, 1419 [59].

[478] Thoms H., Schnupp J., *Ann. Chem.* 1923, **434**, 296 [60].

[479] Butula I., *Croat. Chem. Acta* 1973, **45**, 313 [60].

[480] Bauer H., *J. Org. Chem.* 1961, **26**, 1649 [60].

[481] Hartmann M., Pannizon L., *Helv. Chim. Acta* 1938, **21**, 1692 [60].

[482] Evans R.C., Wiselogle F.Y., *J. Am. Chem. Soc.* 1945, **67**, 60 [60].

[483] Smith V.H., Christensen B.E., *J. Org. Chem.* 1955, **20**, 829 [60].

[484] Behun J.D., Levine R., *J. Org. Chem.* 1961, **26**, 3379 [60].

[485] Armand J., Chekir K., Pinson J., *Can. J. Chem.* 1974, **52**, 3971 [60].

[486] Schatz F., Wagner-Jauregg T., *Helv. Chim. Acta* 1968, **51**, 1919 [60].

[487] Neber P.W., Knoller G., Herbst K., Trissler A., *Ann. Chem.* 1929, **471**, 113 [60].

[488] Lund H., *Acta Chem. Scand.* 1967, **21**, 2525 [60].

[*489*] Marr E.B., Bogert M.T., *J. Am. Chem. Soc.* 1935, **57**, 729 [61].

[*490*] Bugle R.C., Oosteryoung R.A., *J. Org. Chem.* 1979, **44**, 1719 [61].

[*491*] Maffei S., Pietra S., *Gazz. Chim. Ital.* 1958, **88**, 556 [61, 62].

[*492*] Cavagnol J.C., Wiselogle F.Y., *J. Am. Chem. Soc.* 1947, **69**, 795 [61].

[*493*] Broadbent H.S., Allred E.L., Pendleton L., Whittle C.W., *J. Am. Chem. Soc.* 1960, **82**, 189 [61].

[*494*] Christie W., Rohde W., Schultz H.P., *J. Org. Chem.* 1956, **21**, 243 [61].

[*495*] Lund H., Jensen E.T., *Acta Chem. Scand.* 1970, **24**, 1867; 1971, **25**, 2727 [61].

[*496*] Elslager E.F., Worth D.F., Haley N.F., Perricone S.S., *J. Heterocycl. Chem.* 1968, **5**, 609 [61].

[*497*] Eckhard I.F., Fielden R., Summers L.A., *Austral. J. Chem.* 1975, **28**, 1149 [62].

[*498*] Karrer P., Waser P., *Helv. Chim. Acta* 1949, **32**, 409 [62].

[*499*] Butula I., *Ann. Chem.* 1969, **729**, 73 [62].

[*500*] Lacher J.R., Kianpour A., Park J.D., *J. Phys. Chem.* 1956, **60**, 1454 [63].

[*501*] Ashby E.C., Lin J.-J., Goel A.B., *J. Org. Chem.* 1978, **43**, 183 [63].

[*502*] Baltzly R., Phillips A.P., *J. Am. Chem. Soc.* 1946, **68**, 261 [63, 67].

[*503*] Grady G.L., Kuivila H.G., *J. Org. Chem.* 1969, **34**, 2014 [63, 64].

[*504*] Hutchins R.O., Hoke D., Keogh J., Koharski B., *Tetrahedron Lett.* **1969**, 3495 [63, 91].

[*505*] Bell H.M., Vanderslice C.W., Spehar A., *J. Org. Chem.* 1969, **34**, 3923 [63, 67].

[*506*] Krishnamurthy S., Brown H.C., *J. Org. Chem.* 1980, **45**, 849; 1982, **47**, 276 [63, 64].

[*507*] Masamune S., Rossy P.A., Bates G.S., *J. Am. Chem. Soc.* 1973, **95**, 6452 [63, 64].

[*508*] Kuivila H.G., Menapace L.W., *J. Org. Chem.* 1963, **28**, 2165 [63, 64, 65, 210].

[*509*] Greene F.D., Lowry N.N., *J. Org. Chem.* 1967, **32**, 882 [63, 64].

[*510*] Kuivila H.G., *Synthesis* **1970**, 499 [63].

[*511*] Trevoy L.W., Brown W.G., *J. Am. Chem. Soc.* 1949, **71**, 1675 [63, 83, 84].

[*512*] Hofmann K., Orochena S.F., Sax S.M., Jeffrey G.A., *J. Am. Chem. Soc.* 1959, **81**, 992 [64].

[*513*] Gaoni Y., *J. Org. Chem.* 1981, **46**, 4502 [64].

[*514*] Sydnes L., Skattebol L., *Tetrahedron Lett.* **1974**, 3703 [64].

[*515*] Fry A.J., Moore R.H., *J. Org. Chem.* 1968, **33**, 1283 [64].

[*516*] Winstein S., Sonnenberg J., *J. Am. Chem. Soc.* 1961, **83**, 3235, 3240 [64].

[*517*] Hartman W.W., Dreger E.E., *Org. Syn. Coll. Vol.* 1932, **1**, 357 [65].

[*518*] Nagao M., Sato N., Akashi T., Yoshida T., *J. Am. Chem. Soc.* 1966, **88**, 3447 [64, 65].

[519] Auwers von,K., Wissenbach H., *Chem. Ber.* 1923, **56,** 715, 730, 738 [65, 142].

[520] Whalley W.B., *J. Chem. Soc.* **1951,** 3229 [65, 123].

[521] Fuller G., Tatlow J.C., *J. Chem. Soc.* **1961,** 3198 [65, 66].

[522] Schlosser M., Heinz G., Le Van Chan, *Chem. Ber.* 1971, **104,** 1921 [65].

[523] Haszeldine R.N., Osborne J.E., *J. Chem. Soc.* **1956,** 61 [65].

[524] Hudlický M., Lejhancová I., *Collect. Czech. Chem. Commun.* 1965, **30,** 2491 [65].

[525] Bruck P., *Tetrahedron Lett.* **1962,** 449 [65].

[526] Strunk R.J., DiGiacomo P.M., Aso K., Kuivila H.G., *J. Am. Chem. Soc.* 1970, **92,** 2849 [65].

[527] Newman M.S., Cohen G.S., Cunico R.F., Dauernheim L.W., *J. Org. Chem.* 1973, **38,** 2760 [65].

[528] Knunyants I.L., Krasuskaya M.P., Mysov E.I., *Izv. Akad. Nauk SSSR* **1960,** 1412; *Chem. Abstr.* 1961, **55,** 349 [66].

[529] Hudlický M., *Chemistry of Organic Fluorine Compounds*, p. 174, Ellis Horwood, Chichester, 1976 [66].

[530] Mettille F.J., Burton D.J., *Fluorine Chemistry Reviews*, Vol. 1, p. 315 (P. Tarrant, ed.), Marcel Dekker, New York, 1967 [66].

[531] Hatch L.F., McDoland D.W., *J. Am. Chem. Soc.* 1952, **74,** 2911, 3328 [66].

[532] Haszeldine R.N., *J. Chem. Soc.* **1953,** 922 [66].

[533] Gassman P.G., Pape P.G., *J. Org. Chem.* 1964, **29,** 160 [66].

[534] Warner R.M., Leitsch L.C., *J. Labelled Compds* 1965, **1,** 42 [66, 158].

[535] Burton D.J., Johnson R.L., *J. Am. Chem. Soc.* 1964, **86,** 5361; *Tetrahedron Lett.* **1966,** 2681 [66, 67].

[536] Kämmerer H., Horner L., Beck H., *Chem. Ber.* 1958, **91,** 1376 [67, 68, 81, 141, 205].

[537] Eliel E.L., *J. Am. Chem. Soc.* 1949, **71,** 3970 [67].

[538] Matsumura S., Tokura N., *Tetrahedron Lett.* **1969,** 363 [67].

[539] Shapiro S.L., Overberger C.G., *J. Am. Chem. Soc.* 1954, **76,** 97 [67].

[540] Parham W.E., Wright C.D., *J. Org. Chem.* 1957, **22,** 1473 [67].

[541] Bell H.M., Brown H.C., *J. Am. Chem. Soc.* 1966, **88,** 1473 [67].

[542] Brown H.C., Krishnamurthy S., *J. Org. Chem.* 1969, **34,** 3918 [67, 68].

[543] Brooke G.M., Burdon J., Tatlow J.C., *J. Chem. Soc.* **1962,** 3253 [67].

[544] Bažant V., Čapka M., Černý M., Chvalovský V., Kochloefl K., Kraus M., Málek J., *Tetrahedron Lett.* **1968,** 3303 [68, 72, 96, 97, 110, 111, 139, 146, 147, 157, 158, 166, 167, 168].

[545] Rothman L.A., Becker E.I., *J. Org. Chem.* 1959, **24,** 294; 1960, **25,** 2203 [68].

[546] Jošt K., Rudinger J., Šorm F., *Collect. Czech. Chem. Commun.* 1963, **28,** 1706 [68].

[547] Gronowitz S., Raznikiewicz T., *Org. Syn. Coll. Vol.* 1973, **5,** 149 [68].

[548] Erlenmeyer H., Grubenmann W., *Helv. Chim. Acta* 1947, **30**, 297 [68, 142].

[549] Govindachari T.R., Nagarajan K., Rajappa S., *J. Chem. Soc.* **1957**, 2725 [68].

[550] Levitz M., Bogert M.T., *J. Org. Chem.* 1945, **10**, 341 [68].

[551] Neumann F.W., Sommer N.B., Kaslow C.E., Shriner R.L., *Org. Syn. Coll. Vol.* 1955, **3**, 519 [68].

[552] Lutz R.E., Ashburn G., Rowlett R.J., Jr, *J. Am. Chem. Soc.* 1946, **68**, 1322 [68].

[553] Wibaut J.P., *Rec. Trav. Chim. Pays-Bas* 1944, **63**, 141 [68].

[554] Fox B.A., Threlfall T.L., *Org. Syn. Coll. Vol.* 1973, **5**, 346 [68].

[555] Banks R.N., Haszeldine R.N., Lathram J.V., Young I.M., *Chem. & Ind.* **1964**, 835; *J. Chem. Soc.* **1965**, 594 [68].

[556] Marshall J.R., Walker J., *J. Chem. Soc.* **1951**, 1004, 1013 [68].

[557] Brown D.J., Waring P., *Austral. J. Chem.* 1973, **26**, 443 [68].

[558] Stephenson E.F.M., *Org. Syn. Coll. Vol.* 1955, **3**, 475 [69].

[559] Senkus M., *Ind. Eng. Chem.* 1948, **40**, 506 [69, 81].

[560] Nielsen A.T., *J. Org. Chem.* 1962, **27**, 1998 [69].

[561] Grundmann C., *Angew. Chem.* 1950, **62**, 558 [70].

[562] Artemev A.A., Genkina A.B., Malimonova A.B., Trofilkina V.P., Isaenkova M.A., *Zh. Vses. Khim. Obshchestva im. D. I. Mendeleeva* 1965, **10**, 588; *Chem. Abstr.* 1966, **64**, 1975 [70].

[563] McMurry J.E., Melton J., *J. Org. Chem.* 1973, **38**, 4367 [70].

[564] Barltrop J.A., Nicholson J.S., *J. Chem. Soc.* **1951**, 2524 [70, 82].

[565] Aeberli P., Houlihan W.J., *J. Org. Chem.* 1967, **32**, 3211 [70].

[566] Meyers A.I., Sircar J.C., *J. Org. Chem.* 1967, **32**, 4134 [70].

[567] Gilsdorf R.T., Nord F.F., *J. Am. Chem. Soc.* 1952, **74**, 1837 [70, 71].

[568] Brossi A., Van Burik J., Teitel S., *Helv. Chim. Acta* 1968, **51**, 1965 [71].

[569] Hass H.B., Susie A.G., Heider R.L., *J. Org. Chem.* 1950, **15**, 8 [71].

[570] Heinzelman R.V., *Org. Syn. Coll. Vol.* 1963, **4**, 573 [71].

[571] Lindemann A., *Helv. Chim. Acta* 1949, **32**, 69 [71].

[572] Brand K., Steiner J., *Chem. Ber.* 1922, **55**, 875, 886 [71, 72, 73, 95, 96].

[573] Harman R.E., *Org. Syn. Coll. Vol.* 1963, **4**, 148 [72].

[574] Kamm O., *Org. Syn. Coll. Vol.* 1932, **1**, 445 [72].

[575] Corbett J.F., *Chem. Commun.* **1968**, 1257 [72].

[576] Nystrom R.F., Brown W.G., *J. Am. Chem. Soc.* 1948, **70**, 3738 [72, 102, 129, 154, 174].

[577] James B.D., *Chem. & Ind.* **1971**, 227 [72, 98, 156, 167, 168].

[578] Bigelow H.E., Robinson D.B., *Org. Syn. Coll. Vol.* 1955, **3**, 103 [72].

[579] Adams R., Cohen F.L., *Org. Syn. Coll. Vol.* 1932, **1**, 240 [73].

[580] Mendenhall G.D., Smith P.A.S., *Org. Syn. Coll. Vol.* 1973, **5**, 829 [73].

[581] Severin T., Schmitz R., *Chem. Ber.* 1962, **95**, 1417 [73].

[582] West R.W., *J. Chem. Soc.* 1925, **127**, 494 [73, 74].

[583] Dankova T.F., Bokova T.N., Preobrazhenskii N.A., *Zh. Obschei Khim.* 1951, **21**, 787; *Chem. Abstr.* 1951, **45**, 9517 [73, 75, 124].

[584] Kuhn W.E., *Org. Syn. Coll. Vol.* 1943, **2**, 448 [73].

[585] Kock E., *Chem. Ber.* 1887, **20**, 1567 [73].

[586] Hurst W.G., Thorpe J.F., *J. Chem. Soc.* **1915**, 934 [73].

[587] Ayling E.E., Gorvin J.H., Hinkel L.E., *J. Chem. Soc.* **1942**, 755 [73, 74].

[588] Entwistle I.D., Johnstone R.A.W., Povall T.J., *J. Chem. Soc. Perkin I* **1975**, 1300 [74].

[589] Hudlický M., Bell H.M., *J. Fluorine Chem.* 1974, **4**, 19 [74].

[590] Sachs F., Sichel E., *Chem. Ber.* 1904, **37**, 1861 [74, 103, 214].

[591] Parkes G.D., Farthing A.C., *J. Chem. Soc.* **1948**, 1275 [74].

[592] Weil T., Prijs B., Erlenmeyer H., *Helv. Chim. Acta* 1953, **36**, 142 [75].

[593] Hendrickson J.B., *J. Am. Chem. Soc.* 1961, **83**, 1251 [75, 142].

[594] Kornblum N., Iffland D.C., *J. Am. Chem. Soc.* 1949, **71**, 2137 [75, 217].

[595] Newman M.S., Hung W.M., *J. Org. Chem.* 1974, **39**, 1317 [75].

[596] Reychler A., *Chem. Ber.* 1887, **20**, 2463 [76].

[597] Fischer E., *Ann. Chem.* 1878, **190**, 67, 78, [76].

[598] VanderWerf C.A., Heisler R.Y., McEwen W.E., *J. Am. Chem. Soc.* 1954, **76**, 1231 [76].

[599] Cama L.D., Leanza W.J., Beattie T.R., Christensen B.G., *J. Am. Chem. Soc.* 1972, **94**, 1408 [76, 151].

[600] Boyer J.H., *J. Am. Chem. Soc.* 1951, **73**, 5865 [76, 125].

[601] Stanovnik B., Tišler M., Polanc S., Gračner M., *Synthesis*, **1978**, 65 [76].

[602] Bayley H., Standring D.N., Knowles J.R., *Tetrahedron Lett.* **1978**, 3633 [76].

[603] Hassner A., Matthews G.J., Fowler F.W., *J. Am. Chem. Soc.* 1969, **91**, 5046 [76].

[604] Landa S., Mostecký J., *Collect. Czech. Chem. Commun.* 1955, **20**, 430; 1956, **21**, 1177 [77].

[605] Brewster J.J., Osman S.F., Bayer H.O., Hopps H.B., *J. Org. Chem.* 1964, **29**, 121 [77].

[606] Vowinkel E., Buthe I., *Chem. Ber.* 1974, **107**, 1353 [77].

[607] Corey E.J., Carey F.A., Winter R.A.E., *J. Am. Chem. Soc.* 1965, **87**, 934 [77].

[608] Ohloff G., Farnow H., Schade G., *Chem. Ber.* 1956, **89**, 1549 [77, 78].

[609] Hochstein F.A., Brown W.G., *J. Am. Chem. Soc.* 1948, **70**, 3484 [77, 80, 102].

[610] Brewster J.H., Bayer H.O., *J. Org. Chem.* 1964, **29**, 116 [78].

[611] Birch A.J., *J. Chem. Soc.* **1945**, 809 [78].

[612] Bohlmann F., Staffeldt J., Skuballa W., *Chem. Ber.* 1976, **109**, 1586 [78].

[613] Corey E.J., Achiwa K., *J. Org. Chem.* 1969, **34**, 3667 [78].

[614] Vowinkel E., Wolf C., *Chem. Ber.* 1974, **107**, 907 [79].

[615] Vowinkel E., Baese H., *Chem. Ber.* 1974, **107**, 1213 [79].

[616] Musliner W.J., Gates J.W., Jr, *J. Am. Chem. Soc.* 1966, **88**, 4271 [79].

[617] Barth L., Goldschmiedt G., *Chem. Ber.* 1879, **12**, 1244 [79].

[618] Severin T., Ipach I., *Synthesis* **1973**, 796 [79].

[619] Mitsui S., Imaizumi S., Esashi Y., *Bull. Chem. Soc. Jap.* 1970, **43**, 2143 [79].

[620] Folkers K., Adkins H., *J. Am. Chem. Soc.* 1932, **54**, 1145 [79, 153].

[621] Gribble G.W., Leese R.M., Evans B., *Synthesis* **1977**, 172 [79].

[622] Brewster J.H., Bayer H.O., *J. Org. Chem.* 1964, **29**, 105, 110 [79].

[623] Breuer E., *Tetrahedron Lett.* **1967**, 1849 [79].

[624] Ferles M., Lebl M., Šilhánková A., Stern P., Wimmer A., *Collect. Czech. Chem. Commun.* 1975, **40**, 1571 [79].

[625] Marvel C.S., Hager F.D., Caudle E.C., *Org. Syn. Coll. Vol.* 1932, **1**, 224 [79].

[626] Musser D.M., Adkins H., *J. Am. Chem. Soc.* 1938, **60**, 664 [80].

[627] Adkins H., Krsek G., *J. Am. Chem. Soc.* 1948, **70**, 412 [80, 81].

[628] Stork G., *J. Am. Chem. Soc.* 1947, **69**, 576 [80, 82].

[629] Gutsche C.D., Peters H.H., *Org. Syn. Coll. Vol.* 1963, **4**, 887 [81].

[630] Gilman N.W., Sternbach L.H., *Chem. Commun.* **1971**, 465 [81, 93].

[631] Wiesner K., *Private communication* [81].

[632] Woodburn H.M., Stuntz C.F., *J. Am. Chem. Soc.* 1950, **72**, 1361 [81].

[633] Bailey W.J., Marktscheffel F., *J. Org. Chem.* 1960, **25**, 1797 [81].

[634] Andrus D.W., Johnson J.R., *Org. Syn. Coll. Vol.* 1955, **3**, 794 [82].

[635] Birch A.J., Subba Rao G.S.R., *Austral. J. Chem.* 1970, **23**, 1641 [82].

[636] Hallsworth A.S., Henbest H.B., Wrigley T.I., *J. Chem. Soc.* **1957**, 1969 [82].

[637] Hartung W.H., Simonoff R., *Org. Reactions* 1953, **7**, 263 [82, 86, 93].

[638] Reimann E., *Ann. Chem.* 1971, **750**, 109, 126 [82].

[639] Olah G.A., Prakash G.K.S., Narang S.C., *Synthesis* **1978**, 825 [82].

[640] Reist E.J., Bartuska V.J., Goodman L., *J. Org. Chem.* 1964, **29**, 3725 [82].

[641] Shriner R.L., Ruby P.R., *Org. Syn. Coll. Vol.* 1963, **4**, 798 [82].

[642] Doyle T.W., *Can. J. Chem.* 1977, **55**, 2714 [82].

[643] Thoms H., Siebeling W., *Chem. Ber.* 1911, **44**, 2134 [82].

[644] Wilds A.L., Nelson N.A., *J. Am. Chem. Soc.* 1953, **75**, 5360, 5366 [82].

[645] Soffer M.D., Bellis M.P., Gellerson H.E., Stewart R.A., *Org. Syn. Coll. Vol.* 1963, **4**, 903 [83].

[646] Fieser L.F., *J. Am. Chem. Soc.* 1953, **75**, 4395 [83, 126].

[647] Kupchan S.M., Maruyama M., *J. Org. Chem.* 1971, **36**, 1187 [83].

[648] Kochi J.K., Singleton D.M., Andrews L.J., *Tetrahedron* 1968, **24**, 3503 [83].

[649] Fujisawa T., Sugimoto K., Ohta H., *Chem. Lett.* **1974**, 883 [83, 84].

[650] Cole W., Julian P.L., *J. Org. Chem.* 1954, **19**, 131 [83, 126].

[651] Brown M., Piszkiewicz L.W., *J. Org. Chem.* 1967, **32**, 2013 [83, 119].

[652] McQuillin F.J., Ord W.O., *J. Chem. Soc.* **1959**, 3169 [83, 84].

[653] Eliel E.L., Delmonte D.W., *J. Am. Chem. Soc.* 1956, **78,** 3226 [83].

[654] Eliel E.L., Rerick M.N., *J. Am. Chem. Soc.* 1960, **82,** 1362 [83, 84].

[655] Elsenbaumer R.L., Mosher H.S., Morrison J.D., Tomaszewski J.E., *J. Org. Chem.* 1981, **46,** 4034 [83, 84].

[656] Hutchins R.O., Taffer I.M., Burgoyne W., *J. Org. Chem.* 1981, **46,** 5214 [83, 84].

[657] Brown H.C., Yoon N.M., *Chem. Commun.* **1968,** 1549 [83].

[658] Brown H.C., Yoon N.M., *J. Am. Chem. Soc.* 1968, **90,** 2686 [83, 84].

[659] Brown H.C., Ikegami S., Kawakami J.H., *J. Org. Chem.* 1970, **35,** 3243 [84].

[660] Zaidlewicz M., Uzarewicz A., Sarnowski R., *Synthesis* **1979,** 62 [84].

[661] Criegee R., Zogel H., *Chem. Ber.* 1951, **84,** 215 [84].

[662] Paget H., *J. Chem. Soc.* **1938,** 829 [84, 85].

[663] Richter F., Presting W., *Chem. Ber.* 1931, **64,** 878 [84, 85].

[664] Wallach O., *Ann. Chem.* 1912, **392,** 49, 60 [84, 85].

[665] Bruyn de P., *Ann. Chim.* 1945, **20,** 551 [84].

[666] Windaus A., Linsert O., *Ann. Chem.* 1928, **465,** 148 [84].

[667] Pappas J.J., Keaveney W.P., Berger M., Rush R.V., *J. Org. Chem.* 1968, **33,** 787 [85].

[668] Kametani T., Ogasawara K., *Chem. & Ind.* **1968,** 1772 [85].

[669] Burfield D.R., *J. Org. Chem.* 1982, **47,** 3821 [85].

[670] Pryde E.H., Anders D.E., Teeter H.M., Cowan J.C., *J. Org. Chem.* 1960, **25,** 618 [85, 86].

[671] Helferich B., Dommer W., *Chem. Ber.* 1920, **53,** 2004, 2009 [85, 86].

[672] Hoffmann F.W., Ess R.J., Simmons T.C., Hanzel R.S., *J. Am. Chem. Soc.* 1956, **78,** 6414 [86, 87, 217].

[673] Truce W.E., Perry F.M., *J. Org. Chem.* 1965, **30,** 1316 [86, 87, 88].

[674] Birch A.J., Walker K.A.M., *Tetrahedron Lett.* **1967,** 1935 [86].

[675] Mukaiyama T., Narasaka K., Maekawa K., Furusato M., *Bull. Chem. Soc. Jap.* 1971, **44,** 2285; *Chem. Lett.* **1973,** 291 [86].

[676] Welch S.C., Loh J.-P., *J. Org. Chem.* 1981, **46,** 4072 [86].

[677] Brink M., *Synthesis* **1975,** 807 [86, 87].

[678] Nayak U.G., Brown R.K., *Can. J. Chem.* 1966, **44,** 591 [86].

[679] Brown E.D., Iqbal S.M., Owen L.N., *J. Chem. Soc. C* **1966,** 415 [87].

[680] Strating J., Backer H.J., *Rec. Trav. Chim. Pays-Bas* 1950, **69,** 638 [87, 89, 90, 91].

[681] Arnold R.C., Lien A.P., Alm R.M., *J. Am. Chem. Soc.* 1950, **72,** 731 [87].

[682] D'Amico J.J., *J. Org. Chem.* 1961, **26,** 3436 [87].

[683] Riegel B., Lappin G.R., Adelson B.H., Jackson R.I., Albisetti C.P., Dodson R.M., Baker R.H., *J. Am. Chem. Soc.* 1946, **68,** 1264 [88].

[684] Ogura K., Yamashita M., Tsuchihashi G., *Synthesis* **1975,** 385 [88].

[685] Drabowicz J., Oae S., *Synthesis* **1977,** 404 [88].

[686] Truce W.E., Simms J.A., *J. Am. Chem. Soc.* 1956, **78,** 2756 [88].

[687] Bordwell F.G., McKellin W.H., *J. Am. Chem. Soc.* 1951, **73**, 2251 [88, 89].

[688] Fehnel E.A., Carmack M., *J. Am. Chem. Soc.* 1948, **70**, 1813 [88, 120].

[689] Gardner J.N., Kaiser S., Krubiner A., Lucas H., *Can. J. Chem.* 1973, **51**, 1419 [88, 89].

[690] Cronyn M.W., Zavarin E., *J. Org. Chem.* 1954, **19**, 139 [89].

[691] Dabby R.E., Kenyon J., Mason R.F., *J. Chem. Soc.* **1952**, 4881 [89].

[692] Wertheim E., *Org. Syn. Coll. Vol.* 1943, **2**, 471 [90].

[693] Field L., Grunwald F.A., *J. Org. Chem.* 1951, **16**, 946 [90].

[694] Ullmann F., Pasdermadjian G., *Chem. Ber.* 1901, **34**, 1151 [90].

[695] Whitmore F.C., Hamilton F.H., *Org. Syn. Coll. Vol.* 1932, **1**, 479 [90].

[696] Adams R., Marvel C.S., *Org. Syn. Coll. Vol.* 1932, **1**, 504 [90].

[697] Sheppard W.A., *Org. Syn. Coll. Vol.* 1973, **5**, 843 [90].

[698] Kenner G.W., Murray M.A., *J. Chem. Soc.* 1949, S1, 178 [90, 91].

[699] Krishmamurthy S., Brown H.C., *J. Org. Chem.* 1976, **41**, 3064 [91].

[700] Kočovský P., Černý V., *Collect. Czech. Chem. Commun.* 1979, **44**, 246 [91, 213].

[701] Closson W.D., Wriede P., Bank S., *J. Am. Chem. Soc.* 1966, **88**, 1581 [91].

[702] Reber F., Reichstein T., *Helv. Chim. Acta* 1945, **28**, 1164, 1170 [91].

[703] Fischer E., *Chem. Ber.* 1915, **48**, 93 [91, 92].

[704] Klamann D., Hofbauer G., *Chem. Ber.* 1953, **86**, 1246 [91, 92].

[705] Ji S., Gortler L.B., Waring A., Battisti A., Bank S., Closson W.D., Wriede P., *J. Am. Chem. Soc.* 1967, **89**, 5311 [91, 92, 212].

[706] Campbell K.N., Sommers A.H., Campbell B.K., *Org. Syn. Coll. Vol.* 1955, **3**, 148 [92].

[707] Sabatier P., Mailhe A., *Compt. Rend.* 1907, **144**, 784 [92].

[708] Doldouras G.A., Kollonitsch J., *J. Am. Chem. Soc.* 1978, **100**, 341 [92].

[709] King T.J., *J. Chem. Soc.* **1951**, 898 [92].

[710] Coulter J.M., Lewis J.W., Lynch P.P., *Tetrahedron* 1968, **24**, 4489 [92].

[711] Schmitt J., Panouse J.J., Hallot A., Pluchet H., Comoy P., Cornu P.J., *Bull. Soc. Chim. Fr.* **1963**, 816 [92].

[712] Borch R.F., Bernstein M.D., Durst H.D., *J. Am. Chem. Soc.* 1971, **93**, 2897 [92].

[713] Benkeser R.A., Lambert R.F., Ryan P.W., Stoffey D.G., *J. Am. Chem. Soc.* 1958, **80**, 6573 [93].

[714] Birch A.J., *J. Chem. Soc.* **1946**, 593 [93].

[715] Waser E.B.H., Mollering H., *Org. Syn. Coll. Vol.* 1932, **1**, 499 [93].

[716] Birkofer L., *Chem. Ber.* 1942, **75**, 429 [93].

[717] Dahn H., Zoller P., *Helv. Chim. Acta* 1952, **35**, 1348 [93].

[718] Dahn H., Solms U., Zoller P., *Helv. Chim. Acta* 1952, **35**, 2117 [93].

[719] Marchand B., *Chem. Ber.* 1962, **95**, 577 [93].

[720] Emmert B., *Chem. Ber.* 1909, **42**, 1507 [93].

[721] Brasen W.R., Hauser C.R., *Org. Syn. Coll. Vol.* 1963, **4**, 508 [93].

[722] Sugasawa S., Ushioda S., Fujisawa T., *Tetrahedron* 1959, **5**, 48 [93, 108].

[723] Martin E.L., *Org. Syn. Coll. Vol.* 1943, **2**, 501 [93].

[724] Boyer J.H., Buriks R.S., *Org. Syn. Coll. Vol.* 1973, **5**, 1067 [93].

[725] Metayer M., *Bull. Soc. Chim. Fr.* **1950**, 1048 [94].

[726] Schultz E.M., Bicking J.B., *J. Am. Chem. Soc.* 1953, **75**, 1128 [94].

[727] Clemo G.R., Raper R., Vipond H.J., *J. Chem. Soc.* **1949**, 2095 [94].

[728] Shih D.H., Ratcliffe R.W., *J. Med. Chem.* 1981, **24**, 639 [94, 151].

[729] Anantharamaiah G.M., Sivanandaiah K.M., *J. Chem. Soc. Perkin I* **1977**, 490 [94].

[730] Haas H.J., *Chem. Ber.* 1961, **94**, 2442 [94].

[731] Hanna C., Schueler F.W., *J. Am. Chem. Soc.* 1952, **74**, 3693 [94].

[732] Schueler F.W., Hanna C., *J. Am. Chem. Soc.* 1951, **73**, 4996 [94].

[733] Hatt H.H., *Org. Syn. Coll. Vol.* 1943, **2**, 211 [94].

[734] Backer H.J., *Rec. Trav. Chim. Pays-Bas* 1912, **31**, 142 [94].

[735] Ingersoll A.W., Bircher L.J., Brubaker M.M., *Org. Syn. Coll. Vol.* 1932, **1**, 485 [94].

[736] Taylor E.C., Crovetti A.J., Boyer N.E., *J. Am. Chem. Soc.* 1957, **79**, 3549 [94, 95].

[737] Berson J.A., Cohen T., *J. Org. Chem.* 1955, **20**, 1461 [94, 95].

[738] Brown H.C., Subba Rao B.C., *J. Am. Chem. Soc.* 1956, **78**, 2582 [94, 139, 141, 142, 155, 160, 209].

[739] Rayburn C.H., Harlan W.R., Hanmer H.R., *J. Am. Chem. Soc.* 1950, **72**, 1721 [95].

[740] Skita A., *Chem. Ber.* 1912, **45**, 3312 [95, 96].

[741] Guither W.D., Clark D.G., Castle R.N., *J. Heterocycl. Chem.* 1965, **2**, 67 [95].

[742] Jardine I., McQuillin F.J., *Chem. Commun.* **1920**, 626 [95, 96].

[743] van Tamelen E.E., Dewey R.S., Timmons R.J., *J. Am. Chem. Soc.* 1961, **83**, 3725 [95].

[744] Cartwright R.A., Tatlow J.C., *J. Chem. Soc.* **1953**, 1994 [95].

[745] Ruggli P., Hölzle K., *Helv. Chim. Acta* 1943, **26**, 1190 [95].

[746] Thiele J., *Ann. Chem.* 1892, **270**, 1, 44 [96].

[747] Schmitt Y., Langlois M., Perrin C., Callet G., *Bull. Soc. Chim. Fr.* **1969**, 2004 [96].

[748] Oae S., Tsujimoto N., Nakanishi A., *Bull. Chem. Soc. Jap.* 1973, **46**, 535 [96].

[749] Carothers W.H., Adams R., *J. Am. Chem. Soc.* 1934, **46**, 1675 [96, 99].

[750] Nystrom R.F., Chaikin S.W., Brown W.G., *J. Am. Chem. Soc.* 1949, **71**, 3245 [96, 97, 98, 99, 100, 107, 108, 110, 155].

[751] Chaikin S.W., Brown W.G., *J. Am. Chem. Soc.* 1949, **71**, 122 [96, 97, 98, 100, 102, 107, 108, 110, 120, 146].

[752] Krishnamurthy S., *J. Org. Chem.* 1981, **46**, 4628 [96, 97].

[753] Fujisawa T., Sugimoto K., Ohta H., *J. Org. Chem.* 1976, **41,** 1667 [96, 97, 99].

[754] Clarke H.T., Dreger E.E., *Org. Syn. Coll. Vol.* 1932, **1,** 304 [96, 97].

[755] Posner G.H., Runquist A.W., Chapdelaine M.J., *J. Org. Chem.* 1977, **42,** 1202 [97, 99, 100, 102, 107, 217].

[756] Fung N.Y.M., Mayo de, P., Schauble J.H., Weedon A.C., *J. Org. Chem.* 1978, **43,** 3977 [97, 99, 100, 107].

[757] Hutchins R.O., Kandasamy D., *J. Am. Chem. Soc.* 1973, **95,** 6131 [96, 97, 100, 107, 108, 110, 117].

[758] Clemmensen E., *Chem. Ber.* 1913, **46,** 1837; 1914, **47,** 51, 681 [97, 101, 108, 113, 118, 126, 213].

[759] Herr C.H., Whitmore F.C., Schiessler R.W., *J. Am. Chem. Soc.* 1945, **67,** 2061 [97, 101, 108, 113].

[760] Brannock K.C., *J. Am. Chem. Soc.* 1959, **81,** 3379 [98].

[761] Hennis H.E., Trapp W.B., *J. Org. Chem.* 1961, **26,** 4678 [98].

[762] Goetz R.W., Orchin M., *J. Am. Chem. Soc.* 1963, **85,** 2782 [98].

[763] Rylander P.N., Steele D.R., *Tetrahedron Lett.* **1969,** 1579 [98, 102].

[764] Krishnamurthy S., Brown H.C., *J. Org. Chem.* 1977, **42,** 1197 [98, 102, 120, 121].

[765] Young W.G., Hartung W.H., Crossley F.S., *J. Am. Chem. Soc.* 1936, **58,** 100 [98, 99, 102].

[766] Law H.D., *J. Chem. Soc.* 1912, **101,** 1016 [98, 102, 120, 121].

[767] Skita A., *Chem. Ber.* 1915, **48,** 1486 [99, 101, 119].

[768] Adams R., McKenzie S., Jr, Loewe S., *J. Am. Chem. Soc.*1948, **70,** 664 [99].

[769] Cook P.L., *J. Org. Chem.* 1962, **27,** 3873 [100, 107, 108, 110, 111, 119, 121].

[770] Brown B.R., White A.M.S., *J. Chem. Soc.* **1957,** 3755 [99, 100, 111, 113].

[771] Raber, D.J., Guido W.C., *J. Org. Chem.* 1976, **41,** 690 [99, 100, 102, 107, 108, 110, 111, 146, 155].

[772] Fry J.L., Orfanopoulos M., Adlington M.G., Dittman W.P., Jr, Silverman S.B., *J. Org. Chem.* 1978, **43,** 374 [99, 100, 101, 108].

[773] Hall S.S., Bartels A.P., Engman A.M., *J. Org. Chem.* 1972, **37,** 760 [99, 101].

[774] Lock G., Stach K., *Chem. Ber.* 1943, **76,** 1252 [99, 101, 103, 113].

[775] Mulholland T.P.C., Ward G., *J. Chem. Soc.* **1954,** 4676 [101].

[776] Burnette L.W., Johns I.B., Holdren R.F., Hixon R.M., *Ind. Eng. Chem.* 1948, **40,** 502 [101].

[777] West C.T., Donnelly S.J., Kooistra D.A., Doyle M.P., *J. Org. Chem.* 1973, **38,** 2675 [101, 113, 143, 144].

[778] Schwarz R., Hering H., *Org. Syn. Coll. Vol.* 1963, **4,** 203 [101].

[779] Jorgenson M.J., *Tetrahedron Lett.* **1962,** 559 [102, 121].

[780] Hutchins R.O., Kandasamy D., *J. Org. Chem.* 1975, **40,** 2530 [102, 120, 121, 122].

[781] Mincione E., *J. Org. Chem.* 1978, **43**, 1829 [102, 121].

[782] Tuley W.F., Adams R., *J. Am. Chem. Soc.* 1925, **47**, 3061 [102].

[783] Likhosherstov M.V., Arsenyuk A.A., Zeberg E.F., Karitskaya I.V., *Zh. Obshchei Khim.* 1950, **20**, 627; *Chem. Abstr.* 1950, **44**, 7823 [102].

[784] Heacock R.A., Hutzinger O., *Can. J. Chem.* 1964, **42**, 514 [102].

[785] Hutchins R.O., Natale N.R., *J. Org. Chem.* 1978, **43**, 2299 [103, 106, 108, 121, 134].

[786] Kabalka G.W., Summers S.T., *J. Org. Chem.* 1981, **46**, 1217 [103, 106, 108, 113, 121, 122].

[787] Nystrom R.B., *J. Am. Chem. Soc.* 1955, **77**, 2544 [103, 146, 174, 208].

[788] Bader H., Downer J.D., Driver P., *J. Chem. Soc.* **1950**, 2775 [103].

[789] Woodward R.B., *Org. Syn. Coll. Vol.* 1955, **3**, 453 [103].

[790] Buck J.S., Ide W.S., *Org. Syn. Coll. Vol.* 1943, **2**, 130 [103].

[791] Remers W.A., Roth R.H., Weiss M.J., *J. Org. Chem.* 1965, **30**, 4381 [103].

[792] Eliel E.L., Badding V.G., Rerick M.N., *J. Am. Chem. Soc.* 1962, **84**, 2371 [103, 130].

[793] Eliel E.L., Nowak B.E., Daignault R.A., Badding V.G., *J. Org. Chem.* 1965, **30**, 2441 [104].

[794] Eliel E.L., Nowak B.E., Daignault R.A., *J. Org. Chem.* 1965, **30**, 2448 [104, 105].

[795] Eliel E.L., Doyle T.W., Daignault R.A., Newman B.C., *J. Am. Chem. Soc.* 1966, **88**, 1828 [104, 105, 130, 131].

[796] Wolfrom M.L., Karabinos J.V., *J. Am. Chem. Soc.* 1944, **66**, 909 [104, 105, 130].

[797] Pettit G.R., van Tamelen E.E., *Org. Reactions* 1962, **12**, 356 [104, 105, 130, 131].

[798] Gutierrez C.G., Stringham R.A., Nitasaka T., Glasscock K.G., *J. Org. Chem.* 1980, **45**, 3393 [104, 105, 131].

[799] Degani I., Fochi R., *J. Chem. Soc. Perkin I* **1976**, 1886; **1978**, 1133 [104, 105, 108].

[800] Hill A.J., Nason E.H., *J. Am. Chem. Soc.* 1924, **46**, 2241 [105].

[801] Eliel E.L., Daignault R.A., *J. Org. Chem.* 1965, **30**, 2450 [105].

[802] Campbell K.N., Sommers A.H., Campbell B.K., *J. Am. Chem. Soc.* 1944, **66**, 82 [105].

[803] Winans C.F., Adkins H., *J. Am. Chem. Soc.* 1932, **54**, 306 [105, 106, 135, 136].

[804] Allen C.F.H., van Allan J., *Org. Syn. Coll. Vol.* 1955, **3**, 827 [105].

[805] Bumgardner C.L., Lawton E.L., Carver J.G., *J. Org. Chem.* 1972, **37**, 407 [105].

[806] Fischer O., *Ann. Chem.* 1887, **241**, 328 [105].

[807] Contento M., Savoia D., Trombini C., Umani-Ronchi A., *Synthesis* **1979**, 30 [105, 120, 121].

[808] Karrer P., Schick E., *Helv. Chim. Acta* 1943, **26**, 800 [106].

[809] Smith D.R., Maienthal M., Tipton J., *J. Org. Chem.* 1952, **17,** 294 [106, 132, 133].

[810] Lycan W.H., Puntambeker S.V., Marvel C.S., *Org. Syn. Coll. Vol.* 1943, **2,** 318 [106].

[811] Caglioti L., *Tetrahedron* 1966, **22,** 487 [106, 113, 118, 172].

[812] Caglioti L., Magi M., *Tetrahedron* 1963, **19,** 1127 [106, 113, 134].

[813] Hutchins R.O., Maryanoff B.E., Milewski C.A., *J. Am. Chem. Soc.* 1971, **93,** 1793 [106, 108, 134].

[814] Breitner E., Roginski E., Rylander P.N., *J. Org. Chem.* 1959, **24,** 1855 [107, 109, 110, 119].

[815] Heusler K., Wieland P., Meystre C.H., *Org. Syn. Coll. Vol.* 1973, **5,** 692 [107].

[816] Eliel E.L., Nasipuri D., *J. Org. Chem.* 1965, **30,** 3809 [107, 115, 116, 117].

[817] Santaniello E., Ponti F., Manzocchi A., *Synthesis* **1978,** 891 [107, 108, 110].

[818] Brown H.C., Mandal A.K., *J. Org. Chem.* 1977, **42,** 2996 [107, 108, 111].

[819] Thoms H., Mannich C., *Chem. Ber.* 1903, **36,** 2544 [107, 108].

[820] Whitmore F.C., Ottenbacher T., *Org. Syn. Coll. Vol.* 1943, **2,** 317 [107].

[821] Truett W.L., Moulton W.N., *J. Am. Chem. Soc.* 1951, **73,** 5913 [107, 108, 110].

[822] Yamashita S., *J. Organometal. Chem.* 1968, **11,** 377 [107, 108, 111].

[823] Newman M.S., Walborsky H.M., *J. Am. Chem. Soc.* 1950, **72,** 4296 [108, 162].

[824] Tafel J., *Chem. Ber.* 1909, **42,** 3146 [108].

[825] Schreibmann P., *Tetrahedron Lett.* **1970,** 4271 [109, 118].

[826] Freifelder M., *J. Org. Chem.* 1964, **29,** 2895 [110, 112, 113].

[827] Tyman J.H.P., *J. Appl. Chem.* 1970, **20,** 179 [110].

[828] Cram D.J., Abd Elhafer F.A., *J. Am. Chem. Soc.* 1952, **74,** 5828 [110, 112].

[829] Cutler R.A., Stenger R.J., Suter C.M., *J. Am. Chem. Soc.* 1952, **74,** 5475 [110].

[830] Ferles M., Attia A., *Collect. Czech. Chem. Commun.* 1973, **38,** 611 [111].

[831] Seebach D., Daum H., *Chem. Ber.* 1974, **107,** 1748 [111].

[832] Brown H.C., Pai G.G., *J. Org. Chem.* 1982, **47,** 1606 [111].

[833] Kabuto K., Imuta M., Kempner E.S., Ziffer H., *J. Org. Chem.* 1978, **43,** 2357 [111, 118].

[834] Imuta M., Ziffer H., *J. Org. Chem.* 1978, **43,** 3530 [111, 127, 218].

[835] Prelog V., *Pure Appl. Chem.* 1964, **9,** 119 [111].

[836] Howe R., Moore R.H., *J. Med. Chem.* 1971, **14,** 287 [112].

[837] Brown H.C., Varma V., *J. Am. Chem. Soc.* 1966, **88,** 2871 [112, 114, 115, 116, 117].

[838] Zagumenny A., *Chem. Ber.* 1881, **14**, 1402 [112].

[839] Gomberg M., Bachmann W.E., *J. Am. Chem. Soc.* 1927, **49**, 236 [112].

[840] Plattner P.A., Fürst A., Keller W., *Helv. Chim. Acta* 1949, **32**, 2464 [112].

[841] Gribble G.W., Kelly W.J., Emery S.E., *Synthesis* **1978**, 763 [113].

[842] Klages A., Allendorff P., *Chem. Ber.* 1898, **31**, 998 [113].

[843] Read R.R., Wood J., Jr, *Org. Syn. Coll. Vol.* 1955, **3**, 444 [113].

[844] Whalley W.B., *J. Chem. Soc.* **1951**, 665 [113].

[845] Cram D.J., Sahyun M.R.V., Knox G.R., *J. Am. Chem. Soc.* 1962, **84**, 1734 [113, 118, 134].

[846] Hardegger E., Plattner P.A., Blank F., *Helv. Chim. Acta* 1944, **27**, 793 [113, 143].

[847] Helg R., Schinz H., Eliel E.L., *Helv. Chim. Acta* 1952, **35**, 2406 [113, 114, 161].

[848] Eliel E.L., Ro R.S., *J. Am. Chem. Soc.* 1957, **79**, 5992 [114].

[849] Haubenstock H., Eliel E.L., *J. Am. Chem. Soc.* 1962, **84**, 2368 [114, 116].

[850] Eliel E.L., Senda Y., *Tetrahedron* 1970, **26**, 2411 [114, 115, 116].

[851] Eliel E.L., Rerick M.N., *J. Am. Chem. Soc.* 1960, **82**, 1367 [114].

[852] Brown H.C., Deck H.R., *J. Am. Chem. Soc.*1965, **87**, 5620 [114, 115, 116, 117].

[853] Eliel E.L., Martin R.J.L., Nasipuri D., *Org. Syn. Coll. Vol.* 1973, **5**, 175 [114].

[854] Richer J.C., *J. Org. Chem.* 1965, **30**, 324 [115, 116].

[855] Kovacs G., Galambos G., Juvancz Z., *Synthesis* **1977**, 171 [115].

[856] Haubenstock H., Eliel E.L., *J. Am. Chem. Soc.*1962, **84**, 2363 [116].

[857] Krishnamurthy S., Brown H.C., *J. Am. Chem. Soc.* 1976, **98**, 3383 [116, 117].

[858] Coulombeau A., Rassat A., *Chem. Commun.* **1968**, 1587 [116, 117].

[859] Jackman L.M., Macbeth A.K., Mills J.A., *J. Chem. Soc.* **1949**, 2641 [117].

[860] Huffman J.W., Charles J.T., *J. Am. Chem. Soc.* 1968, **90**, 6486 [115].

[861] Browne P.A., Kirk D.N., *J. Chem. Soc. C* **1969**, 1653 [117].

[862] Throop L., Tökés L., *J. Am. Chem. Soc.* 1967, **89**, 4789 [118].

[863] Toda M., Hirata Y., Yamamura S., *Chem. Commun.* **1969**, 919 [118].

[864] Leonard N.J., Figueras J., *J. Am. Chem. Soc.* 1952, **74**, 917 [118, 126].

[865] Sondheimer F., Velasco M., Rosenkranz G., *J. Am. Chem. Soc.* 1955, **77**, 192 [119].

[866] Shepherd D.A., Donia R.A., Campbell J.A., Johnson B.A., *J. Am. Chem. Soc.* 1955, **77**, 1212 [119].

[867] Adams R., Kern J.W., Shriner R.L., *Org. Syn. Coll. Vol.* 1932, **1**, 101 [119].

[868] Augustine R.L., *J. Org. Chem.* 1958, **23**, 1853 [119, 120].

[869] Djerassi C., Gutzwiller J., *J. Am. Chem. Soc.* 1966, **88**, 4537 [119].

[870] Sakai K., Watanabe K., *Bull. Chem. Soc. Jap.* 1967, **40**, 1548 [119].

[871] Ashby E.C., Lin J.J., *Tetrahedron Lett.* **1976**, 3865 [120, 121].

[872] Wolf H.R., Zink M.P., *Helv. Chim. Acta* 1973, **56**, 1062 [120].

[873] Ashby E.C., Lin J.J., *Tetrahedron Lett.* **1975**, 4453 [120, 121].

[874] Semmelhack M.F., Stauffer R.D., *J. Org. Chem.* 1975, **40**, 3619 [120, 121].

[875] Collman J.P., Finke R.G., Matlock P.L., Wahren R., Komoto R.G., Brauman J. I., *J. Am. Chem. Soc.* 1978, **100**, 1978 [120].

[876] Fetizon M., Gore J., *Tetrahedron Lett.* **1966**, 471 [120].

[877] Davis B.R., Woodgate P.D., *J. Chem. Soc.* **1966**, 2006 [120].

[878] Kergomard A., Renaud M.F., Veschambre H., *J. Org. Chem.* 1982, **47**, 792 [120].

[879] Brown H.C., Hess H.M., *J. Org. Chem.* 1969, **34**, 2206 [120].

[880] Gannon W.F., House H.O., *Org. Syn. Coll. Vol.* 1973, **5**, 294 [120].

[881] Meek J.S., Lorenzi F.J., Cristol S.J., *J. Am. Chem. Soc.* 1949, **71**, 1830 [120, 121].

[882] Hochstein F.A., *J. Am. Chem. Soc.* 1949, **71**, 305 [120, 121].

[883] Wilson K.E., Seidner R.T., Masamune S., *Chem. Commun.* **1970**, 213 [120].

[884] Kim S., Moon Y.C., Ahn K.H., *J. Org. Chem.* 1982, **47**, 3311 [120, 121].

[885] Hutchins R.O., Natale N.R., Taffer I.M., *Chem. Commun.* **1978**, 1088 [120].

[886] Macbeth A.K., Mills J.A., *J. Chem. Soc.* **1949**, 2646 [121].

[887] Mozingo R., Adkins H., *J. Am. Chem. Soc.* 1938, **60**, 669 [121].

[888] Hayes N.F., *Synthesis* **1975**, 702 [121].

[889] Kabalka G.W., Yang D.T.C., Baker J.D., *J. Org. Chem.* 1976, **41**, 574 [121].

[890] Biemann K., Büchi G., Walker B.H., *J. Am. Chem. Soc.* 1957, **79**, 5558 [121].

[891] Fischer R., Lardelli G., Jerger O., *Helv. Chim. Acta* 1951, **34**, 1577 [121].

[892] Harries C., Eschenbach G., *Chem. Ber.* 1896, **29**, 380 [122].

[893] Cohen N., Lopresti R.J., Neukom C., Saucy G., *J. Org. Chem.* 1980, **45**, 582 [122].

[894] Schlenk H., Lamp B., *J. Am. Chem. Soc.* 1951, **73**, 5493 [122].

[895] Hudlický M., Lejhancová I., *Collect. Czech. Chem. Commun.* 1966, **31**, 1416 [122].

[896] Huffmann J.W., *J. Org. Chem.* 1959, **24**, 1759 [123].

[897] Wharton P.S., Dunny S., Krebbs L.S., *J. Org. Chem.* 1965, **29**, 958 [123].

[898] Inhoffen H.H., Kolling G., Koch G., Nebel I., *Chem. Ber.* 1951, **84**, 361 [123].

[899] McGuckin W.F., Mason H.L., *J. Am. Chem. Soc.* 1955, **77**, 1822 [123].

[*900*] Julian P.L., Cole W., Magnani A., Meyer E.W., *J. Am. Chem. Soc.* 1945, **67**, 1728 [123].

[*901*] Shechter H., Ley D.L., Zeldin L., *J. Am. Chem. Soc.* 1952, **74**, 3664 [123].

[*902*] Levai L., Fodor G., Ritvay-Emandity K., Fuchs O., Hajos A., *Chem. Ber.* 1960, **93**, 387 [123].

[*903*] Simpson J.C.E., Atkinson C.M., Schofield K., Stephenson O., *J. Chem. Soc.* **1945**, 646 [124].

[*904*] Oelschlager H., *Chem. Ber.* 1956, **89**, 2025 [124].

[*905*] McMurry J.E., Melton J., *J. Am. Chem. Soc.* 1971, **93**, 5309; *J. Org. Chem.* 1973, **38**, 4367 [124].

[*906*] Gruber W., Renner H., *Monatsh.* 1950, **81**, 751 [124, 125].

[*907*] Birkofer L., *Chem. Ber.* 1947, **80**, 83 [124, 160].

[*908*] Wolfrom M.L., Brown R.L., *J. Am. Chem. Soc.* 1943, **65**, 1516 [124].

[*909*] Mamoli L., *Chem. Ber.* 1938, **71**, 2696 [125].

[*910*] Simon H., Rambeck B., Hashimoto H., Gunther H., Nohynek G., Neumann H., *Angew. Chem. Internat. Ed. Eng.* 1974, **13**, 609 [125].

[*911*] Guetté J.P., Spassky N., Boucherot D., *Bull. Soc. Chim. Fr.* **1972**, 4217 [125].

[*912*] Blomquist A.T., Goldstein A., *Org. Syn. Coll. Vol.* 1963, **4**, 216 [125].

[*913*] Prelog V., Schenker K., Günthard H.H., *Helv. Chim. Acta* 1952, **35**, 1598 [125, 126].

[*914*] Cope A.C., Barthel J.W., Smith R.D., *Org. Syn. Coll. Vol.* 1963, **4**, 218 [125].

[*915*] Smith W.T., Jr, *J. Am. Chem. Soc.* 1951, **73**, 1883 [125].

[*916*] Reusch W., LeMahieu R., *J. Am. Chem. Soc.* 1964, **86**, 3068 [125, 127].

[*917*] Ho T.-L., Wong C.M., *Synthesis* **1975**, 161 [125, 215].

[*918*] Rosenfeld R.S., *J. Am. Chem. Soc.* 1957, **79**, 5540 [126, 151].

[*919*] Wharton P.S., Bohlen D.H., *J. Org. Chem.* 1961, **26**, 3615 [126].

[*920*] Fehnel E.A., *J. Am. Chem. Soc.* 1949, **71**, 1063 [126].

[*921*] Pechmann, von, H., Dahl F., *Chem. Ber.* 1890, **23**, 2421 [126].

[*922*] Rubin M.B., Ben-Bassat J.M., *Tetrahedron Lett.* **1971**, 3403 [126, 127].

[*923*] Seibert W., *Chem. Ber.* 1947, **80**, 494 [127, 134].

[*924*] Cusack N.J., Davis B.R., *Chem. Ind. (London)* **1964**, 1426; *J. Org. Chem.* 1965, **30**, 2062 [127].

[*925*] Wenkert E., Kariv E., *Chem. Commun.* **1965**, 570 [127].

[*926*] Bonnet M., Geneste P., Rodriguez M., *J. Org. Chem.* 1980, **45**, 40 [127].

[*927*] Buchanan J.G. S.C., Davis B.R., *J. Chem. Soc. C* **1967**, 1340 [127, 128].

[*928*] Rosenblatt E.F., *J. Am. Chem. Soc.* 1940, **62**, 1092 [129].

[*929*] Fatiadi A.J., Sager W.F., *Org. Syn. Coll. Vol.* 1973, **5**, 595 [129].

[930] Blasczak L.C., McMurry J.E., *J. Org. Chem.* 1974, **39**, 258 [129].

[931] Boyland E., Manson D., *J. Chem. Soc.* **1951**, 1837 [129].

[932] Clar E., *Chem. Ber.* 1939, **72**, 1645 [129].

[933] Howard W.L., Brown J.H., Jr, *J. Org. Chem.* 1961, **26**, 1026 [130, 131].

[934] Daignault R.A., Eliel E.L., *Org. Syn. Coll. Vol.* 1973, **5**, 303 [130].

[935] Eliel E.L., Krishnamurthy S., *J. Org. Chem.* 1965, **30**, 848 [130].

[936] Eliel E.L., Pilato L.A., Badding V.G., *J. Am. Chem. Soc.* 1962, **84**, 2377 [130, 131].

[937] Truce W.E., Roberts F.E., *J. Org. Chem.* 1953, **28**, 961 [131].

[938] Georgian V., Harrison R., Gubich N., *J. Am. Chem. Soc.* 1959, **81**, 5834 [132].

[939] Nishiwaki Z., Fujiyama F., *Synthesis* **1972**, 569 [132].

[940] Freifelder M., Smart W.D., Stone G.R., *J. Org. Chem.* 1962, **27**, 2209 [132].

[941] Anziani P., Cornubert R., *Bull. Soc. Chim. Fr.* **1948**, 857 [132, 133].

[942] Rausser R., Finckenor D., Weber L., Hershberg E.B., Oliveto E.P., *J. Org. Chem.* 1966, **31**, 1346 [132].

[943] Rausser R., Weber L., Hershberg E.B., Oliveto E.P., *J. Org. Chem.* 1966, **31**, 1342 [132].

[944] Lloyd D., McDougall R.H., Wason F.T., *J. Chem. Soc.* **1965**, 822 [132, 133].

[945] Terentev A.P., Gusar N.I., *Zh. Obshchei Khim.* 1965, **35**, 125; *Chem. Abstr.* 1965, **62**, 13068 [132, 133].

[946] Prelog V., Fausy El-Neweihy M., Häflinger O., *Helv. Chim. Acta* 1950, **33**, 365 [132].

[947] Ricart G., *Bull. Soc. Chim. Fr.* **1974**, 615 [133].

[948] Feuer H., Braunstein D.M., *J. Org. Chem.* 1969, **34**, 1817 [133].

[949] Murphy J.G., *J. Org. Chem.* 1961, **26**, 3104 [133].

[950] Goodwin R.C., Bailey J.R., *J. Am. Chem. Soc.* 1925, **47**, 167 [133].

[951] Armand J., Boulares L., *Bull. Soc. Chim. Fr.* **1975**, 366 [133].

[952] Norton T.R., Benson A.A., Seibert R.A., Bergstrom F.W., *J. Am. Chem. Soc.* 1946, **68**, 1330 [133].

[953] Emerson W.S., *Org. Reactions* 1948, **4**, 174 [134, 135].

[954] Schellenberg K.A., *J. Org. Chem.* 1963, **28**, 3259 [134, 135].

[955] Borch R.F., Durst H.D., *J. Am. Chem. Soc.* 1969, **91**, 3996 [134, 135, 136].

[956] Winans C.F., *J. Am. Chem. Soc.* 1939, **61**, 3566 [135].

[957] Hancock E.M., Cope A.C., *Org. Syn. Coll. Vol.* 1955, **3**, 501 [135].

[958] Mignonac G., *Compt. Rend.* 1920, **171**, 1148; 1921, **172**, 223 [135].

[959] Balcom D.M., Noller C.R., *Org. Syn. Coll. Vol.* 1963, **4**, 603 [135].

[960] Robinson J.C., Jr, Snyder H.R., *Org. Syn. Coll. Vol.* 1955, **3**, 717 [135, 136].

[961] Ingersoll A.W., Brown J.H., Kim C.K., Beauchamp W.D., Jennings G., *J. Am. Chem. Soc.* 1936, **58**, 1808 [136].

[962] Eschweiler W., *Chem. Ber.* 1905, **38**, 880 [136].

[963] Muraki M., Mukaiyama T., *Chem. Lett.* **1974**, 1447 [137, 139].

[964] Bedenbaugh A.O., Bedenbaugh J.H., Bergin W.A., Adkins K.D., *J. Am. Chem. Soc.* 1970, **92**, 5774 [137].

[965] Tafel J., Friedrichs G., *Chem. Ber.* 1904, **37**, 3187 [137].

[966] Guyer A., Bieler A., Sommaruga M., *Helv. Chim. Acta* 1955, **38**, 976 [137].

[967] Ryashentseva M.A., Minachev K.M., Belanova E.P., *Izv. Akad. Nauk SSSR* **1974**, 1906; *Chem. Abstr.* 1974, **81**, 135391 [137].

[968] Nystrom R.F., Brown W.G., *J. Am. Chem. Soc.* 1947, **69**, 2548 [137, 138, 139, 140, 141, 145].

[969] Černý M., Málek J., Čapka M., Chvalovský V., *Collect. Czech. Chem. Commun.* 1969, **34**, 1025 [137, 139, 146, 147, 154, 156].

[970] Černý M., Málek J., *Collect. Czech. Chem. Commun.* 1971, **36**, 2394 [137, 139].

[971] Yoon N.M., Pak C.S., Brown H.C., Krishnamurthy S., Stocky T.P., *J. Org. Chem.* 1973, **38**, 2786 [138, 139, 141, 143, 155, 163, 209].

[972] Burton H., Ingold C.K., *J. Chem. Soc.* **1929**, 2022 [138].

[973] Farmer E.H., Hughes L.A., *J. Chem. Soc.* **1934**, 1929 [138].

[974] Castro C.E., Stephens R.D., Moje S., *J. Am. Chem. Soc.* 1966, **88**, 4964 [138, 140, 141, 157].

[975] Benedict G.E., Russell R.R., *J. Am. Chem. Soc.* 1951, **73**, 5444 [138].

[976] Brown H.C., Heim P., Yoon N.M., *J. Am. Chem. Soc.* 1970, **92**, 1637 [139].

[977] Brown H.C., Subba Rao B.C., *J. Am. Chem. Soc.* 1960, **82**, 681 [139, 141, 142, 155, 174].

[978] Blackwood R.H., Hess G.B., Larrabee C.E., Pilgrim E.J., *J. Am. Chem. Soc.* 1958, **80**, 6244 [139].

[979] Coleman G.H., Johnson H.L., *Org. Syn. Coll. Vol.* 1955, **3**, 60 [139].

[980] Černý M., Málek J., *Tetrahedron Lett.* **1969**, 1739 [139].

[981] Černý M., Málek J., *Collect. Czech. Chem. Commun.* 1970, **35**, 1216 [139].

[982] Benkeser R.A., Foley K.M., Gaul J.M., Li G.S., *J. Am. Chem. Soc.* 1970, **92**, 3232 [139, 141].

[983] Burgstahler A.W., Bithos Z.J., *Org. Syn. Coll. Vol.* 1973, **5**, 591 [140].

[984] Kuehne M.E., Lambert B.F., *J. Am. Chem. Soc.* 1959, **81**, 4278: *Org. Syn. Coll. Vol.* 1973, **5**, 400 [140, 168].

[985] Johnson W.S., Gutsche C.D., Offenhauser R.D., *J. Am. Chem. Soc.* 1946, **68**, 1648 [140].

[986] Eliel E.L., Hoover T.E., *J. Org. Chem.* 1959, **24**, 938 [140].

[987] May E.L., Mosettig E., *J. Am. Chem. Soc.* 1948, **70**, 1077 [140].

[988] Dang T.P., Kagan H.B., *Chem. Commun.* **1971**, 481 [141].

[989] Page G.A., Tarbell D.S., *Org. Syn. Coll. Vol.* 1963, **4**, 136 [141].

[990] Badger G.M., *J. Chem. Soc.* **1948**, 999 [141].

[991] Ingersoll A.W., *Org. Syn. Coll. Vol.* 1932, **1**, 311 [140, 141].

[992] Paal C., Hartmann W., *Chem. Ber.* 1909, **42**, 3930; 1918, **51**, 640 [141].

[993] Nystrom R.F., *J. Am. Chem. Soc.* 1959, **81,** 610 [141, 146, 155, 159].

[994] Yoon N.M., Brown H.C., *J. Am. Chem. Soc.* 1968, **90,** 2927 [141, 155, 159, 160, 167, 168, 174].

[995] Gaux B., Le Henaff P., *Bull. Soc. Chim. Fr.* **1974,** 505 [141].

[996] Rosenmund K.W., Zetzsche F., *Chem. Ber.* 1918, **51,** 578 [141, 142].

[997] Liebermann C., *Chem. Ber.* 1895, **28,** 134 [142].

[998] Moppelt C.E., Sutherland J.K., *J. Chem. Soc. C* **1968,** 3040 [142].

[999] Allen C.F.H., MacKay D.D., *Org. Syn. Coll. Vol.* 1943, **2,** 580 [143].

[1000] Mohrig J.R., Vreede P.J., Schultz S.C., Fierke C.A., *J. Org. Chem.* 1981, **46,** 4655 [143].

[1001] Rosenmund K.W., Zymalkowski F., Engels P., *Chem. Ber.* 1951, **84,** 711 [143].

[1002] Tuynenburg M.G., van der Ven B., De Jonge A.P., *Nature* 1962, **194,** 995 [143].

[1003] Steinkopf W., Wolfram A., *Ann. Chem.* 1923, **430,** 113 [143].

[1004] Ray F.E., Weisburger E.K., Weisburger J.H., *J. Org. Chem.* 1948, **13,** 655 [143].

[1005] Beyerman H.C., Boeke P., *Rec. Trav. Chim. Pays-Bas* 1959, **78,** 648 [143].

[1006] Martin E.L., *Org. Syn. Coll. Vol.* 1943, **2,** 499 [143, 144].

[1007] Durham L.J., McLeod D.J., Cason J., *Org. Syn. Coll. Vol.* 1963, **4,** 510 [144].

[1008] Weygand C., Meusel W., *Chem. Ber.* 1943, **76,** 503 [144, 145].

[1009] Rosenmund K.W., *Chem. Ber.* 1918, **51,** 585 [144, 145].

[1010] Barnes R.P., *Org. Syn. Coll. Vol.* 1955, **3,** 551 [144].

[1011] Brown H.C., Subba Rao B.C., *J. Am. Chem. Soc.* 1958, **80,** 5377 [145, 207, 208].

[1012] Horner L., Röder H., *Chem. Ber.* 1970, **103,** 2984 [145].

[1013] Sroog C.E., Woodburn H.M., *Org. Syn. Coll. Vol.* 1963, **4,** 271 [146].

[1014] Santaniello E., Farachi C., Manzocchi A., *Synthesis* **1979,** 912 [146].

[1015] Austin P.R., Bousquet E.W., Lazier W.A., *J. Am. Chem. Soc.* 1937, **59,** 864 [146, 147].

[1016] Morand P., Kayser M., *Chem. Commun.* **1976,** 314 [146].

[1017] Bloomfield J.J., Lee S.L., *J. Org. Chem.* 1967, **32,** 3919 [147].

[1018] Bailey D., Johnson R., *J. Org. Chem.* 1970, **35,** 3574 [147].

[1019] Makhlouf M.A., Rickborn B., *J. Org. Chem.* 1981, **46,** 4810 [147].

[1020] Wislicenus J., *Chem. Ber.* 1884, **17,** 2178 [147].

[1021] Reissert A., *Chem. Ber.* 1913, **46,** 1484 [147].

[1022] Zakharin L.I., Gavrilenko V.V., Maslin D.N., Khorlina I.M., *Tetrahedron Lett.* **1963,** 2087 [148, 149].

[1023] Zakharin L.I., Khorlina I.M., *Tetrahedron Lett.* **1962,** 619 [148, 149].

[1024] Winterfeld, E., *Synthesis* **1975,** 617 [148, 149].

[1025] Greenwald R.B., Evans D.H., *J. Org. Chem.* 1976, **41,** 1470 [148].

[1026] Weissman P.M., Brown H.C., *J. Org. Chem.* 1966, **31,** 283 [149].

[1027] Muraki M., Mukaiyama T., *Chem. Lett.* **1975,** 215 [149].

[*1028*] Arth G.E., *J. Am. Chem. Soc.* 1953, **75**, 2413 [149].

[*1029*] Wolfrom M.L., Wood H.B., *J. Am. Chem. Soc.* 1951, **73**, 2933 [149].

[*1030*] Wolfrom M.L., Anno K., *J. Am. Chem. Soc.* 1952, **74**, 5583 [149].

[*1031*] Sperber N., Zaugg H.E., Sandstrom W.M., *J. Am. Chem. Soc.* 1947, **69**, 915 [149].

[*1032*] Pettit G.R., Piatok D.M., *J. Org. Chem.* 1962, **27**, 2127 [149].

[*1033*] Baldwin S.W., Haut S.A., *J. Org. Chem.* 1975, **40**, 3885 [150].

[*1034*] Kraus G.A., Frazier K.A., Roth B.D., Taschner M.J., Neunschwander K., *J. Org. Chem.* 1981, **46**, 2417 [150].

[*1035*] Zymalkowski F., Schuster T., Scherer H., *Arch. Pharm.* 1969, **302**, 272; *Chem. Abstr.* 1969, **71**, 12277 [150].

[*1036*] Letsinger R.L., Jamison J.D., Hussey A.S., *J. Org. Chem.* 1961, **26**, 97 [150].

[*1037*] Beels C.M.D., Abu-Rabie M.S., Murray-Rust P., Murray-Rust J., *Chem. Commun.* **1979**, 665 [150].

[*1038*] Uhle F.C., McEwen C.M., Jr, Schröter H., Yuan C., Baker B.W., *J. Am. Chem. Soc.* 1960, **82**, 1200, 1206 [150].

[*1039*] Bowman R.E., *J. Chem. Soc.* **1950**, 325 [150, 151].

[*1040*] Bergmann M., Zervas L., *Chem. Ber.* 1932, **65**, 1192, 1201 [151].

[*1041*] Wunsch E., Zwick A., *Hoppe-Seylers Z. Physiol. Chem.* 1963, **333**, 108 [151].

[*1042*] Kovacs J., Rodin R.L., *J. Org. Chem.* 1968, **33**, 2418 [151].

[*1043*] Kuromizu K., Meienhofer J., *J. Am. Chem. Soc.* 1974, **96**, 4978 [151].

[*1044*] Carpenter F.H., Gish D.T., *J. Am. Chem. Soc.* 1952, **74**, 3818 [151].

[*1045*] Jackson A.E., Johnstone R.A.W., *Synthesis* **1976**, 685 [151].

[*1046*] House H.O., Carlson R.G., *J. Org. Chem.* 1964, **29**, 74 [151].

[*1047*] Fieser L.F., *J. Am. Chem. Soc.* 1953, **75**, 4377 [151].

[*1048*] Hansley V.L., *J. Am. Chem. Soc.* 1935, **57**, 2303 [152, 211].

[*1049*] Allinger N.L., *Org. Syn. Coll. Vol.* 1963, **4**, 840 [152].

[*1050*] Prelog V., Frenkiel L., Kobelt M., Barman P., *Helv. Chim. Acta* 1947, **30**, 1741 [152].

[*1051*] Blomquist A.T., Burge R.E., Sucsy A.C., *J. Am. Chem. Soc.* 1952, **74**, 3636 [152].

[*1052*] Bouveault L., Blanc G., *Bull. Soc. Chim. Fr.* 1904, [3], **31**, 666, 1203; *Compt. Rend.* 1903, **136**, 1676 [152].

[*1053*] Ford S.G., Marvel C.S., *Org. Syn. Coll. Vol.* 1943, **2**, 372 [152].

[*1054*] Manske R.H., *Org. Syn. Coll. Vol.* 1943, **2**, 154 [152].

[*1055*] Adkins H., Gillespie R.H., *Org. Syn. Coll. Vol.* 1955, **3**, 671 [152, 153].

[*1056*] Adkins H., *Org. Reactions* 1954, **8**, 1 [153, 157].

[*1057*] Adkins H., Folkers K., *J. Am. Chem. Soc.* 1931, **53**, 1095 [153, 157].

[*1058*] Lazier W.A., Hill J.W., Amend W.Y., *Org. Syn. Coll. Vol.* 1943, **2**, 325 [153].

[*1059*] Moffet R.B., *Org. Syn. Coll. Vol.* 1963, **4**, 834 [154].

[*1060*] Brown H.C., Narasimhan S., *J. Org. Chem.* 1982, **47**, 1604 [155, 156].

[1061] Brown M.S., Rapoport H., *J. Org. Chem.* 1963, **28**, 3261 [155, 156, 158].

[1062] Santaniello E., Ferraboschi P., Sozzani P., *J. Org. Chem.* 1981, **46**, 4584 [155].

[1063] Brown H.C., Narasinham S., Yong Moon Choi, *J. Org. Chem.* 1982, **47**, 4702 [155, 156].

[1064] Brown H.C., Yong Moon Choi, Narasinham S., *J. Org. Chem.* 1982, **47**, 3153 [156, 167, 168].

[1065] Firestone R.A., *Tetrahedron Lett.* **1967**, 2629 [156].

[1066] Villani F.J., King M.S., Papa D., *J. Org. Chem.* 1953, **18**, 1578 [156].

[1067] Paal C., Gerum J., *Chem. Ber.* 1908, **41**, 2273, 2278 [140, 156].

[1068] McCullan W.R., Connor R., *J. Am. Chem. Soc.* 1941, **63**, 484 [157].

[1069] Marshall J.A., Carroll R.D., *J. Org. Chem.* 1965, **30**, 2748 [157, 175].

[1070] Kadin S.B., *J. Org. Chem.* 1966, **31**, 620 [157, 175].

[1071] Pereyre M., Colin G., Valade J., *Tetrahedron Lett.* **1967**, 4805 [157].

[1072] Zurqiyah A., Castro C.E., *Org. Syn. Coll. Vol.* 1973, **5**, 993 [157].

[1073] Snyder E.I., *J. Org. Chem.* 1967, **32**, 3531 [157, 158].

[1074] Schmidt O., *Chem. Ber.* 1931, **64**, 2051 [157].

[1075] de Benneville P.L., Connor R., *J. Am. Chem. Soc.* 1940, **62**, 283 [158].

[1076] Dornow A., Bartsch W., *Chem. Ber.* 1954, **87**, 633 [158].

[1077] Paul R., Joseph N., *Bull. Soc. Chim. Fr.* **1952**, 550 [158].

[1078] Bouveault L., Blanc G., *Bull. Soc. Chim. Fr.* 1904 [3], **31**, 1209; *Compt. Rend.* 1903, **137**, 328 [158].

[1079] Bates E.B., Jones E.R.H., Whiting M.C., *J. Chem. Soc.* **1954**, 1854 [159].

[1080] Richards E.M., Tebby J.C., Ward R.S., Williams D.H., *J. Chem. Soc. C* **1969**, 1542 [159].

[1081] Weizmann A., *J. Am. Chem. Soc.* 1949, **71**, 4154 [159].

[1082] Hardegger E., Montavon R.M., *Helv. Chim. Acta* 1946, **29**, 1203 [159].

[1083] Feuer H., Kucera T.J., *J. Am. Chem. Soc.* 1955, **77**, 5740 [160].

[1084] Pechman, von, H., *Chem. Ber.* 1895, **28**, 1847 [160].

[1085] Darapsky A., Prabhakar M., *Chem. Ber.* 1912, **45**, 1654, 2622 [160].

[1086] Moore A.T., Rydon H.N., *Org. Syn. Coll. Vol.* 1973, **5**, 586 [160].

[1087] Adkins H., Wojcik B., Covert L.W., *J. Am. Chem. Soc.* 1942, **55**, 1669 [160, 161].

[1088] Nerdel F., Frank D., Barth G., *Chem. Ber.* 1969, **102**, 395 [161].

[1089] Deol B.S., Ridley D.D., Simpson G.W., *Austral. J. Chem.* 1976, **29**, 2459 [162, 170].

[1090] Nutaitis C.F., Schultz R.A., Obaza J., Smith F.X., *J. Org. Chem.* 1980, **45**, 4606 [162].

[1091] Howe, R., McQuillin F.J., Temple R.W., *J. Chem. Soc.* **1959**, 363 [162].

[1092] Kraus G.A., Frazier K., *J. Org. Chem.* 1980, **45**, 4262 [162].

[1093] Schmid L., Swoboda W., Wichtl M., *Monatsh.* 1952, **83**, 185 [163].

[*1094*] Hartung W.H., Beaujon J.H.R., Cocolas G., *Org. Syn. Coll. Vol.* 1973, **5**, 376 [163].

[*1095*] Putochin N.J., *Chem. Ber.* 1923, **56**, 2211 [163].

[*1096*] Zambito A.J., Howe E.E., *Org. Syn. Coll. Vol.* 1973, **5**, 373 [163].

[*1097*] Chang Y.-T., Hartung W.H., *J. Am. Chem. Soc.* 1953, **75**, 89 [163].

[*1098*] Claus C.J., Morgenthau J.L., Jr, *J. Am. Chem. Soc.* 1951, **73**, 5005 [163].

[*1099*] Eliel E.L., Daignault R.A., *J. Org. Chem.* 1964, **29**, 1630 [163].

[*1100*] Bublitz D.E., *J. Org. Chem.* 1967, **32**, 1630 [163, 164].

[*1101*] Wolfrom M.L., Karabinos J.V., *J. Am. Chem. Soc.* 1946, **68**, 724, 1455 [164, 205].

[*1102*] Baddiley J., *J. Chem. Soc.* **1950**, 3693 [164].

[*1103*] Weygand F., Eberhardt G., Linden H., Schäfer F., Eigen I., *Angew. Chem.* 1953, **65**, 525 [164].

[*1104*] Micović V.M., Mihailović M.L., *J. Org. Chem.* 1953, **18**, 1190 [164, 165, 166, 167].

[*1105*] Walborsky H.M., Baum M., Loncrini D.F., *J. Am. Chem. Soc.* 1955, **77**, 3637 [164].

[*1106*] Brown H.C., Tsukamoto A., *J. Am. Chem. Soc.* 1961, **83**, 2016, 4549 [165].

[*1107*] Málek J., Černý M., *Synthesis* **1972**, 217 [165].

[*1108*] Ramegowda N.S., Modi M.N., Koul A.K., Bora J.M., Narang C.K., Mathur N.K., *Tetrahedron* 1973, **29**, 3985 [165].

[*1109*] Birch A.J., Cymerman-Craig J., Slayton M., *Austral. J. Chem.* 1955, **8**, 512 [165, 166, 171].

[*1110*] Gilman H., Jones R.G., *J. Am. Chem. Soc.* 1948, **70**, 1281 [166].

[*1111*] Brown H.C., Kim S.C., *Synthesis* **1977**, 635 [166].

[*1112*] Weygand F., Frauendorfer E., *Chem. Ber.* 1970, **103**, 2437 [166].

[*1113*] Wojcik B., Adkins H., *J. Am. Chem. Soc.* 1934, **56**, 2419 [166, 167].

[*1114*] Sauer J.C., Adkins H., *J. Am. Chem. Soc.* 1938, **60**, 402 [167].

[*1115*] Wilson C.V., Stenberg J.F., *Org. Syn. Coll. Vol.* 1963, **4**, 564 [167].

[*1116*] Cope A.C., Ciganek E., *Org. Syn. Coll. Vol.* 1963, **4**, 339 [167].

[*1117*] Schindlbauer H. *Monatsh.* 1969, **100**, 1413 [167, 170].

[*1118*] Umino N., Iwakuma T., Itoh N., *Tetrahedron Lett.* **1976**, 763 [167, 168].

[*1119*] Brown H.C., Heim P., *J. Am. Chem. Soc.* 1964, **86**, 3566; *J. Org. Chem.* 1973, **38**, 912 [167, 168].

[*1120*] Baille T.B., Tafel J., *Chem. Ber.* 1899, **32**, 68 [167, 168].

[*1121*] Borch R.F., *Tetrahedron Lett.* **1968**, 61 [167, 168].

[*1122*] Galinovsky F., Stern E., *Chem. Ber.* 1943, **76**, 1034; 1944, **77**, 132 [168].

[*1123*] Moffet R.B., *Org. Syn. Coll. Vol.* 1963, **4**, 354 [168].

[*1124*] Tafel J., Stern M., *Chem. Ber.* 1900, **33**, 2224 [168, 169].

[*1125*] Kondo Y., Witkop B., *J. Org. Chem.* 1968, **33**, 206 [168, 169].

[*1126*] Merkel W., Mania D., Bormann D., *Ann. Chem.* **1979**, 461 [169].

[1127] Lukeš R., Smetáčková M., *Collect. Czech. Chem. Commun.* 1933, **5,** 61 [169].

[1128] Lukeš R., Ferles M., *Collect. Czech. Chem. Commun.* 1951, **16,** 252 [169].

[1129] Butula L., Kolbah D., Butula I., *Croat. Chem. Acta* 1972, **44,** 481 [169].

[1130] Dunet A., Rollet R., Willemart A., *Bull. Soc. Chim. Fr.* **1950,** 877 [169].

[1131] Herbert R.M., Shemin D., *Org. Syn. Coll. Vol.* 1943, **2,** 491 [169].

[1132] Shamma M., Rosenstock P.D., *J. Org. Chem.* 1961, **26,** 718 [170].

[1133] Holík M., Tesařová A., Ferles M., *Collect. Czech. Chem. Commun.* 1967, **32,** 1730 [170].

[1134] Lukeš R., Černý M., *Collect. Czech. Chem. Commun.* 1959, **24,** 1287 [170].

[1135] Kornet M.J., Poo An Thio, Sip Ie Tan, *J. Org. Chem.* 1968, **33,** 3637 [171].

[1136] Gribble G.W., Jasinski J.M., Pellicone J.T., Panetta J.A., *Synthesis* **1978,** 766 [171].

[1137] Brovet D., *Archiv Kem.* 1948, **20,** 70 [171, 172].

[1138] Cronyn M.W., Goodrich J.E., *J. Am. Chem. Soc.* 1952, **74,** 3936 [171].

[1139] Kornfeld E.C., *J. Org. Chem.* 1951, **16,** 131 [171].

[1140] Sundberg R.J., Walters C.P., Bloom J.D., *J. Org. Chem.* 1981, **46,** 3730 [171].

[1141] Karady S., Amato J.S., Weinstock L.M., Sletzinger M., *Tetrahedron Lett.* **1978,** 403 [172].

[1142] Staudinger H., *Chem. Ber.* 1908, **41,** 2217 [172].

[1143] Williams J.W., Witten C.H., Krynitsky J.A., *Org. Syn. Coll. Vol.* 1955, **3,** 818 [172].

[1144] Braun, von, J., Rudolph W., Kröper H., Pinkernelle W., *Chem. Ber.* 1934, **67,** 269, 1735 [172].

[1145] Nicolaus B.I.R., Mariani L., Gallo G., Testa E., *J. Org. Chem.* 1961, **26,** 2253 [172].

[1146] Newman M.S., Caflisch E.G., Jr, *J. Am. Chem. Soc.* 1958, **80,** 862 [172].

[1147] Robert J.D., *J. Am. Chem. Soc.* 1951, **73,** 2959 [172].

[1148] Sprecher M., Feldkimel M., Wilchek M., *J. Org. Chem.* 1961, **26,** 3664 [172].

[1149] Hesse G., Schroedel R., *Ann. Chem.* 1957, **607,** 24 [173].

[1150] Brown H.C., Garg C.P., *J. Am. Chem. Soc.* 1964, **86,** 1085 [172, 173].

[1151] Miller A.E.G., Biss J.W., Schwartzman L.H., *J. Org. Chem.* 1959, **24,** 627 [173, 208].

[1152] Gardner T.S., Smith F.A., Wenis E., Lee J., *J. Org. Chem.* 1951, **16,** 1121 [173].

[1153] Fuson R.C., Emmons W.D., Tull R., *J. Org. Chem.* 1951, **16,** 648 [173, 175].

[1154] Backeberg O.G., Staskun B., *J. Chem. Soc.* **1962,** 3961 [173].

[*1155*] Staskun B., Backeberg O.G., *J. Chem. Soc.* **1964,** 5880 [173].

[*1156*] van Es T., Staskun B., *J. Chem. Soc.* **1965,** 5775 [173].

[*1157*] Freifelder M., *J. Am. Chem. Soc.* 1960, **82,** 2386 [173].

[*1158*] Albert A., Magrath D., *J. Chem. Soc.* **1944,** 678 [173].

[*1159*] Schreifels J.A., Maybury D.C., Swartz W.E., Jr, *J. Org. Chem.* 1981, **46,** 1263 [174].

[*1160*] Whitemore F.C., *J. Am. Chem. Soc.* 1944, **66,** 725 [174].

[*1161*] Biggs B.S., Bishop W.S., *Org. Syn. Coll. Vol.* 1955, **3,** 229 [174].

[*1162*] Greenfield H., *Ind. Eng. Chem. Prod. Res. Develop.* 1967, **6,** 142 [174].

[*1163*] Suter C.M., Moffett E.W., *J. Am. Chem. Soc.* 1934, **56,** 487 [175].

[*1164*] Schnider O., Hellerbach J., *Helv. Chim. Acta* 1950, **33,** 1437 [175].

[*1165*] Bergmann E.D., Ikan R., *J. Am. Chem. Soc.* 1956, **78,** 1482 [175].

[*1166*] Osborn M.E., Pegues J.F., Paquette L.A., *J. Org. Chem.* 1980, **45,** 167 [175].

[*1167*] Profitt A., Watt D.S., Corey E.J., *J. Org. Chem.* 1975, **40,** 127 [175].

[*1168*] Toda F., Iida K., *Chem. Lett.* **1976,** 695 [175].

[*1169*] Kornblum N., Fishbein L., *J. Am. Chem. Soc.* 1955, **77,** 6266 [175].

[*1170*] Burger A., Hornbaker E.D., *J. Am. Chem. Soc.* 1952, **74,** 5514 [175].

[*1171*] Fields M., Walz D.E., Rothchild S., *J. Am. Chem. Soc.* 1951, **73,** 1000 [175].

[*1172*] Ferris J.P., Sanchez R.A., Mancuso R.W., *Org. Syn. Coll. Vol.* 1973, **5,** 32 [175].

[*1173*] Zartman W.H., Adkins H., *J. Am. Chem. Soc.* 1932, **54,** 3398 [176].

[*1174*] Becker W.E., Cox S.E., *J. Am. Chem. Soc.* 1960, **82,** 6264 [176].

[*1175*] Bordwell F.G., Douglas M.L., *J. Am. Chem. Soc.* 1966, **88,** 993 [176].

BIBLIOGRAPHY

REVIEWS IN *ORGANIC REACTIONS*

1. The Clemmensen reduction. Martin, E.L., *Org. Reactions*, **1**, 155 (1942).
2. The Meerwein-Ponndorf-Verley reduction (reduction with aluminum alkoxides). Wilds, A.L., *Org. Reactions* **2**, 178 (1944).
3. Replacement of the aromatic primary amino group by hydrogen. Kornblum, N., *Org. Reactions* **2**, 262 (1944).
4. The preparation of amines by reductive alkylation. Emerson, W.S., *Org. Reactions* **4**, 174 (1948).
5. The Rosenmund reduction. Mosettig, E., Mozingo, R., *Org. Reactions* **4**, 362 (1948).
6. The Wolff-Kishner reduction. Todd, D., *Org. Reactions* **4**, 378 (1948).
7. The Leuckart reaction. Moore, M.L., *Org. Reactions* **5**, 301 (1949).
8. Reductions by lithium aluminum hydride. Brown, W.G., *Org. Reactions* **6**, 469 (1951).
9. Hydrogenolysis of benzyl groups attached to oxygen, nitrogen, or sulfur. Hartung, W.H., Simonoff, R., *Org. Reactions* **7**, 263 (1953).
10. Catalytic hydrogenation of esters to alcohols. Adkins H., *Org. Reactions* **8**, 1 (1954).
11. The synthesis of aldehydes from carboxylic acids. Mosettig, E., *Org. Reactions* **8**, 218 (1954).
12. Desulfurization with Raney nickel. Pettit, G.R., van Tamelen, E.E., *Org. Reactions* **12**, 356 (1962).
13. The Zinin reduction of nitroarenes. Porter, H.K., *Org. Reactions* **20**, 455 (1973).
14. Clemmensen reduction of ketones in anhydrous organic solvents. Vedejs, E., *Org. Reactions* **22**, 401 (1975).
15. Reduction and related reactions of α,β-unsaturated compounds with metals in liquid ammonia. Caine, D., *Org. Reactions* **23**, 1, (1976).
17. Homogeneous hydrogenation catalysts in organic synthesis. Birch, A.J., Williamson, D.H., *Org. Reactions* **24**, 1 (1976).

MONOGRAPHS

1. Hudlický, M., *Reduction and Oxidation* (in Czech), Academia, Prague 1953.
2. Micović, V.M., Mihailović, M.L., *Lithium Aluminum Hydride in Organic Chemistry*, Naućna Knjiga, Beograd, 1955.
3. Gaylord, N.G., *Reduction with Complex Metal Hydrides*, Wiley-Interscience, New York, 1956.
4. Rudinger, J., Ferles, M., *Lithium Aluminum Hydride and Kindred Reagents in Organic Chemistry* (in Czech). Academia, Prague, 1956.
5. Augustine, R.L., *Catalytic Hydrogenation*, Marcel Dekker, New York, 1965.
6. Zymalkovsky F., *Katalytische Hydrierungen im Organisch-Chemischen Laboratorium* (in German), F. Enke, Stuttgart, 1965.

7. Rylander, P.N., *Catalytic Hydrogenation over Platinum Metals*, Academic Press, New York, 1967.
8. Augustine, R.L., *Reduction*, Marcel Dekker, New York, 1968.
9. Freifelder, M., *Practical Catalytic Hydrogenation*, Wiley-Interscience, New York, 1971.
10. House, H.O., *Modern Synthetic Reactions*, W.A. Benjamin, Menlo Park, Calif., 1972.
11. James, B.R., *Homogeneous Hydrogenation*, John Wiley and Sons, New York, 1973.
12. Brown, H.C., *Organic Syntheses via Boranes*, Wiley-Interscience, New York, 1975.
13. Doyle, M.P., West, C.T. (eds), *Benchmark Papers in Organic Chemistry* Vol. VI, *Stereoselective Reductions*, Academic Press, New York, 1976.
14. McQuillin, F.J., *Homogeneous Hydrogenation in Organic Chemistry*, D. Reidel, Boston, 1976.
15. Pizey, J.S., *Lithium Aluminium Hydride*, Ellis Horwood, Chichester, 1977.
16. Freifelder, M., *Catalytic Hydrogenations in Organic Synthesis. Procedures and Commentary*, Wiley-Interscience, New York, 1978.
17. Hajos, A., *Complex Hydrides and Related Reducing Agents in Organic Synthesis*, Elsevier, New York, 1979.
18. Rylander, P.N., *Catalytic Hydrogenation in Organic Syntheses*, Academic Press, New York, 1979.
19. Houben-Weyl, *Methoden der Organischen Chemie* (E. Müller, ed.), Vols IV/Ic and IV/Id, G. Thieme, Stuttgart, 1980, 1981.

CHAPTERS IN *SYNTHETIC REAGENTS* MONOGRAPHS

Pizey, J.S., Ellis Horwood, Chichester, England.
Volume 1: Lithium Aluminium Hydride (pp. 101–294), 1974.
Volume 2: Raney Nickel (pp. 175–311), 1974.
Volume 3: Diborane (pp. 1–191), 1977.

AUTHOR INDEX

Author Index

SUBJECT INDEX

Numbers in *italics* refer to Correlation Tables (pp. 178-200), **bold-face** numbers to Procedures (pp. 201-218). The index lists reagents, types of reactions and types of compounds. Of individual compounds, only those shown in Schemes and Procedures are listed.

A

Acenapthene, from acenaphthylene, 51, *179*
Acenaphthylene, reduction, 51, *179*
Acetalization, protection of keto esters, 162
Acetals, see *Aldehyde acetals* or *Ketals*
Acetals, from ortho esters, 163, *198*
Acetates, from aldehyde diacetates, 105
Acetic acid
  solvent
    in hydrogenation, 10, 11, 46, 47
    in reductions with zinc, 29
  of chromous salts, 30
Acetic anhydride
  solvent in reductions with zinc, 28
Acetone, reduction, 108
  reductive amination, 135
  solvent of chromous salts, 30
Acetophenone
  from diazoacetophenone, 124
  reduction, 109-111, **212, 213**
  reductive amination, 136, **218**
Acetylacetone, reduction, 127
Acetylene, reduction to ethylene, 1
Acetylenes
  See also *Alkynes*
Acetylenes
  reduction to
    cis-alkenes, 6, 7, 9, 43, 44, *178*
    trans-alkenes, 30, 44, *178*
Acetylenic acids, reduction, 138, 141, *195*
Acetylenic halogen, replacement by hydrogen, 66
Acetylenic ketones, reduction, 122, *191*
Acetylenylcyclohexanol, reduction, 78
Acetylpyridines, reduction, 110, 111
Acid anhydrides
  from diacylperoxides, 85
  reduction, 146-148, *196*
  stoichiometry in hydride reductions, 19
Acid anhydrides, cyclic, reduction, 146, 147, *196*
Acid derivatives, reduction, 141-175, *196*, *197*
Acid-esters, reduction to ester alcohols, 163, *198*
Acidic quenching after hydride reductions, 22, **207**
Acidity in hydrogenation, 10, 11
Acids
  See *Carboxylic acids*

Acids, mineral
  decomposition of lithium aluminum hydride, 22
Acridine, reduction, 59
Activated carbon, *Activated charcoal*, catalyst support, 6
Activation energy of hydrogenation, 4
Activators of catalysts, 5, 10, 11
Activity of Raney nickel catalysts, 8
Acyl chlorides
  reduction to alcohols, 145, 146, *196*
    aldehydes, 16, 144, 145, *196*, **208**
  stoichiometry in hydride reductions, 19
Acyloin condensation of esters, 151, 152, **211**
Acyloins
  from diketones, 35, 126, 127, *193*
    esters, 151, 152, *197*, **211**
  reduction to hydrocarbons, 126, 127, *193*
    ketones, 30, 31, 124-127, *193*, **215**
    vicinal diols, 125, *193*
Acylsaccharins, reduction to aldehydes, 165
Adam's catalyst, 5, 6, 46, 47
Adipate, monoethyl, reduction, 163, **209**
Adkins' catalysts, 9, 47
Advances in Catalysis, review of hydrogenation, 55, 58
Alane
  addition to alkenes, 15
  cleavage of
    cyclic ethers, 81, *185*
    tetrahydrofuran, 81
  hydrogenolysis of enamines, 92, *188*
  mechanism of reduction, 19
  preparation, 13, 14, 83, 159, **206**
  reduction of acyl chlorides, 146, *196*
    aldehyde acetals, 103, 104, *190*
    amides, 167, 168, 170, *199*
    aromatic aldehydes, 99, 100, *189*
    aromatic ketones, 113, *191*
    cinnamaldehyde, 102, *189*
    enamines, 92, *188*
    epoxides, 83, *186*
    ester acids, 138, 163, *198*
    esters, 154, 155, *197*
    free carboxyl preferentially, 163
    halo acids, 141, *196*
    halo esters, 159, *198*
    ketals, 130, *194*
    monothioketals, 130, 131, *194*
    nitriles, 174, *200*, **208**
    nitro aldehydes, 103, *190*

Explosion
  of borane-THF complex, 14
  in handling hydrides and complex hydrides, 20

F

Ferric chloride
  catalyst activator, 5
  in hydrogenation of aromatic aldehydes, 99
  reduction of aldehydes, 96–98
  reductions with iron, 73
Ferromagnetic catalysts, Raney nickel, 8
Ferrous sulfate
  catalyst activator, 10
    in reductions with iron, 29, 73
  reduction of nitro aldehydes, 103, *190*
    nitro compounds to amines, 31, 73, *182*
  removal of peroxides from ethers, 85
Filtration, in catalytic hydrogenations, 12
Fire hazard
  in catalytic hydrogenation, 13
    handling hydrides and complex hydrides, 20
Fluorene
  proton donor in reduction with metals, 43
Fluorides
  See also *Halides*
Fluorides, reduction, 63
Fluorine, hydrogenolysis, 66–68, 81, 159
Fluoroalkanes, reduction, 63
Fluorobenzene, reduction, 68
Fluorosuccinic acid, from difluoromaleic acid, 142
Formaldehyde
  reduction of
    aromatic aldehydes, 100, *189*
    aromatic ketones, 110, *191*
    reductive amination, 36, 134, 136
Formic acid
  hydrogen donor, in catalytic hydrogen transfer, 13, 73
  properties, 36
  reduction of aromatic aldehydes, 100, *189*
    enamines, 92, *188*
    ketimines, 132
    ketoximes, 133
    nitriles, 173
    nitro compounds, 71, 73, 74, *182*
    pyridine, 56, 180
    pyridinium salts, 56
  reductive amination, 134–136
  solvent for reduction with zinc, 29
Furan, reduction, 53, 80, *180*

G

Geminal halides
  from trigeminal halides, 32, 65
  reduction, 32, 33, 35, 64, *181*
Glucose
  in biochemical reduction, **218**
  reducing agent, 36
  reduction of nitroketones, 124, *192*

Glycols
  See *Diols*
Grignard reagents
  reduction of aromatic ketones, 110
    of halides, 27

H

Half-hydrogenation times, 39, 40
Halides
  See also *Bromides, Chlorides, Fluorides* and *Iodides*
Halides
  reduction, 6, 27, 29–30, 32, 37, 62–69, *181*
  stoichiometry in hydride reductions, 19
Halo acids, reduction, 141, 142, *196*
Halo alcohols
  from halo acids, 141, *196*
    halo esters, 159, *198*
    halo ketones, 122, 123, *192*
  reduction, 81
Halo aldehydes, reduction, 103
Halo alkanes, reduction, 63–65, *181*
Halo alkenes, reduction, 66, 67, *181*
Halo alkynes, reduction, 66, *181*
Halo amines, from halo nitro compounds, 74, 75, *182*
Halo aromatics, reduction, 67–69, *181*
Halo azides
  reduction to alkenes, 76, *183*
    aziridines, 76, *183*
Halo cholestanones, reduction, 123
Halo cycloalkanes, reduction, 63–67
Halo esters, reduction, 159, *198*
Halogen compounds
  See *Halides*
Halogens
  replacement by hydrogen, 16, 36, 62–69, *181*
    in acids, 141, *196*
    alcohols, 81, *184*
    esters, 159, *198*
    ketones, 123, *192*
Halohydrins, reduction to alcohols, to alkenes, 81, *184*
Halo ketones
  reduction to alkenes, 123, *192*
    halo alcohols, 122, 123, *192*
    ketones, 31, 35, 123, *192*
Halonitro compounds, reduction, 74, 75, *182*
Halophenols, reduction, 81
Halopyridines, reduction, 68
Halopyrimidines, reduction, 68
Handling of hydrides and complex hydrides, 20–22
Heat of hydrogenation, 8, 39, 46
Hemiacetals of aldehydes, 148, 149
Henbest reduction, 117
Heptadecanone, synthesis, 151
Heptanal, reduction, 97
Heptanol, from heptanal, 97
Heptyl nonyl ketone, synthesis, 151
Heterocycles, aromatic
  reduction, 5, 53–62, *180*
Heterocycles, aromatic, halogen derivatives, reduction, 68, 69
Hexaamocalcium, reduction of benzene, 48

Methanol
  proton donor in reductions with alkali metals, 26, 48
  solvent in hydrogenations, 11
  in reductions with chromous salts, 30
    with sodium, 26
Methoxybenzaldehyde, reduction, 100
B-Methoxy-9-borabicyclo[3.3.1]nonane
  in reduction of esters to alcohols, 155
4-Methoxycarbonyl-3-indolylmethyl-acetami-domalonic acid from its dibenzyl ester, 150
Methylaldimines from carboxylic acids, 137
Methylamine
  from ammonium chloride and formaldehyde, 36, 136
  solvent in electroreduction, 48, **210**
  dissolving metal reduction, 26, 48
Methylaminoethylbenzene, from acetophenone, 136
Methylanisole, from anisaldehyde, 100
Methyl benzyl ketoxime, from 2-nitro-1-phenylpropene, 71
Methylbutene, hydrogenation heat and rate, 5
Methyl caproate, reduction, 149
Methyl cinnamate, reduction, 158
Methylcinnoline, reduction, 60
Methylcyclohexane
  from 1-methylcyclohexene, **210**
  toluene, 47
Methylcyclohexanols, from methylcyclohexene oxide, 84
Methylcyclohexanone oxime, reduction, 133
Methylcyclohexene, reduction, **210**
Methylcyclohexene oxide, reduction, 84
Methylcyclohexylamine, from methylcyclohexanone oxime, 133
Methylene group, by reduction of carbonyl, 34, 130, 131, 143, 144
Methylenemethylcyclohexane, hydrogenation, 40
Methyl ethyl ketone, reduction, 108, **209**
Methyl group, from aldehyde group, 97, 100–102, 106
Methyl hexanoate, reduction, 149
Methylhexanol, from methylisoamyl ketone, 108
Methylhexanone, reduction, 108
Methyl hexyl ketone, reduction, 108
Methyl hydrocinnamate, from methyl cinnamate, 158
Methylindole
  from indolealdehyde, 106
    methylcinnoline, 60
  reduction, 57
Methylindoline, from methylindole, 57
Methyl isoamyl ketone, reduction, 108
Methyl isopropyl ketone, from acetylacetone, 127
Methyl ketones, from trichloromethyl ketones, 123
Methyl ketopiperidine, reduction, 119
Methyl laurate, reduction, **211**
Methyl mercaptan, reducing agent, 32
Methylnaphthalene, from naphthaldehyde, 106
Methyl nonyl ketone, reduction, 108
Methyloctahydroindole, 57

Methyl octyl ketone, reduction, 108
Methylphenylcarbinol
  from acetophenone, 111
  reduction, 79
N-methylpicolinium halides, reduction, 56
N-methylpipecoline, from N-methylpicolinium halides, 56
Methylpiperidine, from methylketopiperidine, 119
Methyl propyl ketone, reduction, 108
Methylstyrylcarbinol, from benzalacetone, 121
Methyl tert-butyl ketone, reduction, 108
N-methyltetrahydropicolines from N-methylpicolinium halides, 56
Methyltetrahydrothiophene, from oxotetrahydrothiopyran, 119
Methyl thiolbenzoate, reduction, 164
Microorganisms, in biochemical reductions, 37, 108, 111, 120, 142
Mixing, effect in hydrogenation, 12
Molecular sieves, removal of peroxides, 85
Molybdenum sulfide
  hydrogenation catalyst, 4, 9
  hydrogenation of
    alcohols to hydrocarbons, 76, 77, *184*
    aromatic ketones to hydrocarbons, 112, *191*
Molybdenum trichloride
  preparation, 31, 88
  reduction of sulfoxides to sulfides, 88, *187*
Molybdenyl chloride
  reduction to molybdenum trichloride, 88
Monoethyl adipate, reduction with borane, **209**
Monoisocampheylborane, synthesis, 15
Monothioketals
  reduction to ethers, sulfides, 130, 131, *194*
Mossy zinc
  preparation of zinc amalgam, 28, **213**

N

Naphthaldehyde, reduction, 106
Naphthalene
  reduction, 26, 50, 51, *179*
  resonance energy, 46
Naphthalene rings
  reduction in alkyl aryl ethers, 82, 83
Naphthoic acids, partial reduction, 140
Naphthols, reduction, 80, 81, **211**
Naphthylamines, reduction, 93
NB-Enanthrane, 15, 16, 122
  reduction of acetylenic ketones, 122
Nef's reaction, 70
Nic catalysts, 9
  hydrogenation of alkenes, 39
    alkynes, 44, 45
Nickel
  See also *Raney nickel*
Nickel
  catalyst in hydrogenation, 4–6, 8, 9, **205**
    hydrogen transfer, 35, 36
  cathode in electroreduction, 24
    of alkynes, 45
  desulfurization of
    disulfides, 87, *187*
    mercaptals, 104, *190*

Sodium borohydride—cont.
  reduction of—cont.
    olefinic—cont.
      amines, 92, *188*
      esters, 157, 158, *197*
      ketones, 120, 121, *191*
      nitriles, 175, *200*
      nitro compounds, 70, *182*
    organometallics, 176
    quinazolines, 61
    quinones, 129, *193*
    quinoxalines, 61
    palladium chloride, 7
    pyridinium salts, 56
    rhodium chloride, 7
    saccharides, 27, 149
    sulfonyl esters, 91 *187*
    sulfonyl hydrazides to hydrocarbons, 172, *200*
    thioamides to amines, 171, *200*
    toluenesulfonyl hydrazones of
      aldehydes, 106
      ketones, 108, 118, 134
    vinyl halides to alkenes, 66, 67, *181*
  reductive amination, 135
    with acids, 171
  solubilities, *21*
  stoichiometry, 19
Sodium chloride
  in reductions with iron, 29, 73
Sodium cyanoborohydride
  handling, 20
  preparation, 15
  reduction of
    acetylenic ketones, 122, *191*
    arenesulfonylhydrazides, 172, *200*
    arenesulfonylhydrazones of
      aldehydes, 106, *190*
      ketones, 108, 121, 134, *190*
    cinnamaldehyde, 102, *189*
    enamines, 92, *188*
    indole, 56, 57, *180*
    isoquinoline, 58, *180*
    keto esters, 162, *198*
    olefinic amines, 92, *188*
      ketones to olefinic alcohols, 120, 121, *191*
    quinoline, 58, *180*
    thioamides to amines, 171, *200*
  reductive amination, 135, 136
  solubilities, *21*
  toxicity, 21
Sodium diethylphosphorotellurate
  reducing agent, 35
  reduction of epoxides, 83, *186*
Sodium diisobutylaluminum hydride
  reduction of esters to aldehydes, 148, 149, *197*
Sodium dithionite
  See *Sodium hydrosulfite*
Sodium fluoroborate
  preparation of sodium borohydride, 14
Sodium hydride
  preparation of sodium borohydride, 14
  reduction of
    aldehydes, 96, *189*
    nickel acetate to nickel, 9, 43, 107

Sodium hydrosulfite
  reducing agent, 33
  reduction of
    aliphatic aldehydes, 96, 97, *189*
    aliphatic ketones, 107, 108, *191*
    aromatic aldehydes, 99, 100, *189*
    aromatic ketones, 110, 111, *191*
    arylarsonic acids, 176
    arylstibonic acids, 176
    azides, 76, *183*
    azo compounds to amines, 96, *188*
    hydrazo compounds, 95, *188*
    ketones, 107, 108, *191*
    nitro alcohols, 81, *184*
    nitro compounds to amines, 73, *182*, **216**
    nitro phenols, 81
    nitroso compounds, 75, *183*
    nitroso phenols, 81
Sodium hydroxide
  in halogen hydrogenolysis, 11
  hydride reductions, 167
Sodium hypophosphite
  reduction of
    nitriles to aldehydes, 173, *200*
    sulfinic acids to disulfides, 89, *187*
Sodium hyposulfite
  See *Sodium hydrosulfite*
Sodium iodide
  reducing agent, 32
  reduction of
    sulfonyl esters, 91, **213**
    sulfoxides, 88
Sodium naphthalene
  preparation, 91, **212**
  reduction of
    sulfonamides, 91, 92, *187*, **212**
    sulfonyl esters, 91, *187*
Sodium potassium tartrate
  quenching after reduction with hydrides, 174
Sodium stannite
  reduction of diazonium compounds, 75, *183*
Sodium sulfide
  reduction of
    nitro alcohols, 81, *184*
    nitro amines, 93
    nitro compounds, 73, *182*
    sulfonyl chlorides to sulfinic acids, 90, *187*
Sodium sulfite
  reducing agent, 33
  reduction of
    diazonium compounds to hydrazines, 76, *183*
    geminal dihalides, 64, *181*
    hydroperoxides, 85, *186*
    sulfonyl chlorides to sulfinic acids, 90, *187*
    trigeminal halides, 65, **216**
  removal of peroxides, 85
Sodium tetrahydridoborate
  See *Sodium borohydride*
Sodium thiosulfate
  reduction of nitro compounds to oximes, 70, *182*
Sodium toluenesulfinate, reduction, 89
Sodium trimethoxyborohydride
  preparation, 15